INTERMEDIATE ALGEBRA

André L. Yandl, *Seattle University*, Elmar Zemgalis, *Highline College*, and Henry S. Mar, *Seattle Community College*.

Numerous specific features make this book useful to the college freshman who is not science oriented, as well as the student who intends to go on to trigonometry and college algebra. Open sentences are introduced early and the dependence of the solution set of an open sentence on its replacement set is carefully explained. Thus, students having used this book will be prepared to handle equations and inequalities.

Real numbers are introduced in a natural way as decimals *after* the decimal representation of rational numbers has been presented. A full chapter is devoted to coordinate geometry because beginning students are often deficient in this area. Equations, inequalities, and systems of equations or inequalities are regarded as open sentences. There is also optional material on such topics as linear programming, the pigeonhole principle, and mathematics of investments.

The authors believe that the development of manipulative skills, while necessary, should not be the goal of the course and emphasis should be given to understanding. Therefore, their mode of presentation is precise, but not excessively formal. Also, their organization has been motivated by both mathematical *and* pedagogic reasons. For example, such a topic as elementary logic is first introduced to develop elementary set theory, and, in fact, is used throughout the book.

Challenging exercises are provided for the more ambitious student and simpler exercises, correlated to examples in the text, are provided for the beginning student.

INTERMEDIATE ALGEBRA

INTERMEDIATE ALGEBRA

André L. Yandl
SEATTLE UNIVERSITY

Elmar Zemgalis
HIGHLINE COLLEGE

Henry S. Mar
SEATTLE COMMUNITY COLLEGE

HOLT, RINEHART AND WINSTON, INC.
New York   Chicago   San Francisco   Atlanta
Dallas   Montreal   Toronto   London   Sydney

PREFACE
TO
INSTRUCTORS

In recent years an attempt has been made to modernize courses in Trigonometry and College Algebra. However, most Intermediate Algebra books with which we are familiar do not adequately prepare students for such modern courses. We have written this text in an attempt to remedy this situation.

This text would also be appropriate for students who do not wish to take further courses in mathematics. It should enhance their general education and give them some idea of the spirit of the "modern approach" in mathematics.

The only prerequisite for this text is one year of high school algebra. However, the book is sufficiently self-contained so that a serious student who has been out of school for some time and who is returning to complete his education should find it accessible without undue effort.

The book is designed for a one-quarter or a one-semester

course. We have included several topics that are not usually covered in Intermediate Algebra texts. This will give each instructor a variety of choices for his own course.

Since there are more sections than class hours in a single quarter (semester), if an instructor wishes to cover the whole text, he will (by the pigeonhole principle) have to cover more than one section on certain days. We have designed this text with this thought in mind. Some sections are very simple and can be combined with others. There are many examples, and most of the exercises contain instructions for the student indicating which examples he should refer to as models for his solutions. In this manner, an instructor may assign exercises on a section that he has not yet covered in class; he may, if he wishes, ask the student to study certain sections on his own and cover them only briefly in class; and most important of all, the student will get the necessary drill and the satisfaction that result from solving many exercises.

Since we feel that the text is very flexible, we do not attempt to give an outline for instructors to follow. However, Chapters 1 and 2 and Section 6.2 should be covered carefully, since the ideas and notation introduced there are used extensively throughout the text.

In spirit we have attempted to make this text *precise without being formal.* Hence, we purposely avoided using the structure "axiom, definition, theorem, proof." This structure is very appealing to a mathematically inclined student, but to most others it may appear somewhat artificial. In order to keep a proper balance between theory and applications, we had to sacrifice some of the mathematical formalism. This decision was based upon our teaching experience in our respective schools over the past few years. We have used, without compunction, the words "law," "property," "rule," although many mathematicians would have preferred the terms axiom, definition, function, and so on. For example, we have defined an operation as a rule, although many would have chosen to define it as a function. We feel that this will not constitute a weakness in the approach, since it is offset by the fact that great care has been given to stress that mathematics is precise, useful, and interesting.

We wish to express our gratitude to Professor Joseph Dorsett of St. Petersburg Junior College and to Professor Victor Klee of the University of Washington for their careful review of the first draft of the manuscript and their valuable suggestions.

A preliminary edition of the text was tested in classrooms of several junior colleges and colleges. We owe many thanks to Professors R. Plagge of Highline College, E. Salisbury of Yakima Junior College, D. DesBrisay of Western Washington State College, and F. Smedley of Seattle University for sending us valuable suggestions based on the reaction of their students to the preliminary edition. These suggestions helped us improve the text greatly.

We also wish to express our appreciation to Mrs. Cecelia Zemgalis who typed most of the final manuscript. Finally, we want to thank Mr. Walter Bishop, mathematics editor of Holt, Rinehart and Winston, whose support was instrumental in the successful completion of this text.

Seattle, Washington
January 1969

André L. Yandl
Elmar Zemgalis
Henry S. Mar

PREFACE
TO
STUDENTS

Algebra is basically a generalization of arithmetic. The main difference is that in arithmetic one deals with specific numbers, whereas in algebra one is mainly interested in stating general laws concerning the operations of arithmetic (addition, subtraction, multiplication, and others), using symbols. This is why a beginning student of algebra often finds the subject difficult; he (or she) is not familiar with the symbolism. Mathematical symbolism is really nothing but a special kind of shorthand. Imagine, for instance, how difficult a first course in shorthand may appear to a young girl who begins her secretarial training. However, as soon as she has mastered the usage of each symbol, it becomes natural for her, and she soon increases her speed and efficiency in taking and transcribing shorthand. Similarly, you may find an introductory course in algebra somewhat puzzling at first, but as soon as you have become familiar with the meaning of the symbols, your speed and comprehension will be improved. You will discover that with the help of algebra you will be able to solve some difficult problems of arithmetic in a simple and clear fashion. This fact alone will constitute a very satisfying experience for you. You will find that when you study this book it is profitable to have paper and pencil in hand. You should outline or summarize what you read, study the examples, and work some exercises, until you thoroughly understand the concepts.

If you feel somewhat discouraged because some of your friends have learned algebra at an earlier stage in their schooling, give some thought to the following quotation:

"Very late in life, when he was studying
geometry, someone said to Lacydes, 'Is it
then a time for you to be learning now?'
'If it is not,' he replied, 'when will it be?' "

— *Diogenes Laertius*, circa A.D. 200

Seattle, Washington
January 1969

André L. Yandl
Elmar Zemgalis
Henry S. Mar

CONTENTS

INTRODUCTION TO ELEMENTARY LOGIC

○

1.1 INTRODUCTORY REMARKS

Today, mathematics occupies a prominent position among the various sciences. It cannot be denied that the usefulness of mathematics largely stems from its application to problem-solving. Typically, when faced with a given problem, one proceeds from known facts and uses correct reasoning to arrive at a conclusion that provides a solution to the problem.

logic The laws of formal correct reasoning are treated in a branch of mathematics known as *logic*. Obviously, one needs some basic knowledge of logic to understand and use mathematics.

The present chapter provides an introduction to elementary logic.

Exercise 1.1

1. Read the Preface to Students of this book.
2. Read the Preface to Instructors of this book.

1.2 SENTENCES, STATEMENTS, PROPOSITIONS

In everyday language one finds sentences, which consist of words. It is well known that the same word may have different meanings in different sentences. For example:

> Rome is the *capital* of Italy.
> He invested his *capital* well.

We shall strive to define some of the common mathematical terms (that is, assign a precise meaning to them) so as to remove ambiguities.

Two types of sentences will be of particular interest to us: statements and propositions.

statement A *statement* is a sentence that asserts or denies some property relative to a specific object or class of objects.

proposition A *proposition* is a statement to which one and only one of the attributes "true," "false" can be assigned.

Example 1. Who was first to discover the North American continent?
Smoked oysters are nice.
$2 + 3 = 5$.
All swans are white.
Every statement is a proposition.
The earth is flat.

The first sentence is not a statement, since it does not assert or deny a specific property. The second sentence is a statement, but not a proposition, because some may consider it true and others false. The last four sentences are propositions, the third being true and the last three being false.

Exercise 1.2

Study Example 1 before doing Exercises 1–19. Which sentences are statements and which are propositions? Justify your answers.

1. $1 + 1 = 2$.
2. Who was first to discover the Pacific Ocean?
3. Broiled lobsters are nice.
4. Every statement is a sentence.
5. Boys like girls.
6. A whale is a mammal.
7. Summers are hot.
8. Every proposition is a statement.

9. When do you plan to go?
10. This is my lucky day.
11. A pound of lead weighs as much as a pound of feathers.
12. Sea urchins are echinoderms.
13. This set of problems is difficult.
14. The symbol 3 is a numeral.
15. Are oranges round?
16. It is raining today.
17. Some men are not good.
18. The moon is made of green cheese.
19. The sky is blue.

1.3 NEGATION, CONJUNCTION, DISJUNCTION

Our major interest will eventually focus on mathematical propositions. However, since we have not yet developed enough algebra, initially we shall often present simple everyday propositions as illustrations.

For notational convenience the lower-case letters p, q, r, $\cdots$ will be used to denote propositions.

Given the propositions p, q, one often desires to form new statements using the connectives *or, and*. These, of course, are common words, and one may tend to think that there would be little room for ambiguity. To see the difficulty, consider the statement:

The sun shines or it rains.

Would you regard this statement as true when it is raining on a sunny day?

Again, consider the statement:

You study and you pass this course.

Assume for a moment that this statement is true. Is it possible for a student to pass this course without studying?

It is conceivable that different persons may provide different answers to these questions. Since we want the resulting statements to be propositions, we must be able to assign to each the attribute *true* or *false*.

We thus give the following:

negation **Definitions.** The *negation* of a proposition p (read "not p" and abbreviate $\sim p$) is a proposition that is false if p is true and true if p is false.

conjunction   The *conjunction* of the propositions p, q (read "p and q" and abbreviate $p \wedge q$) is a proposition that is true when both p and q are true and false otherwise.

disjunction   The *disjunction* of the propositions p, q (read "p or q" and abbreviate $p \vee q$) is a proposition that is false if both p and q are false, and true otherwise.

We can summarize the foregoing definitions in the following

truth table *truth tables:*

Negation:

p	$\sim p$
T	F
F	T

Conjunction:

p	q	$p \wedge q$ (p and q)
T	T	T
T	F	F
F	T	F
F	F	F

Disjunction:

p	q	$p \vee q$ (p or q)
T	T	T
T	F	T
F	T	T
F	F	F

We remark that in the last two tables, we have indeed considered all possible combinations of the truth values for p and q. To see this, we recall the following *fundamental principle of counting.*

If a thing can be done in m different ways and, after it has been done in any of these ways, another thing can be done in any of n different ways, then the two things can be done, one after the other, in mn different ways. Thus, there are two ways of giving a truth value to p (true, false). When this has been done, there are two ways of assigning a truth value to q, thus there are $2 \times 2 = 4$ ways of obtaining combination of truth values for p and q. This is why the last two tables have four rows.

If we combine propositions p, q, r and want to consider all possibilities of assigning a truth value to each, we have a table with $2 \cdot 2 \cdot 2 = 8$ rows. We illustrate this with the following *tree diagram:*

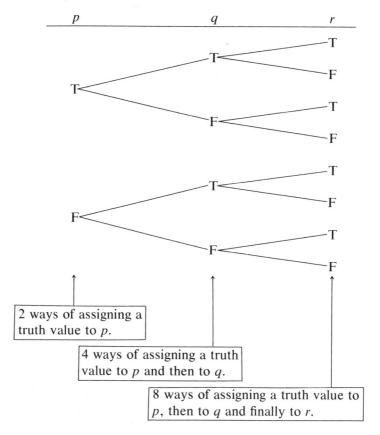

2 ways of assigning a truth value to p.

4 ways of assigning a truth value to p and then to q.

8 ways of assigning a truth value to p, then to q and finally to r.

Example 1. "7 is not an even integer" is the negation of "7 is an even integer." Note that the first proposition is true while the second is false.

Example 2. "Elmar Zemgalis is a bad president of the United States" is not the negation of "Elmar Zemgalis is a good president of the United States"; both propositions are false, because he is not a president of the United States.

Example 3. Verify that if p, q, r represent propositions, then for each of the eight combinations of truth values for p, q, r, the propositions $p \wedge (q \vee r)$, $(p \wedge q) \vee (p \wedge r)$ have the same truth value.

Solution: We construct the following truth table:

p	q	r	$q \lor r$	$p \land (q \lor r)$	$p \land q$	$p \land r$	$(p \land q) \lor (p \land r)$
T	T	T	T	T	T	T	T
T	T	F	T	T	T	F	T
T	F	T	T	T	F	T	T
T	F	F	F	F	F	F	F
F	T	T	T	F	F	F	F
F	T	F	T	F	F	F	F
F	F	T	T	F	F	F	F
F	F	F	F	F	F	F	F

Comparison of columns 5 and 8 gives the desired verification.

Example 4. We have assigned truth values to the following propositions in accordance with the definitions and well-known facts.

Paris is a city or the sun rises in the west.   (T)
Paris is a city or the sun rises in the east.   (T)
The earth is flat or the sun rises in the west.   (F)
Paris is a city and the sun rises in the west.   (F)
Paris is a city and the sun rises in the east.   (T)

Example 5. Negate the proposition: "All swans are white."

Solution: Note that the given proposition is false, since some black swans have been found in Australia. Hence, the negation of the proposition must be true. It can be expressed as "It is false that all swans are white," or equivalently, "Some swans are not white." To verify the truth of the last proposition it is sufficient to produce one swan that is not white.

Example 6. Negate the proposition "Some normal dog has five legs."

Solution: First note that the given proposition is false, since a dog with five legs would not be considered normal. The negation of the given proposition can be stated as "All normal dogs do not have five legs," or equivalently, "No normal dog has five legs."

Remark. According to the definition of negation, if p is a proposition, any proposition whose truth value is opposite that of p can be considered as a negation of p. For example, "California is north of Washington" is false, while "$1 + 1 = 2$" is true; thus the second proposition can be considered to be a negation of the other. However, it is common practice to consider a proposition to be the negation of another proposition provided the two propositions are related in some sense and also have opposite truth values.

Exercise 1.3

Study Examples 1 and 2 before doing Exercises 1–6. In each problem write the negation of the given proposition. Also state the truth value of each proposition and the truth value of its negation.

1. All apples are not fruits.
2. Oregon is north of Washington.
3. $2 + 2$ is equal to 4.
4. This text is a chemistry book.
5. Some square is not round.
6. Las Vegas is on the Pacific Coast.

Study Example 3 before working Exercises 7–12. Assume that p, q, r represent propositions.

7. Construct a truth table and show that the propositions $p, \sim(\sim p)$ have the same truth value.

8. Verify that the propositions $p \vee (q \vee r)$, $(p \vee q) \vee r$ have the same truth value by completing the truth table.

p	q	r	$q \vee r$	$p \vee (q \vee r)$	$p \vee q$	$(p \vee q) \vee r$
T	T	T	T			
T	T	F	T			
T	F	T	T			
T	F	F	F			
F	T	T	T			
F	T	F	T			
F	F	T	T			
F	F	F	F			

9. Verify that the propositions $p \wedge (q \wedge r)$, $(p \wedge q) \wedge r$ have the same truth value by completing the truth table.

p	q	r	$q \wedge r$	$p \wedge (q \wedge r)$	$p \wedge q$	$(p \wedge q) \wedge r$
T	T	T	T			
T	T	F	F			
T	F	T	F			
T	F	F	F			
F	T	T	T			
F	T	F	F			
F	F	T	F			
F	F	F	F			

10. Verify that the propositions $p \vee (q \wedge r)$, $(p \vee q) \wedge (p \vee r)$ have the same truth value by completing the truth table.

p	q	r	$q \wedge r$	$p \vee (q \wedge r)$	$p \vee q$	$p \vee r$	$(p \vee q) \wedge (p \vee r)$
T	T	T					
T	T	F					
T	F	T					
T	F	F					
F	T	T					
F	T	F					
F	F	T					
F	F	F					

11. Verify that the propositions $\sim(p \wedge q)$, $(\sim p) \vee (\sim q)$ have the same truth value by completing the truth table.

p	q	$p \wedge q$	$\sim(p \wedge q)$	$\sim p$	$\sim q$	$\sim p \vee \sim q$
T	T					
T	F					
F	T					
F	F					

12. Verify that the propositions $\sim(p \vee q)$, $\sim p \wedge \sim q$ have the same truth value by constructing a truth table.

Study Example 4 before doing Exercises 13–18. Assign a truth value to each of the following propositions in accordance with the definitions and well-known facts.

13. Seattle is a city and $1 + 1$ is not equal to 2.
14. Seattle is a city or $1 + 1$ is not equal to 2.
15. Seattle is a city and $1 + 1$ is equal to 2.
16. Seattle is a city or $1 + 1$ is equal to 2.
17. The moon is purple or $1 + 1$ is equal to 2.
18. The moon is purple or $1 + 1$ is not equal to 2.

Review Examples 5 and 6 before working Exercises 19–24. Assuming the following statements are propositions, state the negation of each.

19. All men are good.
20. All girls are beautiful.
21. Some lions are hungry.
22. Some dancing is fun.
23. All birds fly.
24. Some cats are cross-eyed.

1.4 IMPLICATION

Given propositions p and q, it is possible to form the statement "If p, then q" (abbreviated "$p \rightarrow q$"). For example, one often hears a statement such as the following:

"If I get a raise, then I will buy a new car."

Again, when p and q are propositions, and therefore each has a definite truth value (it is either true or false), we would like the statement "If p, then q" to be a proposition. Thus we must have a way to assign a definite truth value to that statement for all possible combinations of truth values of p, q. The definition we are about to give might seem strange at first, but it is almost universally accepted by mathematicians. Thus, we make a few remarks concerning the choice of definition. First, in everyday language, if $p \rightarrow q$ is a proposition, we more or less expect the propositions p, q to be related in some way. In mathematics, we do not require this. The reason for this practice is that it is possible for the propositions p, q not to seem related today, although some future discovery may prove that they are.

For example, suppose p denotes the proposition "John Smith has been bitten by an infectious mosquito" and q denotes "John Smith will have yellow fever." Then prior to 1881, the propositions p, q would have been thought to be unrelated. However, in 1900, the Yellow Fever Commission of the United States Army conducted experiments that proved conclusively that they were related.

Further, suppose that a man made the statement: "If it does not rain tomorrow, then I will play tennis." Most of us would agree that the man made a false statement if it did not rain the next day and yet he failed to play tennis. However, we would hesitate to call him a liar if any one of the following situations occurred:

It did not rain and he played tennis.
It did rain and he did not play tennis.
It rained and he decided to play tennis anyway.

The definition given below agrees closely with this usage of "If p, then q" in ordinary parlance.

implication **Definition.** If p and q are propositions, the proposition "*if p, then q*" is false when p is true and q is false and true in the other three cases.

We illustrate this in the following truth table:

p	q	If p, then q
T	T	T
T	F	F
F	T	T
F	F	T

The proposition "If p, then q" is often stated in one of the following three ways:

(1) p implies q.
(2) p is sufficient for q.
(3) q is necessary for p.

It is also abbreviated $p \rightarrow q$.

hypothesis   In the foregoing p, q are called the *hypothesis* and *con-*
conclusion *clusion*, respectively.

A proposition that has been proved to be true is called a
theorem *theorem*. Most theorems are of the form "If p, then q." The preceding truth table suggests one valid method to prove (that is, to
establish the truth of) a theorem. If we wish to prove that the
proposition $p \rightarrow q$ is true, we first assume that p (the hypothesis)
is true. Using this assumption and any other established rules,
we proceed to show that q (the conclusion) must also be true.
Obviously, we need not be concerned about the case of p being
false, since then the implication is true by definition.

Example 1. "If Paris is the capital of France, then $2 + 2 = 4$"
is true, since both the hypothesis and conclusion are true.

"Canada is part of the United States implies $2 + 3 = 5$" is
true, since the hypothesis is false.

"The earth is flat $\rightarrow 2 + 3 = 4$" is true, since the hypothesis
is false.

"Rome is the capital of Italy is sufficient for $2 + 2 = 5$"
is false, since the hypothesis is true while the conclusion is false.

Example 2. Show that $[(p \rightarrow q) \wedge (q \rightarrow r)] \rightarrow (p \rightarrow r)$ is true.

Solution: Note that in the given proposition, the hypothesis is in
brackets (namely, $(p \rightarrow q) \wedge (q \rightarrow r)$), and the conclusion is
$p \rightarrow r$. As suggested earlier, we begin by assuming the truth of
the hypothesis and seek to prove the truth of the conclusion.
We first remark that since the hypothesis is true, both proposi-
tions $p \rightarrow q, q \rightarrow r$ are true. To show that the conclusion $p \rightarrow r$
is true, we consider two cases.

CASE 1. p is false. Then $p \rightarrow r$ is automatically true by definition, regardless of the truth value of r.

CASE 2. p is true. As stated earlier, $p \rightarrow q$ is true; therefore,
q must be true. But then, recalling that $q \rightarrow r$ is also true, we
conclude that r is true. We have shown that the truth of p implies that of r. Hence, $p \rightarrow r$ is true, and the proof is complete.

As an exercise the student may give a different solution for

Example 2, making use of a truth table in which all eight possible combinations of truth values for p, q, r are displayed and observing in each case the truth value for the proposition $[(p \rightarrow q) \wedge (q \rightarrow r)] \rightarrow (p \rightarrow r)$.

Question: Why is $[(p \rightarrow q) \wedge (q \rightarrow r)] \rightarrow (p \rightarrow r)$ a theorem?

Example 3. Using a truth table, show that for each of the four possible combinations of truth values for p, q, the propositions $p \rightarrow q$ and $(\sim q) \rightarrow (\sim p)$ have the same truth value.

Solution:

p	q	$p \rightarrow q$	$\sim q$	$\sim p$	$\sim q \rightarrow \sim p$
T	T	T	F	F	T
T	F	F	T	F	F
F	T	T	F	T	T
F	F	T	T	T	T

Comparing columns 3 and 6, we conclude that $p \rightarrow q$, $(\sim q) \rightarrow (\sim p)$ have the same truth values in all four cases.

The foregoing example suggests that if one wants to prove the truth of $p \rightarrow q$, he is logically justified in proving the truth of $(\sim q) \rightarrow (\sim p)$ instead.

Exercise 1.4

Study Example 1 before doing Exercises 1–6. Justify the assigned truth value for each of the given propositions.

    1. "If London is the capital of England, then $2 + 2 = 4$" is true.

    2. "$2 + 1 = 4$ implies Mexico is a part of France" is true.

    3. "Water never freezes $\rightarrow 2 + 2 = 4$" is true.

    4. "$2 + 2 = 4$ is sufficient for triangles to have four sides" is false.

    5. "Tigers never eat" is necessary for "circles are round" is false.

    6. "$2 + 2 = 5 \rightarrow 1 = 2$" is true.

In Exercises 7–18 assume p, q, r are true and determine the truth value of each of the given propositions.

   7. $[\sim(\sim p)] \to (p \lor \sim p)$.
   8. $[\sim(p \to q)] \to (p \land \sim q)$.
   9. $[(p \to q) \land p] \to q$.
   10. $[(p \to q) \land (\sim q)] \to \sim p$.
   11. $[(\sim q) \land (p \to q)] \to p$.
   12. $[(\sim p) \land (p \lor q)] \to q$.
   13. $(p \land \sim q) \to \sim(p \to q)$.
   14. $[(p \land q) \to r] \to [p \to (q \to r)]$.
   15. $[p \lor (p \to q)] \to q$.
   16. $(\sim p \lor q) \to (p \to q)$.
   17. $(\sim p) \to [(\sim p) \lor (\sim q)]$.
   18. $[(p) \lor (\sim q)] \to \sim p$.

   19. Use a truth table to show that the proposition in Example 2 of the preceding section is true for all truth values of p, q, r.

   20. Use a truth table to determine the truth value of $p \lor \sim p$.

   21. Use a truth table to determine the truth value of $p \land \sim p$.

   22. Show that the propositions $p \to q$, $(\sim p) \lor q$ have the same truth value for each of the four possible combinations of truth values for p, q.

   23. Why is the following "proof" invalid?

Theorem. $1 = 2$.

Proof: If $1 = 2$, then $2 = 1$. Adding, we get $3 = 3$. Hence $1 = 2$ is a true proposition.

   24. In each of the propositions given in Exercises 1–6 state which part in the hypothesis and which is the conclusion.

Study Example 3 before working Exercises 25–30. In these, write a proposition that has the same truth value as the given proposition.

   25. $p \to \sim q$.
   26. If Jim is a brother of Susan, then Susan is a sister of Jim.
   27. If b is an even integer, then $2 + b$ is an even integer.
   28. If the diagonals of a parallelogram are mutually perpendicular, then the parallelogram is a rectangle.
   29. If two lines have no point in common, then they do not coincide.
   30. $(\sim q) \to p$.

1.5 CONVERSE, CONTRAPOSITIVE

<u>*contrapositive*</u> The implication $(\sim q) \to (\sim p)$ is called the *contrapositive* of
<u>*converse*</u> $p \to q$. The implication $q \to p$ is called the *converse* of $p \to q$.

In Example 3 of the preceding section we have shown that an implication and its contrapositive have the same truth values.

The following example shows that this is not necessarily the case for an implication and its converse.

Example 1. The implication "If $2 = 3$, then $5 = 5$" is true, since the hypothesis is false. On the other hand, its converse — which is "If $5 = 5$, then $2 = 3$" — is false, since the hypothesis is true while the conclusion is false.

Example 2. State the converse and the contraposition of
$$\text{"If } -1 = 1, \text{ then } (-1)^2 = 1.\text{"}$$

Note first that this implication is true. (Why?) The converse is "If $(-1)^2 = 1$, then $-1 = 1$," which is false, since the hypothesis is true while the conclusion is false. The contrapositive "If $(-1)^2 \neq 1$, then $-1 \neq 1$" is true, since its hypothesis is false.

Exercise 1.5

Study Examples 1 and 2 before working Exercises 1–10. State the converse and the contrapositive of each of the following propositions. Also state the truth value of each proposition, its converse, and its contrapositive.

1. If $1 = 2$, then $3 = 3$.
2. If Paris is in France, then $1 + 1 = 2$.
3. If $2 + 2 \neq 5$, then $1 = 2$.
4. If $p \wedge q$, then p (assume p, q are both true).
5. If $p \vee \sim p$, then p (assume p is true).
6. If $p \wedge \sim p$, then p (assume p is false).
7. $p \rightarrow \sim q$ (assume that p is false and q is true).
8. $p \rightarrow (q \rightarrow r)$ (assume p, q, r are true).
9. $2 + 1 = 3$ is sufficient for $5 + 1 = 7$.
10. "The earth is flat" is necessary for "$1 + 3 = 5$."
11. Give an example of a true implication whose converse is also true.
12. Give an example of a false implication whose converse is true.
13. Give an example of a true implication whose converse is false.
14. Is it possible to give an example of a false implication whose converse is also false?

REVIEW EXERCISES

1. State the negation of each of the following propositions:
(a) All automobiles have four wheels.
(b) Mars is a planet.
(c) Every proposition is a statement.
(d) Some counting number is odd.
(e) Mt. Everest is higher than Mt. Rainier.

2. (a) Construct a truth table to show that the propositions $\sim(p \wedge q)$, $(\sim p) \vee (\sim q)$ have the same truth value.
(b) State the negation of the proposition "Every proposition is a statement and some counting number is odd."

3. Suppose q is a false proposition and $p \rightarrow q$ is a true proposition. Construct a truth table to show that p is false.

4. Suppose $p \rightarrow q$ is a false proposition. Construct a truth table to show that then $p \rightarrow \sim q$ is a true proposition.

5. State the converse and the contrapositive of the proposition "If Mars is a planet, then all automobiles have four wheels."

6. Without using a truth table show that the proposition $p \rightarrow (q \rightarrow p)$ is true. [*Hint:* Assume that p is true. What can you conclude about the truth value of $q \rightarrow p$?]

7. Suppose p is a false proposition. Determine the truth value of each of the following propositions: **(a)** $(p \rightarrow q) \rightarrow p$, **(b)** $(p \rightarrow \sim q) \rightarrow p$, **(c)** $(p \rightarrow q) \rightarrow \sim p$, **(d)** $(\sim p \rightarrow p) \rightarrow q$, **(e)** $q \rightarrow \sim p$.

Assign a truth value to each of the following propositions in accordance with the definitions and well-known facts.

8. Mars is a planet and some counting number is odd.
9. Jupiter is a planet and $2 + 3$ is not equal to 5.
10. Jupiter is a planet and $2 + 3$ is equal to 5.
11. Mars is a planet or the earth is flat.
12. Jupiter is not a planet or Seattle is not a city.

In Exercises 13–26 assume that both p, q are true and determine the truth value of each of the given propositions.

13. $[\sim(\sim p)] \rightarrow q$.   **14.** $(\sim q) \rightarrow p$.
15. $p \rightarrow \sim q$.   **16.** $\sim(\sim p \vee q)$.
17. $(p \rightarrow q) \rightarrow (q \rightarrow p)$.   **18.** $\sim(p \wedge q)$.
19. $(\sim p) \wedge q$.   **20.** $(\sim p) \vee (\sim q)$.
21. $p \rightarrow [(\sim p) \vee q]$.   **22.** $p \rightarrow [(\sim p) \vee (\sim q)]$.
23. $(\sim p) \vee (\sim q) \vee q$.   **24.** $p \wedge q \wedge (\sim p)$.
25. $(p \rightarrow q) \wedge (q \rightarrow p)$.   **26.** $(p \rightarrow q) \wedge [(\sim q) \rightarrow p]$.

CHAPTER

2

SETS

○

2.1 SETS

A set is a collection of objects. In order to describe a set, we must be able to state a precise property that an object must have in order to be a member of that set. For example, "the set of all healthy people" is not very well described; one doctor may declare a certain person healthy, yet another could find him ill. On the other hand, we can very well talk about the set consisting of the first ten presidents of the United States.

In general, mathematicians use capital letters, $A, B, C, \cdots$, $S, T, \cdots$, to denote sets, and small letters, $a, b, c, \cdots$, to denote members of sets. If S is a set and a is an object, the sentence "a is a member of S" is abbreviated $a \in S$. Note that this sentence is a proposition, since it is either true or false. Similarly, *member* $x \notin T$ shall mean "x is not an element of the set T." We agree *element* that the terms *member* and *element* are synonymous in this context.

If a set has only a few members, one can describe it by displaying all of its members, listing each member only once, separating consecutive ones with commas, and putting the list within braces.

Example 1. The set whose members are the first four counting numbers is described as follows:

$$\{1, 2, 3, 4\}.$$

Example 2. The set whose members are the vowels of the English alphabet is denoted

$$\{a, e, i, o, u, y\}.$$

empty set   It is convenient to consider a collection with no members at all a set. It is called the *empty set*. Observe that we said *the* empty set rather than an empty set. The reason is that it can logically be shown that there is one and only one empty set. A very descriptive notation for the empty set is { }. This notation is certainly consistent with the method described above. Another very common symbol to denote the empty set is ∅, which we shall primarily use in this text.

Exercise 2.1

Study Examples 1 and 2 before working Exercises 1–9. In these, some objects are described. In each case describe the set whose members are these objects by listing all the members of each set.

1. The first eight counting numbers.

2. The first counting number.

3. The last four letters of the alphabet.

4. All states of the United States whose names begin with the letter A.

5. All of the counting numbers between 10 and 20.

6. All of the letters of the alphabet between h and p.

7. All of the counting numbers between 2 and 3.

8. The letters of the alphabet between b and c.

9. The presidents of the United States who succeeded John F. Kennedy.

10. Let A denote the set $\{1, 3, 5\}$. If each element of a set B is in A and we know that $1 \notin B$, $3 \notin B$, and $5 \notin B$, describe B.

11. Let A denote the set $\{a, b, c, d, e\}$ and suppose that each member of a set C is in A. Further suppose that each member of C is a consonant. Describe the set C.

2.2 VARIABLE, REPLACEMENT SET, OPEN SENTENCE

As we pointed out earlier, in algebra we often use letters to represent numbers. For example, the reader may have seen statements such as these:

$$x + 3 = 3x - 1,$$

$$2 + y = 3t,$$

and $$x \cdot y = y \cdot x.$$

At first, it may seem strange to add a letter to a number.

However, a moment of reflection should reveal that even in a statement such as

$$2 + 3 = 3 \cdot 2 - 1,$$

the 2, 3, and 1 are symbols that denote some specific numbers.

On the other hand, in

$$x + 3 = 3 \cdot x - 1,$$

variable x may be regarded as a symbol that does not denote a specific number if it is understood that it may be replaced by any one member from a certain set of numbers. In general, we call such a symbol a *variable*. In some respects, a variable is a placeholder. For example, instead of

$$x + 3 = 3 \cdot x - 1,$$

we could have written

$$\underline{\quad} + 3 = 3 \underline{\quad} - 1,$$

where the lines indicate the place to write numbers. Of course, we agree that once we write a number in one blank, we must also write the same number in the other blank.

Suppose you were asked, "Is the statement, $x + 3 = 3 \cdot x - 1$, true?" This, of course, you could not answer. Does this mean that the statement is false? No, since it is neither true nor false.

open It is an *open sentence*. An open sentence is a sentence that in-
sentence volves at least one variable.

In any discussion in which a variable is involved, we must have a set so that we may replace the variable by the name of
replacement some member of that set. This set is called the *replacement set*
set for the variable.

For example, in the sentence

$$x + 3 = 5 - x$$

the replacement set may be the set of counting numbers. If we replace x by 1, we get

$$1 + 3 = 5 - 1,$$

which is a true statement.

On the other hand, if we replace x by 2, we obtain

$$2 + 3 = 5 - 2,$$

which is false.

We say that 1 is a solution of the sentence $x + 3 = 5 - x$, while 2 is not a solution.

For the sake of simplicity, we shall at first consider only open sentences with one variable. In each case we must have a replacement set for this variable. A member of the replacement *solution* set will be called a *solution* if, whenever the variable is replaced throughout by the name of that member, a true proposition is obtained. It is now clear that whether or not an open sentence has a solution depends on both the open sentence and the replacement set. For example, consider the open sentence, $2 \cdot x = 1$. It has no solution if the replacement set is the collection of counting numbers. On the other hand, it has one solution (namely $\frac{1}{2}$) if the replacement set is the set of fractions.

To further explore these ideas, consider the following examples.

Example 1. Suppose we were interested in singling out the residents of the State of Washington who are Certified Public Accountants. We may consider the open sentence:

x is a Certified Public Accountant.

From the context we assert that the replacement set for the foregoing open sentence is the set of all residents of the State of Washington. If we replace x by the name of a member of that set, we shall obtain a proposition (that is, a statement that is either true or false). If it is a true statement, then the resident whose name we have used is a solution of the open sentence and should be singled out. It is clear that after each name from the list of residents of the State of Washington has been used as a substitute for x, the collection of solutions of the open sentence is the set of residents we wanted to single out.

Note that if we tried to single out the collection of all CPA's who reside in the United States, the open sentence would be the same, namely:

x is a Certified Public Accountant.

However, here the replacement set is larger. It is the set of residents of the United States. It is likely that the set of solutions in this new situation is larger than the previous one.

Example 2. Consider the open sentence $x + 2 = 8 - 2x$, with replacement set $\{1, 2, 3\}$. We replace x by 1 and we get $1 + 2 = 8 - 2$, which is false. Now, replacing x by 2, we obtain $2 + 2$

$= 8 - 4$, which is true; replacing x by 3, we obtain $3 + 2 = 8 - 6$, which is false. Hence, the solution set is $\{2\}$.

On the other hand, if the replacement set is $\{1, 3, 5, 7\}$, we obtain a false statement regardless which of the numbers $1, 3, 5, 7$ we use to replace x in $x + 2 = 8 - 2x$. Thus, the solution set is the empty set.

Thus, we see again that the solution set of an open sentence depends very strongly on its replacement set.

In summary, we have introduced the following:

variable   A *variable* is a symbol in a sentence (called an open sentence) that may be replaced by names of members of some specified set (called the *replacement set for the variable*). Any member of the replacement set having the property that the open sentence becomes a true statement when the name of that member is *solution* substituted for the variable is called a *solution* of the open *solution set* sentence. The collection of all solutions is called the *solution set* of the sentence relative to the given replacement set.

Exercise 2.2

Study Examples 1 and 2 before working Exercises 1–6.

1. Find the solution set of the open sentence

$$x - 1 = 3 - x$$

if the replacement set for the variable x is (a) $\{1, 2\}$, (b) $\{1, 2, 3\}$.

2. Find the solution set of the open sentence

$$2x + 1 = 4 - x$$

if the replacement set for the variable x is (a) $\{2, 4, 6\}$, (b) $\{1, 2, 4, 6\}$.

3. List the solutions of the open sentence

$$7y = 1$$

if the replacement set for the variable y is (a) $\{1, \frac{1}{2}, \frac{1}{3}\}$, (b) $\{1, \frac{1}{2}, \frac{1}{3}, \frac{1}{7}\}$.

4. Find the solution set of the open sentence

$$2u - 1 = u + 3$$

if the replacement set for the variable u is (a) $\{\frac{1}{2}, 2, 1\}$, (b) $\{\frac{1}{2}, 3, 5, 2, 1\}$.

5. Describe how one can single out the residents of the State of Washington who are medical doctors (M.D.'s).

6. Describe how one can single out the residents of the United States who are medical doctors (M.D.'s).

2.3 SET NOTATION

Suppose that an open sentence and a certain replacement set are given. For convenience, let the sentence and replacement set be denoted by $p(x)$ (read "p of x") and A, respectively. Then the solution set of that open sentence is usually described as follows:

$$\{x \in A \mid p(x)\}$$

($\{x \mid p(x)\}$ if it is clear from context what the replacement set is). This last expression is often read "the set of all x's in A that satisfy the sentence p of x."

Example 1. Consider the sentence $(x - 1) \cdot (x - 2) = 0$ with replacement set $\{0, 1, 2, 3, 4\} = A$. Clearly, if we replace x by 1 we obtain a true statement, namely, $(1 - 1) \cdot (1 - 2) = 0$. Thus, 1 is a solution of the given open sentence.

Similarly, if we replace x by 2, we get the true statement

$$(2 - 1) \cdot (2 - 2) = 0.$$

Hence, 2 is also a solution. However, if x is replaced by 0, 3, or 4, we obtain false statements. Thus, the solution set of the sentence $(x - 1) \cdot (x - 2) = 0$ is $\{1, 2\}$. Specifically, $\{x \in A \mid (x - 1) \cdot (x - 2) = 0\}$ and $\{1, 2\}$ are different notations for the same set.

It now seems appropriate to discuss the meaning of equality. When we write $a = b$, we mean that a and b are symbols denoting the same object. Note that the statement $a = b$ is a proposition. Its negation is commonly denoted by $a \neq b$ (read "a is not equal to b").

For example, $2 = 5 - 3$ is true, while $6 = 13 + 1$ is false. Hence, $2 \neq 5 - 3$ is false, while $6 \neq 13 + 1$ is true. Further, from Example 1, we can write

$$\{x \in A \mid (x - 1) \cdot (x - 2) = 0\} = \{1, 2\}.$$

In particular, if A and B denote sets, then the statement $A = B$ means that A and B denote the same set. Thus, if $A = B$ is true, then $x \in A$ implies $x \in B$, and conversely, $y \in B$ implies $y \in A$. Also, if $x \in A$ implies $x \in B$ and $y \in B$ implies $y \in A$, then $A = B$.

Example 2. Describe the set $\{x \mid x \neq x\}$.

Solution: Note that regardless which set we use as replacement set for the open sentence $x \neq x$, if we replace x by the name of one of its members, we obtain a false statement. Thus, the open sentence $x \neq x$ has no solution, and its solution set is the empty set. Thus, we write

$$\{x \mid x \neq x\} = \{\quad\} = \varnothing.$$

Example 3. Describe the set $\{x \mid (x - \frac{1}{2}) \cdot (x - 2) \cdot (x + 1) = 0\}$, if the replacement set is

(a) the set of counting numbers,

(b) the set of all integers (positive, negative, and zero),

(c) the set of all ratios of integers $\left(\text{excluding } \frac{a}{0}, \text{ which is not defined}\right)$.

Solution: (a) If we consider the open sentence $(x - \frac{1}{2})(x - 2)(x + 1) = 0$, and replace x by 2, we obtain the true statement $(2 - \frac{1}{2})(2 - 2)(2 + 1) = 0$. Thus, 2 is a solution. However, no other replacement from the set of counting numbers yields a true statement. Thus for part (a), the given set is $\{2\}$.

Similarly for part (b), we obtain the set $\{-1, 2\}$, and for part (c), the set $\{-1, \frac{1}{2}, 2\}$.

Once more we have seen that the solution set of an open sentence depends very strongly on its replacement set.

Observe that in the present section we have used the letter x as a variable in an open sentence. Although this is common practice, any other symbol (preferably one easy to read) could have served the same purpose. For example, it is true that

$$\{x \mid (x - 1) \cdot (x - 2) = 0\} = \{y \mid (y - 1) \cdot (y - 2) = 0\}$$
$$= \{t \mid (t - 1) \cdot (t - 2) = 0\}$$
$$= \{? \mid (? - 1) \cdot (? - 2) = 0\}.$$

The only requirements are the following:

Whatever symbol we use between the first brace and the vertical bar must be used again in the open sentence between the vertical bar and the second brace. For example, it would be incorrect to write $\{x \mid (y - 1) \cdot (y - 2) = 0\}$. Furthermore, we must be careful not to use a symbol that is already the name of a specific object. Thus, for example, the symbols 4 or π should never be used as variables.

Exercise 2.3

Study Example 1 before working Exercises 1–6. In each of these, find the solution set of the given open sentence. The letters x, u, y, v are variables while a, b, A, B represent specific numbers.

1. The sentence "$(x - 2) \cdot (x - 3) = 0$" with replacement set $\{0, 1, 2, 3, 4\}$.

2. The sentence "$(x - 2) \cdot (x - 3) = 0$" with replacement set $\{0, 1, 3, 4\}$.

3. The sentence "$(x - 2) \cdot (x - 5) = 0$" with replacement set $\{0, 1, 2, 3, 4\}$.

4. The sentence "$(u - 3) \cdot (u - 1) = 0$" with replacement set $\{0, 2\}$.

5. The sentence "$(y - a) \cdot (y - b) = 0$" with replacement set $\{A, B\}$.

6. The sentence "$(v - a) \cdot (v - b) = 0$" with replacement set $\{a, B\}$.

Study Examples 2 and 3 before working Exercises 7–16. Describe each of the following sets if the replacement set is (a) the set of counting numbers, (b) the set of all integers (positive, negative, and zero), (c) the set of all ratios of integers (excluding $\frac{a}{0}$, which is not defined).

7. $\{x \mid (x - \frac{1}{3}) \cdot (x + 3) \cdot (x - 1) = 0\}$.

8. $\{y \mid (y + \frac{1}{2}) \cdot (y - 4) \cdot (y + 1) = 0\}$.

9. $\{x \mid 1 \neq 1 + x\}$.

10. $\{u \mid (u - \frac{1}{5}) \cdot (u + 5) \cdot (u - 5) = 0\}$.

11. $\{v \mid (v + 4) \cdot (v + 2) \cdot (v - 7) = 0\}$.

12. $\{z \mid (z + \frac{1}{3}) \cdot (z - \frac{1}{4}) \cdot (z + 6) = 0\}$.

13. $\{x \mid x \cdot (x + \frac{1}{6}) \cdot (x + \frac{1}{9}) = 0\}$.

14. $\{y \mid y$ is a round square$\}$.

15. $\{w \mid (w + \frac{1}{2}) \cdot (w + \frac{1}{3}) \cdot (w + \frac{1}{4}) = 0\}$.

16. $\{t \mid (t + \frac{1}{5})(t - \frac{1}{6})(t + \frac{1}{7}) = 0\}$.

17. Describe the set $\{x \mid (x - a) \cdot (x - b) \cdot (x - c) = 0\}$ if the replacement set is $\{a, B, c, d\}$.

18. Describe the set $\{x \mid (x - a) \cdot (x - b) = 0\}$ if the replacement set is $\{a, b\}$.

19. Describe the set $\{t \mid (t - a) \cdot (t - b) \cdot (t - c) = 0\}$ if the replacement set is $\{A, B, d\}$.

20. Describe the set $\{u \mid (u + a) \cdot (u + b) \cdot (u + c) = 0\}$ if the replacement set is $\{-a, -b, -c, d\}$.

21. Let $A = \{-1, 3, 4\}$. Give an example of an open sentence with a replacement set such that the solution set of the open sentence is A.

2.4 SUBSETS

As we did in the foregoing section, consider an open sentence, such as "x is presently an instructor of mathematics in an American university." Suppose its replacement set is the collection of all living men. First note that a woman mathematics instructor in an American university would not be a solution of the sentence, since she does not belong to the given replacement set. That is, each member of the solution set must be a priori a member of the replacement set. We say that the solution set is a subset of the replacement set.

subset In general, if A and B are sets, and each element of A is also an element of B, we say that A is *a subset* of B. If in addition there is at least one element of B that is not in A, then we say *proper* that A is a *proper subset* of B. The abbreviation for "A is a proper *subset* subset of B" is $A \subset B$, and that for "A is a subset of B" is $A \subseteq B$. Note that if $A = B$, then $A \subseteq B$ is true. Further, $B \subseteq A$ is also true. In fact, the conditions $A \subseteq B$ and $B \subseteq A$ tell us that $A = B$.

We note that in many mathematics books, $A \subset B$ is the abbreviation for "A is a subset of B" whether A is a proper subset or not.

It can be shown, using elementary logic, that the empty set is a subset of any set.*

Example 1. Show that $\{1, 2\} \subset \{0, 1, 2, 3\}$ is a true statement.

Solution: Since $1 \in \{0, 1, 2, 3\}$, and $2 \in \{0, 1, 2, 3\}$ are both true, $\{1, 2\}$ is a subset of $\{0, 1, 2, 3\}$. Further, since $0 \in \{0, 1, 2, 3\}$, but $0 \notin \{1, 2\}$, we conclude that $\{1, 2\}$ is a proper subset of $\{0, 1, 2, 3\}$.

Example 2. List all subsets of the following sets: (a) $\{1\}$, (b) $\{1, 2\}$, (c) $\{1, 2, 3\}$.

Solution: (a) The subsets of the set $\{1\}$ are the empty set $\varnothing$ (which is a subset of any set) and $\{1\}$. Note that there are two subsets.

(b) The subsets of $\{1, 2\}$ are $\varnothing, \{1\}, \{2\}, \{1, 2\}$. There are four subsets.

(c) Clearly, $\varnothing, \{1\}, \{2\}, \{3\}, \{1, 2\}, \{1, 3\}, \{2, 3\}, \{1, 2, 3\}$ are the subsets of $\{1, 2, 3\}$. Note there are eight subsets of $\{1, 2, 3\}$.

* See Exercise 21 below.

It can be shown that if a set has k elements, then it has 2^k subsets. The student may find listing all sixteen subsets of the set $\{a, b, c, d\}$ an interesting exercise. He should proceed to list first the empty set, then all subsets with one member, then all subsets with two members, and so on.

Example 3. Let $S = \{\varnothing, \{\varnothing\}\}$. This set has two distinct elements, namely $\varnothing$ and $\{\varnothing\}$. Thus it must have four subsets, which are $\varnothing, \{\varnothing\}, \{\{\varnothing\}\}, \{\varnothing, \{\varnothing\}\}$. The student should verify that the four sets listed are indeed subsets of S.

Note that in this very special case the elements $\varnothing$ and $\{\varnothing\}$ of S are also subsets of S. In general, however, this need not occur. We must always be careful to differentiate between the phrases "is an element of" and "is a subset of." For example, if $T = \{a, b, c\}$, then we should write $a \in T$ rather than $a \subset T$, and $\{a\} \subset T$ rather than $\{a\} \in T$. In other words, in true statements, we should have symbols denoting sets on each side of $\subset$. Similarly, we should have a symbol denoting a set on the right of $\in$, and on its left a symbol denoting a member of that set.

Exercise 2.4

Study Example 1 before working Exercises 1–8. In these, determine whether the set A is a subset or proper subset of the set B. Justify your answers.

    1. $A = \{1\}, B = \{1, 2\}$.
    2. $A = \{3, 4\}, B = \{3, 4, 5\}$.
    3. $A = \{1, 7, 8\}, B = \{1, 7, 8\}$.
    4. $A = \{a, b, c\}, B = \{a, b, c, d\}$.
    5. $A = \{0\}, B = \{0\}$.
    6. $A = \varnothing, B = \{0\}$.
    7. $A = \{a, b, c\}, B = \{a, b, d\}, c \neq d$.
    8. $A = \varnothing, B = \varnothing$.

Study Examples 2 and 3 before working Exercises 9–20. List all subsets of each of the following sets.

    9. $\{1, 4\}$.   **10.** $\{7\}$.
    11. $\{1, 2, 3, 4\}$.   **12.** $\{0, 1\}$.
    13. $\{0\}$.   **14.** $\{a, b\}$.
    15. $\{a, b, c, d\}$.   **16.** $\{s, 0, 7\}$.
    17. $\{a, \{a\}\}$.   **18.** $\{0, \{0\}\}$.
    19. $\{\{0\}, \{\{0\}\}\}$.   **20.** $\{0, \{0\}, \{\{0\}\}\}$.

    21. Prove that if A is any set, then the proposition $\varnothing \subseteq A$

is true. [*Hint:* Justify that the implication $x \in \varnothing \rightarrow x \in A$ is true.]

22. Let $A = \{1, 2, 3\}$ and let B be the set of all subsets of A. Determine which of the following are true, stating why in each case: (a) $\varnothing \in A$; (b) $\varnothing \subset A$; (c) $\varnothing \subset B$; (d) $\varnothing \in B$; (e) $1 \in A$; (f) $1 \in B$; (g) $\{A\} \in B$; (h) $A \subset B$; (i) $\{1, 2\} \in A$; (j) $\{1, 2\} \in B$.

23. Let $A = \{a, b, c\}$. List all those proper subsets of A which contain a.

2.5 MORE ON OPEN SENTENCES (OPTIONAL)

Suppose the set A is the replacement set for some open sentence $p(x)$. We should like to consider statements such as the following:

> For all x in the set A, $p(x)$ is true.
> For some x in the set A, $p(x)$ is true.

Very often these two sentences are abbreviated:

> For all $x \in A$, $p(x)$.
> For some $x \in A$, $p(x)$.

We agree that the first statement is true, if the solution set of the open sentence $p(x)$ is A, and false otherwise. We also agree that the second statement is true, if the solution set is nonempty.

Example 1. Let A be the set of all residents of the United States. Further, let $p(x)$ be an abbreviation for the open sentence, "x is a citizen of the United States." Then $\sim p(x)$ should be an abbreviation for "x is not a citizen of the United States." Assume that A is the replacement set for both $p(x)$ and $\sim p(x)$. Consider the following two statements:

> (1) For all $x \in A$, $p(x)$.
> (2) For some $x \in A$, $\sim p(x)$.

Using the idea illustrated in Example 5, Section 1.3, we see that statement (2) is the negation of statement (1). Thus, we should expect one of them to be true and the other false.

Note that (1) is false, since the solution set of $p(x)$ is a proper subset of A. As an optional exercise, the student may verify the truth of our last claim by contacting any United States Immigration and Naturalization Office and obtaining the name of one alien residing in the United States. We remark that when this has been done, the truth of (2) has been verified.

In general, if an open sentence $q(x)$ is given with replacement set S, each of the following is a proposition:

(1) For all $x \in S$, $q(x)$.
(2) For some $x \in S$, $\sim q(x)$.
(3) For some $x \in S$, $q(x)$.
(4) For all $x \in S$, $\sim q(x)$.

Question. Why are these propositions?

We further remark that each of the first two propositions is the negation of the other.

Example 2. Let $T = \{1, 2, 3, 4, 5, 6, 7, 8\}$ and let $r(x)$ abbreviate the open sentence "x is less than 4." Then if T is the replacement set, the solution set of $r(x)$ is $\{1, 2, 3\}$. It happens to be a proper subset of T. Thus, "For all $x \in T$, $r(x)$" is false, whereas "For some $x \in T$, $\sim r(x)$" is true, since, for example, $7 \in T$ but 7 is not less than 4.

Exercise 2.5

Study Example 1 before working Exercises 1–2.

1. Let A be the set of all United States citizens, and let $p(x)$ be the abbreviation for the following open sentences:

(a) "x is a college teacher."
(b) "x is a carpenter."
(c) "x is a woman."
(d) "x is an artist."
(e) "x is an alcoholic."

Assume that A is the replacement set for $p(x)$ in each of the foregoing five cases. In each case, what does $\sim p(x)$ abbreviate? Also in each case, determine the truth value of the propositions "For all $x \in A$, $p(x)$" and "For some $x \in A$, $\sim p(x)$."

2. Repeat Exercise 1 with the set A as the set of all animals and $p(x)$ the abbreviation for the following open sentences: (a) "x is a dog." (b) "x is a bird." (c) "x is not a horse." (d) "x is an elephant." (e) "x is a sick cow."

Study Example 2 before working Exercises 3–5.

3. Let A be the set of counting numbers and

$$B = \{x \in A \mid 1 < x < 12\}.$$

Let $p(x)$ abbreviate the open sentence "x is greater than or equal to 9." What is the solution set of $p(x)$ if the replacement set is B? What is the solution set of $\sim p(x)$ with the same replacement set?

4. Repeat Exercise 3 with $B = \{3, 7, 10, 15, 19\}$ and $p(x)$ abbreviating "x is an even integer."

5. Repeat Exercise 3 with B the set of counting numbers and $p(x)$ abbreviating "x is odd and x is between 6 and 14."

6. Repeat Exercise 3 with $B = \{1, 2, 3, 4, 5, 6, 7\}$ and $p(x)$ abbreviating "x is greater than 3 and less than 4."

7. Repeat Exercise 3 with B the set of counting numbers and $p(x)$ abbreviating "x is odd."

2.6 VENN DIAGRAMS AND COMPLEMENTS

Often it is helpful to illustrate pictorially some notions about
Venn diagram sets. A convenient way of doing this is with the aid of a *Venn diagram*.

In a Venn diagram a set is represented by the collection of points inside a simple closed curve, as illustrated in Figure 2.1.

Representation of a set A

Figure 2.1

As another illustration, Figure 2.2 represents three sets A, B, and C, where A and C have no point in common. We see that $B \subset A$.

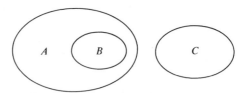

Figure 2.2

We are now ready to introduce the idea of the complement of a set relative to some other set.

Suppose $q(x)$ represents an open sentence whose replacement set is denoted by A. In the preceding section we showed that the solution set of $q(x)$ is in fact a subset of A, which we call B, as illustrated in Figure 2.3. Thus $b \in B$ if and only if $q(b)$ is true.

Now consider the open sentence $\sim q(x)$ (the negation of $q(x)$) with the same replacement set A.

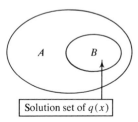

Solution set of $q(x)$

Figure 2.3

It is clear that the solution set of the sentence $\sim q(x)$ is the set of all members of A that do not belong to B. To see this, we note that if $a \in A$ and $\sim q(a)$ is true, then $q(a)$ must be false and, therefore, a is not a solution of $q(x)$. On the other hand, if a is in A and not in B, then $q(a)$ is false and hence $\sim q(a)$ is true.

More generally consider a set S and let T be a subset of S. Then if $q(x)$ denotes the open sentence "$x \in T$" with replacement set S, $\sim q(x)$ denotes "$x \notin T$." We remark that the solution set of the sentence $x \in T$ is indeed T. The solution set of $x \notin T$ (relative to the replacement set S) is called the *complement of T relative to S*. It is often denoted by $S - T$. The idea is represented pictorially in Figure 2.4.

complement

S

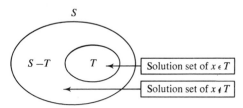

$S - T$ T Solution set of $x \in T$

Solution set of $x \notin T$

Figure 2.4

Formally we give the following:

Definition. If A and B are sets with $B \subseteq A$, the set $\{x \in A \mid x \notin B\}$ is called *the complement of B relative to A* and is denoted $A - B$.

Example 1. Let $A = \{1, 2, 3, 4, 5, 6, 7\}$ and let $q(x)$ be the open sentence

$$(x - 1) \cdot (x - 3) \cdot (x - 7) = 0.$$

Then the solution set of $q(x)$ relative to A is the set $B = \{1, 3, 7\}$ and is clearly a subset of A. The complement of B relative to A is the set $A - B = \{2, 4, 5, 6\}$.

Example 2. Let T be a set. The complement of T relative to T is the empty set $\varnothing$. To see this, note that by definition $T - T = \{x \mid x \notin T$ and $x \in T\}$. Now note that the open sentence "$x \notin T$ and $x \in T$" will always become a false proposition when x is replaced by the name of a member of T. Thus its solution set must be empty.

Question: Let X be a set. We know that $\varnothing \subseteq X$ (since the empty set is a subset of any set). What is the complement of $\varnothing$ relative to X?

Example 3. Let A be the set of all American teenagers and let $B = \{x \in A \mid x$ is an American female teenager$\}$. Then $A - B$ is clearly the set of all American male teenagers.

A concept entirely analogous to the foregoing is introduced as follows. Suppose that A, B are sets. Then the solution set of the open sentence $x \notin B$ using A as replacement set is denoted *set* $A - B$ and is called the *set difference* of A and B. (See Figure 2.5.) *difference* Note that if $B \subseteq A$, then the set difference $A - B$ is simply the complement of B relative to A.

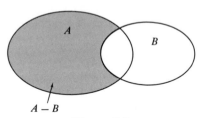

$A - B$

Figure 2.5

Example 4. Let $A = \{1, 2, 3, 4, 5\}$ and $B = \{1, 3\}$. Then $A - B = \{x \in A \mid x \notin B\} = \{2, 4, 5\}$ and $B - A = \{x \in B \mid x \notin A\} = \varnothing$.

Example 5. Let $A = \{1, 3, 5, 7, 9\}$ and $B = \{2, 4, 6\}$. Then $A - B = \{x \in A \mid x \notin B\} = \{1, 3, 5, 7, 9\}$ and $B - A = \{2, 4, 6\}$.

Exercise 2.6

Study Examples 1 and 2 before doing Exercises 1–5. In these, a set A is described and an open sentence $q(x)$ is given. In each case find the solution set B of that open sentence and also the complement of B relative to A.

1. $A = \{2, 4, 6, 8, 10, 12, 14\}$; $(x - 3) \cdot (x - 2) \cdot (x - 12) = 0$.

2. $A = \{-3, -1, 0, 1, 3, 5\}$; $x^2 - 1 = 0$.

3. $A = \{x \in C \mid (x - 2) \cdot (x - 3) \cdot (x - 5) = 0\}$, where C is the set of counting numbers; $(x + 1) \cdot (2x - 4) = 0$.

4. A is the set of odd positive integers; $(x - 2) \cdot (x - 4) = 0$.

5. $A = \{y \in C \mid (y - 1) \cdot (y - 2) \cdot (y - 3) = 0\}$, where C is the set of counting numbers; x is less than 3.

Study Example 3 before doing Exercises 6–10. In these, sets A and B are given. In each case, describe $A - B$.

6. A is the set of counting numbers; $B = \{x \in A \mid x \text{ is odd}\}$.

7. A is the set of authors of this text; $B = \{\text{Zemgalis, Mar}\}$.

8. A is the set of American citizens; $B = \{x \in A \mid x \text{ is a male}\}$.

9. $A = \{1, 2, 3, 4, 5\}$; $B = \varnothing$.

10. $A = \{1, 2, 3, 4\}$; $B = \{y \in C \mid (y - 1) \cdot (y - 2) \cdot (y - 3) \cdot (y - 4) = 0\}$, where C is the set of counting numbers.

11. What can you assert about the sets A and B if $B \subseteq A$ and $A - B = \varnothing$? Justify your answer.

12. What can you assert about the set A if $A - A = A$? Justify your answer.

Read Examples 4 and 5 before doing Exercises 13–17.

13. Let $A = \{-1, 3, 5, 7, 11, 25\}$ and $B = \{3, 5, 25, 36\}$. Find $A - B$ and $B - A$.

14. Let $A = \{-6, -4, -3, 0, 1, 2\}$ and $B = \{0, 2, 7, 10\}$. Find $A - B$ and $B - A$.

15. Let $A = \{1, 3, 7, 8, 13\}$, $B = \varnothing$. Find $A - B$ and $B - A$.

16. Let $A = \{3, 5, 7, 9, 11\}$ and $B = \{2, 4, 6, 8, 10, 12\}$. Find $A - B$ and $B - A$.

17. Suppose A and B are sets. Under what conditions can you be sure that $A - B = A$?

2.7 INTERSECTION AND UNION OF SETS

Recall that in Chapter 1 we combined propositions p, q to obtain the new propositions $p \wedge q$, $p \vee q$. Similarly we now consider open sentences $p(x), q(x)$, and we wish to combine them to obtain new open sentences. For convenience, let U be the re-

universal set placement set for all open sentences under discussion. Frequently, such a set is called the *universal set* for the discussion. A member a of U will be a solution of the sentence $p(x) \land q(x)$ if $p(a) \land q(a)$ is true — that is, if both $p(a), q(a)$ are true, which means that a is a solution of both $p(x), q(x)$.

If A, B denote, respectively, the solution sets of $p(x), q(x)$, then the solution set of the open sentence $p(x) \land q(x)$ is called *intersection* the *intersection* of the sets A, B and is denoted by $A \cap B$.

Formally we have $\{x \mid p(x)\} \cap \{x \mid q(x)\} = \{x \mid p(x) \land q(x)\}$.

We remark that the intersection of any two subsets of U has been defined. To see this, note that if S and T are any two subsets of U, then S is the solution set of "$x \in S$." Similarly, $T = \{x \mid x \in T\}$. Thus, the intersection of the sets S, T is given by

$$S \cap T = \{x \mid x \in S \land x \in T\}.$$

Example 1. Let the universe U be the set of all counting numbers $\{1, 2, 3, \cdots\}$ and suppose $A = \{3, 8, 17, 24\}$ and $B = \{1, 8, 10, 13, 24, 37\}$. Then

$$A \cap B = \{x \mid x \in A \land x \in B\}$$

$$= \{8, 24\}.$$

To see this note that $8 \in A \land 8 \in B$ is true (since both $8 \in A$, $8 \in B$ are true), and also $24 \in A \land 24 \in B$ is true. Obviously, $8, 24$ are the only solutions of the sentence $x \in A \land x \in B$. Hence, the set $\{8, 24\}$ is the solution set of that sentence.

Example 2. Use a Venn diagram to represent the intersection of two sets A, B.

Solution: In Figure 2.6 the shaded area represents the intersection of two sets A, B.

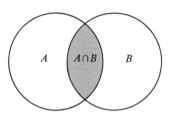

Figure 2.6

In Figure 2.7 we illustrate that the intersection may be the empty set in the case A, B have no elements in common.

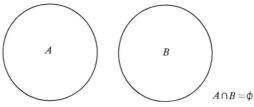

Figure 2.7

Question: What is $A \cap B$ if $A \subseteq B$? (Recall that $A \subseteq B$ means that $A \subset B$ or $A = B$.)

union
Again, let A, B denote, respectively, the solution sets of $p(x), q(x)$ relative to the replacement set U. The *union* of the sets A, B is the solution set of $p(x) \lor q(x)$. It is denoted by $A \cup B$. Symbolically,

$$\{x \mid p(x)\} \cup \{x \mid q(x)\} = \{x \mid p(x) \lor q(x)\}.$$

As we did in the case of intersection, we remark that if $S \subseteq U$ and $T \subseteq U$, then S, T are, respectively, the solution sets of "$x \in S$", "$x \in T$." Thus,

$$S \cup T = \{x \mid x \in S \lor x \in T\}.$$

Example 3. If A, B, U are as in Example 1, then

$$A \cup B = \{x \mid x \in A \lor x \in B\}$$

$$= \{1, 3, 8, 10, 13, 17, 24, 37\}.$$

To see this we note that $1 \in A \lor 1 \in B$ is true since $1 \in B$ is true. Thus 1 is a solution of the open sentence $x \in A \lor x \in B$, and therefore $1 \in A \cup B$. Similarly, we can verify that $3, 8, 10, 13, 17, 24, 37$ are also solutions of the sentence $x \in A \lor x \in B$, and we conclude that $A \cup B$ is the set described above.

Exercise 2.7

Study Examples 1 and 3 before working Exercises 1–10. In each of these, the universe is the set of counting numbers and two sets A and B are given. In each case find (a) $A \cap B$, (b) $A \cup B$, (c) $A - (A \cap B)$, (d) $B - (A \cap B)$.

    1. $A = \{1, 2, 3, 4, 5, 6\}$; $B = \{3, 6, 10, 12\}$.
    2. $A = \{x \mid (x - 1)(x - 2)(x - 4) = 0\}$;
       $B = \{y \mid (y - 4)(y - 5) = 0\}$.

3. $A = \{x \mid x \text{ is odd}\}$; $B = \{x \mid x \text{ is even}\}$.

4. $A = \{x \mid x \text{ is between 3 and 12}\}$;
   $B = \{u \mid u \text{ is between 10 and 17}\}$.

5. $A = \varnothing$; $B = \{1, 3, 5\}$.

6. $A = \varnothing$; $B = \{12, 13, 15, 17\}$.

7. $A = \{x \mid (2x - 1)(3x - 6) = 0\}$;
   $B = \{x \mid (x - 2)(x - 5) = 0\}$.

8. $A = \{x \mid x^2 \text{ is less than 10}\}$; $B = \{x \mid x^3 \text{ is less than 25}\}$.

9. $A = \{t \mid t^4 - 16 = 0\}$; $B = \{2, 3, 4\}$.

10. $A = \{x \mid x \text{ is divisible by 3}\}$; $B = \{1, 2, 3, 4, 5, 6, 7\}$.

Study Example 2 before working Exercises 11–18. In these, A, B, and C are sets. Use a Venn diagram to represent the sets described in terms of A, B, C.

11. $A - (A \cap B)$.

12. $B - (A \cap B)$.

13. $A \cap (B \cup C)$.

14. $(A \cap B) \cup (A \cap C)$.

15. $A \cup (B \cap C)$.

16. $(A \cup B) \cap (A \cup C)$.

17. $A \cap B \cap C$.

18. $A \cup B \cup C$.

19. Let A and B be sets such that (1) $A \cup B = \{a, b, c, d\}$, (2) $A \cap B = \{a, c\}$, (3) $A - B = \{b\}$. Determine A and B.

20. What must be true about the sets A and B, if (a) $A \cup B \subseteq B$, (b) $A \cup B \subseteq A$, (c) $A \cup B \subseteq B$ and $A \cup B \subseteq A$? Explain your conclusions with appropriate Venn diagrams.

21. Among 49 students taking college entrance exams in English and mathematics, 25 passed English and 22 passed mathematics, while 10 students passed both English and mathematics. (a) How many students passed at least one of the exams? (b) How many students failed in both exams? (c) How many students passed one and only one exam?

∗22. Five hundred people were studied and the following data reported: 250 are male, 250 are teachers, 200 are married, 100 are male teachers, 40 males are married, 150 teachers are married, and 30 male teachers are married. Draw an appropriate Venn diagram and determine the number of unmarried females who are not teachers.

∗23. What can you conclude about the sets A and B if $A \cap B = A \cup B$? [*Hint:* Use a Venn diagram.]

REVIEW EXERCISES

In Exercises 1–3, some objects are described. In each case, describe the set whose members are these objects by listing all the members of each set.

1. All the states of the United States whose names begin with the letter W.

2. The first four letters of the alphabet.

3. All Presidents of the United States who preceded George Washington.

4. Find the solutions of the open sentence "x is less than 6" if the replacement set for the variable x is **(a)** $\{7, 8\}$, **(b)** $\{5, 6, 7, 8\}$, **(c)** $\{1, 2, 3, 4, 5, 6, 7, 8\}$.

5. Find the solution set for the open sentence "$3x + 2 = 8$" if the replacement set for the variable x is **(a)** $\{4, 6, 8\}$, **(b)** $\{1, 4, 6, 8\}$, **(c)** $\{2, 4, 6, 8\}$.

6. Find the solution set for the open sentence "$(x - 2)(x - 3) = 2$" if the replacement set is **(a)** $\{1, 2, 3\}$, **(b)** $\{1, 2, 5\}$, **(c)** $\{1, 2, 3, 4, 5\}$.

7. Let S be the set whose members are all the subsets of the set $\{1, 2\}$. **(a)** List all the members of S. **(b)** Determine the truth value of each of the following statements:

$$S \subseteq S, \quad S \subset S, \quad S \in S, \quad \varnothing \in S, \quad \{\varnothing\} \in S,$$

$$1 \in S, \quad \{1\} \in S, \quad \{\{1, 2\}\} \subset S, \quad \{1, 2\} \subset S.$$

Justify your answers.

8. If $S = \{1, 2, 4\}$ and $g(x)$ is the open sentence "$x^2 \in S$," determine the truth value of each of the following propositions: **(a)** for all $x \in S$, $g(x)$, **(b)** for some $x \in S$, $\sim g(x)$, **(c)** for some $x \in S$, $g(x)$, **(d)** for all $x \in S$, $\sim g(x)$.

In Exercises 9–20, $A = \{1, 2, 3, 4, 5\}$, $B = \{1, 3, 5\}$, $C = \{2, 3, 5\}$. Give an explicit description of each set.

9. $A \cup B \cup C$.	**10.** $(A \cap B) \cup C$.
11. $(A - B) \cup C$.	**12.** $(A - C) \cup B$.
13. $(B \cup C) - B$.	**14.** $(B \cup C) - C$.
15. $A \cap B \cap C$.	**16.** $B - \{1, 3\}$.
17. $(B - \{1, 3\}) \cup C$.	**18.** $(C - \{2, 3\}) \cap B$.
19. $(A - \{1, 2\}) \cap B$.	**20.** $(B - \{1, 5\}) \cup C$.

21. Find two subsets A and B of $\{a, b, c\}$ such that $A \neq \varnothing$, $B \neq \varnothing$, $A \cap B = \varnothing$, and $\{a, b, c\} = A \cup B$.

22. Give an example of a set having two members, such that each member of the set is also a subset of the set.

23. Show that if A, B are sets, the proposition "$A \subset B \rightarrow A \subseteq B$" is true.

THE
INTEGERS

3.1 BINARY OPERATIONS

The student, undoubtedly, is familiar with the term "operation" from his earlier training. For example, he has performed arithmetic with the counting numbers, using the operations addition and multiplication, and has learned how to subtract one integer from another.

Consider for a moment the operation subtraction. We make the following observations:

(1) If a and b are integers, we can combine them to get the difference $a - b$ under the well-known rule called subtraction.

(2) If a and b are integers, the difference $a - b$ is an integer.

(3) If we calculate the difference $a - b$ twice and obtain two distinct integers, we conclude that at least one of our calculations must contain an error, since we know that $a - b$ represents one and only one integer.

(4) The sentence "combine the integers 4 and 3 by performing subtraction" is ambiguous, since it does not specify which of the two differences $4 - 3$ or $3 - 4$ is intended. Obviously $4 - 3 \neq 3 - 4$. This ambiguity can be removed if we agree to the interpretation that the integer that was mentioned second (namely 3) is to be subtracted from the integer mentioned first (namely 4).

(5) Unless some definite convention is adopted, the expres-

sion $a - b - c$, where a, b, c are integers, does not make sense. Note that we can combine only two integers at a time under the rule called subtraction, and therefore a combination of three integers at a time is undefined.

The usual convention, which we adopt, is that $a - b - c = (a - b) - c$. Observe that the expression on the right does make sense, since it directs us to combine a and b first and then combine $a - b$ and c.

We are now ready to give the following definition:

binary **Definition.** A *binary operation* on a nonempty set S is a rule that
operation assigns to each pair a, b (selected in that order) of elements in S, one and only one element in S.† If $*$ denotes a binary operation on a set S and a, b are sequentially selected elements, we write $a * b$ to denote the corresponding unique element in S.

Remark. If $*$ is a binary operation on a set S and a, b, c are members of S, then

$$a = b \rightarrow a * c = b * c.$$

Similarly,

$$a = b \rightarrow c * a = c * b.$$

These facts will be used frequently.

It will be shown in the following section that it is possible for $a = b$ to be true and $a * c = c * b$ to be false.

Often when we have a pair of elements a, b of some set and we want to indicate that a is first and b second, we call this pair
ordered an *ordered pair*, and we denote it by (a, b). We also consider
pair (a, a) an ordered pair, thus allowing the selection of the same element twice in this context.

Since all the operations treated in this course will be binary
operation operations, we shall henceforth use the term *operation* instead of binary operation.

Example 1. Let I be the set of integers — that is,

$$I = \{ \cdots, -3, -2, -1, 0, 1, 2, 3, \cdots \}.$$

Subtraction (symbolically "$-$") is indeed an operation on I, in the sense of the foregoing definition. That is, for any two integers a, b (a first, b next), one and only one integer, commonly denoted by $a - b$, is assigned to the ordered pair (a, b).

closed † In some texts the fact that if $a \in S \wedge b \in S$ then $a * b \in S$ is expressed by saying that $*$ is *closed*.

For example, to the ordered pair $(7, 4)$ subtraction assigns the integer 3, since $7 - 4 = 3$.

Exercise

Explain why $+$ (ordinary addition) and $\cdot$ (ordinary multiplication) are operations on the set of counting numbers.

Example 2. Let $T = \{0, 1\}$. The rule $\oplus$ is defined by the following table:

$\oplus$	0	1
0	0	1
1	1	0

To find $x \oplus y$ we locate the row preceded by x and the column headed by y, and where the two intersect we find $x \oplus y$. For example, $1 \oplus 1 = 0$. Note that $\oplus$ is an operation on T.

Example 3. Let $S = \{a, b, c\}$. We define the rule $\odot$ by the following table, to be read as explained in Example 2.

$\odot$	a	b	c
a	a	b	c
b	b	c	b
c	c	b	a

The table shows that $\odot$ is an operation on S. We also see that

$$a \odot b = b \odot a,$$

$$a \odot c = c \odot a,$$

$$b \odot c = c \odot b;$$

hence for every pair x, y in S

$$x \odot y = y \odot x.$$

Note that $c \odot (b \odot b) = c \odot c = a$ and $(c \odot b) \odot b = b \odot b$ $= c$, hence

$$c \odot (b \odot b) \neq (c \odot b) \odot b.$$

Example 4. Let N be the set of counting numbers and define the rule $*$ as follows: If a, b are in N, $a * b = a$. Thus, for example, $2 * 4 = 2$.

Note that $*$ is an operation on N, since it assigns to each ordered pair (a, b) of elements in N, one and only one element in N, namely a.

If we choose the elements a, b, c, in that order in N and compute $(a * b) * c$ and $a * (b * c)$, we find that $(a * b) * c = a * c = a$, and $a * (b * c) = a * b = a$. Thus, $(a * b) * c = a * (b * c)$ for any choice of the three elements a, b, c in N.

On the other hand, it is interesting to note that $1 * 2 \neq 2 * 1$, since $1 * 2 = 1$ but $2 * 1 = 2$.

Example 5. Let A be the set of odd integers and let $+$ denote ordinary addition. Then $+$ is not an operation on A, since, for example, $5 \in A$ (because 5 is odd), $9 \in A$ (since 9 is odd), but $5 + 9 = 14$ is not in A, since 14 is even.

Example 6. Let I be the set of integers and let

$$a * b = \begin{cases} a & \text{if } a \text{ is even,} \\ b & \text{if } b \text{ is even.} \end{cases}$$

For example, $4 * 3 = 4$ (since 4 is even) and $5 * 8 = 8$ (since 8 is even). We see that for any ordered pair (a, b) of integers, $a * b$ is an integer. However, $*$ is not an operation, since $a * b$ is not uniquely defined in the case that both a and b are even. For example, $6 * 8$ can be either 6 or 8.

Thus, if we are given a set S and a rule $*$ that associates with ordered pairs of elements of S some other elements, and we want to show that $*$ is not an operation, we shall have succeeded if we can do any one of the following three things:

(1) Find an ordered pair (a, b) of elements of S for which $*$ does not assign a corresponding element.
(2) Find an ordered pair (a, b) of elements of S for which $a * b$ is not in S.
(3) Find an ordered pair (a, b) of elements of S for which the rule $*$ assigns more than one element.

Exercise 3.1

Read carefully the definition of a binary operation on a nonempty set and also Example 1. Then answer the questions in Exercises 1–6.

1. Let $I = \{\cdots, -3, -2, -1, 0, 1, 2, \cdots\}$. Is multiplication an operation on I?

2. Let I be as in Exercise 1. Is division an operation on I?

3. Let A be the set of even integers. Is addition an operation on A?

4. Let A be as in Exercise 3. Is multiplication an operation on A?

5. Let A be as in Exercise 3. Is subtraction an operation on A?

6. Let A be as in Exercise 3. Is division an operation on A?

Read Examples 2 and 4 before doing Exercises 7–9.

7. Let $S = \{a, b, c\}$. The rule $\oplus$ is defined by the following table:

$\oplus$	a	b	c
a	c	a	b
b	a	b	c
c	b	c	a

Is $\oplus$ an operation on S?

8. Let $T = \{x, y, z\}$. The rule $\odot$ is defined by the following table.

$\odot$	x	y	z
x	y	x	z
y	x	y	z
z	z	z	x

Is $\odot$ an operation on T?

9. Let $A = \{a, b\}$. The rule $\oplus$ is defined as follows:

$\oplus$	a	b
a	a	c
b	c	b

Is $\oplus$ an operation on A?

Read Examples 4 and 5 before doing Exercises 10–12.

10. Let N be the set of counting numbers and define the rule $\oplus$ as follows: If a, b are in N, $a \oplus b = b$. (a) Find $3 \oplus 2$. (b) Is $\oplus$ an operation on N?

11. Let A be the set of odd integers and let $\odot$ be defined as follows: If x, y are in A, then

$$x \odot y = \frac{x + y}{2},$$

where $+$ denotes ordinary addition. (a) Find $3 \odot 5$. (b) Is $\odot$ an operation on A?

12. Let A be as in Exercise 11. Is subtraction an operation on A?

3.2 GENERAL PROPERTIES OF OPERATIONS

Given an operation, it is natural to investigate its properties. Knowledge of those properties will make learning easier and will help to eliminate errors in computation. On the other hand, the absence of certain properties is also significant: it alerts the student to exercise proper care when dealing with the given operation.

Referring to the four operations discussed in Examples 1 to 4 in Section 3.1, let us ask the question: "What are some of the major similarities and dissimilarities among them?" Rather than answering this question, we prefer to discuss properties of operations in general, so that the interested student may be able to satisfy his curiosity after having studied the present section.

It is common knowledge that when two tasks are performed, the ordering of the tasks may affect the end result. For example, if a gourmet broils a steak and then eats it, this may constitute an enjoyable experience for him. He would not think of reversing the ordering of the tasks here. On the other hand, putting on a coat and then a hat may provide the same protection against the weather as would putting on the hat first and then the coat.

Thinking along similar lines, an operation $*$ on a set S may or may not assign the same element to each of the ordered pairs (a, b) and (b, a) of elements in S.

commutative If $a * b = b * a$ for all a, b in S, then we say that the operation $*$ is *commutative*. If $*$ is an operation on S that is not commutative (that is, there are at least two elements a, b in S such *non-* that $a * b \neq b * a$), then $*$ is said to be *noncommutative*. *commutative* Consulting the examples in Section 3.1, we see that in both Example 2 and Example 3 we have commutative operations. On the other hand, Examples 1 and 4 illustrate noncommutative operations. The student should refer to these examples to verify our claims.

Another property that is relevant to the study of operations *associative* is associativity. An operation $*$ on a set S is said to be *associative* if for all triples a, b, c of elements in S, $(a * b) * c = a * (b * c)$. If $*$ is an operation on S that is not associative (that is, there is at least one triple a, b, c of elements in S such that $(a * b) * c$ *non-* $\neq a * (b * c)$), then we say that $*$ is *nonassociative*. *associative* The operations in Examples 2 and 4 in Section 3.1 are associative, while the operations in Examples 1 and 3 in the same section are nonassociative. We urge the student to study these examples again.

From his experience with integers the student should know that $a + 0 = 0 + a = a$ for every integer a. This shows that the number 0 enjoys a rather unique property relative to the operation $+$.

In general, if $*$ is an operation on a set S and if there is an element e in S such that $s * e = e * s = s$ for every s in S, then *identity* e is called the *identity element* relative to the operation $*$.

element We have already mentioned one example of an identity element — namely, the integer 0 is the identity element relative to the *additive* operation $+$ on the set of integers. We often call 0 the *additive* *identity* *identity*.

Note that we say "the identity element," rather than "an identity element." It can be shown that for every operation there is at most one identity element.

Example 1. Show that there is no identity element relative to the operation $*$ in Example 4, Section 3.1.

Solution: Recall that $*$ is an operation defined on the set N of counting numbers by the rule: If a, b are in N, $a * b = a$.

Consider the open sentence $(x * 1 = 1) \wedge (x * 2 = 2)$, with replacement set N.

Since by definition of $*$, $a * 1 = a$ for all $a \in N$, we see that the only solution of $x * 1 = 1$ is 1. Similarly, $b * 2 = b$ for all $b \in N$ and therefore the only solution of $x * 2 = 2$ is 2.

Hence,

$$\{x \mid (x * 1 = 1) \wedge (x * 2 = 2)\} = \{x \mid x * 1 = 1\} \cap \{x \mid x * 2 = 2\}$$

$$= \{1\} \cap \{2\}$$

$$= \varnothing.$$

Therefore, $*$ cannot have an identity e, since otherwise e would be a solution to the open sentence $(x * 1 = 1) \wedge (x * 2 = 2)$. But we know this open sentence has no solution.

Example 2. Consider the operation $\odot$ as defined in Example 3, Section 3.1. From the definition of $\odot$ we see that $x \odot a = a \odot x = x$ for every x in S. Thus a is the identity element relative to the operation $\odot$.

Example 3. Consider the operation in Example 2, Section 3.1. Recall that

$$0 \oplus 0 = 0,$$

$$1 \oplus 1 = 0,$$

$$1 \oplus 0 = 0 \oplus 1 = 1.$$

Thus 0 is the identity element relative to the operation $\oplus$. Also note that for every $x \in \{0, 1\}$ there is a $y \in \{0, 1\}$ such that $x \oplus y = 0$.

The last example illustrates a special case of the following.

Suppose $*$ is an operation on a set S and e is the identity element relative to $*$. If $a \in S$ and there is an element $b \in S$ such
inverse that $a * b = b * a = e$, then b is called an *inverse of a relative to the operation* $*$. If $*$ is the only operation defined on S, we may use the shorter form: "b is an inverse of a."

Example 4. Let $U = \{0, 1, 2\}$ and define the operation $\otimes$ on U as follows:

$\otimes$	0	1	2
0	0	1	2
1	1	0	0
2	2	0	0

The table shows that 0 is the identity element relative to $\otimes$. Note that

$$1 \otimes 1 = 0 \quad \text{and} \quad 1 \otimes 2 = 2 \otimes 1 = 0,$$

also

$$2 \otimes 2 = 0.$$

We have now an unusual situation: the element 1 has two distinct inverses, namely 1 and 2. Further, the element 2 also has two distinct inverses.

We see now why it would be incorrect in general to use the phrase "b is the inverse of a."

If, however, we have an associative operation $*$ on a set S and an element c in S has an inverse b, then it can be shown that c cannot have more than one inverse. Accordingly, in this case it would be correct to use the phrase "b is the inverse of c."

In summary, we have introduced the following: If S is a nonempty set and $*$ is an operation on S, then

(a) $*$ is *commutative* provided $a * b = b * a$ for all elements a, b in S;

(b) $*$ is *associative* if $(a * b) * c = a * (b * c)$ for all a, b, c in S;

(c) $*$ has an *identity* e provided $a * e = e * a = a$ for each $a \in S$.

Further, if $*$ has an identity e, $a \in S$, and there is an element

$b \in S$ such that $a * b = b * a = e$, then b is called an *inverse* of a, which is unique in the case $*$ is associative.

Exercise 3.2

Read the examples and study the summary at the end of Section 3.2 before doing Exercises 1–6.

1. Let S and $\oplus$ be as in Exercise 7 of the preceding section. (a) Is $\oplus$ commutative? (b) Is $\oplus$ associative? (c) Does $\oplus$ have an identity?

2. Repeat Exercise 1 for T and $\odot$ defined as in Exercise 8 of the preceding section.

3. Repeat Exercise 1 for N and $\oplus$ defined as in Exercise 10 of the preceding section.

4. If $\oplus$ of Exercise 1 has an identity, find an inverse of a.

5. If $\odot$ of Exercise 2 has an identity, find an inverse of x. Does z have an inverse?

6. Does the operation $\oplus$ described in Exercise 10 of the preceding section have an identity?

7. Prove that if $*$ is an operation on S, then $*$ has at most one identity element. [*Hint:* Show that if e_1 and e_2 are identity elements, then $e_1 = e_2$.]

8. Let $*$ be an associative operation on S with the identity element e. Prove that each element in S has at most one inverse relative to the operation $*$.

3.3 THE INTEGERS

The student is familiar with the set of counting numbers $\{1, 2, 3, \cdots\}$, also called the set of *natural numbers*. In school arithmetic he learned to add and multiply such numbers. Thus, there are two basic operations defined on that set. Do these operations have the properties listed in the preceding section?

First let us consider addition. If N denotes the set of natural numbers, we recall that $a \in N$ and $b \in N$ implies $a + b = b + a$. For example, $3 + 8 = 8 + 3$. Also for any three natural numbers, a, b, c, it is true that $a + (b + c) = (a + b) + c$. That is, addition is both commutative and associative. However, for no natural number a can we find a natural number e such that $a + e = a$. Hence, there is no identity element for addition. Thus, we wish to extend the set of natural numbers as follows:

natural numbers

We consider a new set, I, of numbers, which we call the set of integers, and which has the following properties:

(1) The set of natural numbers N is a proper subset of I.

(2) There is an addition and a multiplication defined on I such that these operations agree with the addition and multiplication defined on N.

zero

additive inverse

(3) Addition has the four basic properties listed in the summary of the previous section. That is, addition is commutative and associative. Further, it has an identity, which we call the number *zero* and denote by the symbol 0. Finally, for each integer a there is a unique integer denoted by ^-a with the property $a + {}^-a = {}^-a + a = 0$. The integer ^-a is called the *additive inverse* of a.

(4) I is the smallest set of numbers with the foregoing properties. That is, if K is a set of numbers with the same properties, then $I \subseteq K$.

Remark. Using our notation, the additive inverse of b is denoted ^-b. Traditionally, this number is written $-b$. In the present chapter we wish to reserve the symbol "$-$" for the operation subtraction.

Example 1.

$$^-3 + 5 = {}^-3 + (3 + 2) \qquad \text{(since } 5 = 3 + 2\text{)}$$
$$= ({}^-3 + 3) + 2 \qquad \text{(since addition is associative)}$$
$$= 0 + 2 \qquad \text{(because } {}^-3 \text{ is the additive inverse of 3)}$$
$$= 2 \qquad \text{(since 0 is the additive identity).}$$

We are now ready to introduce the following simple but useful theorem:

Theorem. If a is an integer, then $^-({}^-a) = a$.

Proof: We know that $^-a + a = a + {}^-a = 0$, since ^-a is the additive inverse of a. But this last statement also tells us that a is the (unique) additive inverse of ^-a. Thus, $a = {}^-({}^-a)$.

As an example, we can write $^-({}^-1) = 1$, and $^-({}^-({}^-3)) = {}^-3$.

Now that we have stated the properties of addition, we are ready to consider multiplication. If a and b are integers, the *product* unique integer obtained by multiplying them is called the *product* of a and b, and is denoted by $a \cdot b$. Frequently, if no confusion arises, the dot may be omitted and we write ab rather than $a \cdot b$.

Multiplication is commutative and associative. It has an identity, namely the number 1. Note that there are some integers (for example, 2) that fail to have a multiplicative inverse in the set of integers. In fact, all integers, except 1 and $^-1$, fail to have multiplicative inverses. That is, if $a \in I - \{^-1, 1\}$, then there is no integer b such that $ab = 1$.

Since there are two operations on the integers, namely addition and multiplication, it is natural to ask whether or not they are related. The fact that they are is illustrated by the following example.

Example 2. A man has a job that pays $2 an hour. He works 6 hours the first day and 5 hours the second. His employer can calculate his wages in the following two different ways:

First, he may decide that the man worked $(6 + 5)$ hours for a total of 11 hours and, thus, earned $(2 \cdot 11)$ dollars = 22 dollars.

On the other hand, he may calculate that the man earned $(2 \cdot 6)$ dollars = 12 dollars the first day, and $(2 \cdot 5)$ dollars = 10 dollars the second day, for a total of $(12 + 10)$ dollars = 22 dollars.

We see in this example that

$$2 \cdot (6 + 5) = 2 \cdot 6 + 2 \cdot 5.$$

distributive This is a particular case of a general property, called the
property *distributive property of multiplication over addition* (also called the *distributive law*), which we now state.

For any three integers a, b, c (not necessarily distinct), it is true that

$$a \cdot (b + c) = a \cdot b + a \cdot c.$$

Exercise 3.3

Read Example 1 before doing Exercises 1–10. In each of these, simplify the given expression.

1. $^-6 + 7$.   2. $8 + ^-3$.
3. $11 + ^-7$.   4. $^-13 + 15$.
5. $8 + ^-9$.   6. $^-35 + 40$.
7. $125 + ^-13$.   8. $999 + ^-99$.
9. $^-8 + 8$.   10. $^-6 + (20 + ^-12)$.

In each of Exercises 11–20, state the law or definition that is illustrated by the given expression.

11. $x + y = y + x$ for all x, y.

12. $x \odot (y \odot z) = (x \odot y) \odot z$ for all x, y, z.
13. $2 + 6 = 6 + 2$.   14. $3 \cdot 5 = 5 \cdot 3$.
15. $2 \cdot (5 + 6) = 10 + 12$.   16. $6 + 0 = 6$.
17. $8 + {}^-8 = 0$.   18. $(3 + 5) + 2 = 3 + 7$.
19. $5 \cdot 1 = 5$.   20. ${}^-({}^-3) = 3$.

3.4 APPLICATIONS OF THE DISTRIBUTIVE LAW

In the preceding section we stated that for any integers a, b, and c,

$$a(b + c) = ab + ac.$$

Since multiplication is commutative, the distributive law can also be stated in the following way:

$$(b + c)a = ba + ca.$$

In the present section we show how the distributive law can be used to prove some useful results from basic algebra.

Example 1. Note that

$$1 + 1 = 2.$$

Therefore

$$(1 + 1)a = 2a, \quad \text{(why?)}$$

from which we get

$$1 \cdot a + 1 \cdot a = 2a. \quad \text{(why?)}$$

That is,

$$a + a = 2a. \quad \text{(why?)}$$

Similarly, starting with

$$1 + 2 = 3,$$

we get

$$(1 + 2)a = 3a \quad \text{(why?)}$$

and

$$1 \cdot a + 2 \cdot a = 3a. \quad \text{(why?)}$$

Therefore,

$$a + 2a = 3a. \quad \text{(why?)}$$

Since we have shown that $2a = a + a$, we get $a + (a + a) = 3a$.

Because addition is associative, it is customary not to write the parentheses in an expression such as the left side of the foregoing equality. Thus we write

$$a + a + a = 3a.$$

In general, if n is a natural number, we can show that

$$\underbrace{a + a + \cdots + a}_{n \text{ terms}} = na.$$

Recall also that if n is a natural number, and a is a number, then

$$\underbrace{a \cdot a \cdot \cdots \cdot a}_{n \text{ factors}} = a^n.$$

exponent
nth power

In the expression a^n, n is called the *exponent* and a^n is called the *nth power of a*.

Example 2. Show that for any integers a and b,

$$a^2 + 2ab + b^2 = (a + b)^2.$$

Solution:

$$a^2 + 2ab + b^2 = a^2 + ab + ab + b^2 \qquad \text{(since } 2ab = ab + ab)$$
$$= a(a + b) + (a + b)b \qquad \text{(by the distributive property)}$$
$$= (a + b)(a + b) \qquad \text{(by the distributive and commutative property)}$$
$$= (a + b)^2.$$

Theorem. If a is an integer, then $a0 = 0$.

Proof: Since a and 0 are integers and multiplication is an operation, $a \cdot 0$ is an integer, which we call b. We would like to show that $b = 0$.

Recalling that 0 is the additive identity, we know that $x + 0 = x$ for any integer x. In particular,

$$0 + 0 = 0.$$

Thus

$$a(0 + 0) = a0.$$

By the distributive law we get

$$a0 + a0 = a0.$$

Hence, $b + b = b$.

Since b is an integer, it has an additive inverse ^-b. We obtain

$$(b + b) + {}^-b = b + {}^-b.$$

Therefore,

$$b + (b + {}^-b) = 0.$$

That is,

$$b + 0 = 0.$$

Hence,

$$b = 0.$$

Since

$$b = a0,$$

$$a0 = 0.$$

The student should supply the reason for each of the foregoing five steps.

Example 3. Write $a^2 + 3ab + 2b^2$ as a product.

Solution:

$$
\begin{aligned}
a^2 + 3ab + 2b^2 &= a^2 + (ab + 2ab) + 2b^2 \\
&= (a^2 + ab) + (2ab + 2b^2) \\
&= (a^2 + ab) + (2ba + 2b^2) \\
&= a(a + b) + 2b(a + b) \\
&= (a + 2b)(a + b).
\end{aligned}
$$

Can you justify each step of the foregoing example by stating the relevant properties?

Remark. The technique illustrated in Example 3 is a special case *factoring* of a general procedure called *factoring*.

Exercise 3.4

Read Example 2 before working Exercises 1–8. In each of these, write the given expression as a perfect square.

1. $x^2 + 6x + 9$.   2. $x^2 + 8x + 16$.
3. $4x^2 + 12x + 9$.   4. $9y^2 + 24y + 16$.
5. $t^2 + 10t + 25$.   6. $w^2 + 16w + 64$.
7. $16x^2 + 8x + 1$.   8. $25x^2 + 60x + 36$.

Read Example 3 before working Exercises 9–16. In each of these, write the given expression as a product.

9. $x^2 + 4xy + 3y^2$.   10. $2a^2 + 3ab + b^2$.
11. $5x^2 + 6xy + y^2$.   12. $3s^2 + 4st + t^2$.
13. $2x^2 + 3xt + t^2$.   14. $6r^2 + 7rs + s^2$.
15. $8r^2 + 12rt + 4t^2$.   16. $a^2 + 7ab + 10b^2$.

17. Using the fact that $a + a + a = 3a$, prove that $a + a + a + a = 4a$. [*Hint:* Study Example 1 and use the equality $1 + 3 = 4$.]

18. (a) Explain why for any integer a, $1 \cdot a = a$.

(b) Suppose n is a natural number and a is an integer and

$$n \cdot a = \underbrace{a + \cdots + a.}_{n \text{ terms}}$$

Prove that then

$$(n + 1)a = \underbrace{a + \cdots + a.}_{(n + 1) \text{ terms}}$$

3.5 REVIEW OF THE ARITHMETIC OF INTEGERS

Recall that for the set I of integers, there exists a unique multiplicative identiy $1 \in I$. That is, for any $a \in I$, $a \cdot 1 = 1 \cdot a = a$. This fact will be used in the proof of the following theorem:

Theorem. If $a \in I$, then $^{-}a = (^{-}1)a$.

Proof: Since the additive inverse of a is unique, we need only show that $a + (^{-}1)a = 0$ for this equality allows us to conclude that $(^{-}1)a = {^{-}}a$. Now

$$a + (^{-}1)a = 1 \cdot a + (^{-}1)a \quad \text{(since 1 is the multiplicative identity)}$$

$$= (1 + {^{-}}1)a \quad \text{(by the distributive property)}$$

$$= 0 \cdot a \quad \text{(since } {^{-}}1 \text{ is the additive inverse of 1)}$$

$$= a \cdot 0 \quad \text{(why?)}$$

$$= 0 \quad \text{(by the theorem in the preceding section).}$$

As a consequence of this theorem we may write numbers such as $^{-}3$ and $^{-}7$ in the forms $(^{-}1)3$ and $(^{-}1)7$, respectively.

Theorem. For all $a, b \in I$, $a(^{-}b) = {^{-}}(ab) = (^{-}a)b$.

Proof:

$$a(^{-}b) = a \cdot [(^{-}1)b] \quad \text{(by the preceding theorem)}$$

$$= [a \cdot (^{-}1)] \cdot b \quad \text{(since multiplication is associative)}$$

$$= [(^{-}1) \cdot a] \cdot b \quad \text{(because multiplication is commutative)}$$

$$= (^{-}1)(ab) \quad \text{(since multiplication is associative)}$$

$$= {^{-}}(ab) \quad \text{(by the last theorem).}$$

Also

$$a(^-b) = a \cdot [(^-1)b]$$
$$= [a(^-1)] \cdot b$$
$$= [(^-1)a] \cdot b$$
$$= (^-a)b.$$

This completes the proof of the theorem. The student may supply the reasons for each of the last four equalities.

Example 1. $2 \cdot (^-3) = {}^-(2 \cdot 3)$ (by the preceding theorem)

$$= (^-2) \cdot 3 \quad \text{(by the preceding theorem)}$$

$$= {}^-6.$$

Example 2. $(^-a) \cdot (^-b) = {}^-[a \cdot (^-b)]$ (by the preceding theorem)

$$= {}^-[{}^-(a \cdot b)] \text{ (by the preceding theorem)}$$

$$= ab \qquad\quad \text{(by the theorem in Section 3.3).}$$

Thus $(^-3) \cdot (^-7) = 3 \cdot 7$

$$= 21.$$

Beginning students of algebra often have difficulties in "removing parentheses." The next theorem should help to clarify this point.

Theorem. If $a, b \in I$, then $^-(a + b) = {}^-a + {}^-b$.

Proof:

$^-(a + b) = (^-1) \cdot (a + b)$ (by the first theorem in this section)

$$= (^-1)a + (^-1)b \quad \text{(by the distributive property)}$$

$$= {}^-a + {}^-b \qquad\quad \text{(since } (^-1)c = {}^-c \text{ for any } c \in I\text{).}$$

Example 3. If $k \in I$, $^-(11 + k) = {}^-11 + {}^-k$.

Example 4. $^-15 + {}^-99 = {}^-(15 + 99) = {}^-114$.

Example 5. $^-(21 + {}^-25) = {}^-21 + {}^-(^-25)$

$$= {}^-21 + 25$$

$$= 4$$

(since $^-21 + (21 + 4) = (^-21 + 21) + 4 = 0 + 4 = 4$).

Example 6. If $m, n \in I$,

$$^-(^-m + ^-n) = ^-(^-m) + ^-(^-n)$$

$$= m + n.$$

Thus far we have dealt with only two operations on the set I — namely, addition and multiplication. Now we define another operation on I, symbolically denoted by "$-$" and called *subtraction subtraction*, by the rule: If $a, b \in I$, $a - b = a + (^-b)$.

To illustrate,

$$7 - 10 = 7 + (^-10)$$

$$= 7 + (^-7 + ^-3)$$

$$= (7 + ^-7) + ^-3$$

$$= 0 + ^-3$$

$$= ^-3.$$

The following theorem asserts that multiplication is distributive over subtraction.

Theorem. If $a, b, c \in I$, then $a \cdot (b - c) = ab - ac$.

Proof:

$$
\begin{aligned}
a \cdot (b - c) &= a \cdot (b + ^-c) && \text{(by definition of subtraction)} \\
&= ab + a(^-c) && \text{(by the distributive property)} \\
&= ab + ^-(ac) && \text{(since } a(^-c) = ^-(ac)\text{)} \\
&= ab - ac && \text{(by definition of subtraction).}
\end{aligned}
$$

Example 7.

$$12 \cdot 73 - 12 \cdot 71 = 12 \cdot (73 - 71)$$

$$= 12 \cdot 2$$

$$= 24.$$

Exercise 3.5

Read Examples 1 and 2 before doing Exercises 1–8. In each of these, perform the indicated multiplication. Justify each step in your work by stating the property or theorem you are using.

1. $3 \cdot (^-5)$.   2. $6 \cdot (^-7)$.
3. $^-5 \cdot 3$.   4. $(^-3) \cdot (^-6)$.

5. $13 \cdot (^-12)$.
6. $12 \cdot (^-4 + 2)$.
7. $^-3 \cdot (6 + ^-1)$.
8. $^-(5 + 7) \cdot 3$.

Read Examples 3–7 before doing Exercises 9–22. In each of these, simplify the given expression. Justify each step in your work by stating the property or theorem you are using.

9. $^-16 + ^-45$.
10. $^-(^-6 + ^-14)$.
11. $4 + ^-7$.
12. $^-(8 + ^-15)$.
13. $^-(17 + ^-16)$.
14. $12 \cdot (18 + ^-14)$.
15. $^-3 \cdot (^-4 + ^-3)$.
16. $15 \cdot (151 + ^-149)$.
17. $(^-4 + ^-3) \cdot (17 + ^-15)$.
18. $(12 + ^-15) \cdot (^-4 + 2)$.
19. $(16 + ^-14) \cdot (^-5 + 3)$.
20. $(81 + ^-3) \cdot (18 + ^-18)$.
21. $(4 + ^-4) \cdot (7 + ^-7)$
22. $(5 + ^-7) \cdot (19 + ^-11)$.

Perform the indicated subtractions in Exercises 23–30.

23. $5 - 18$.
24. $6 - 7$.
25. $12 - 19$.
26. $1 - 5$.
27. $14 - 59$.
28. $1103 - 9852$.
29. $15 \cdot 421 - 15 \cdot 415$.
30. $31 \cdot 25 - 25 \cdot 27$.

3.6 PRIME NUMBERS

In the present section we single out certain subsets of the set N of natural numbers. A natural number m is said to be *even* if $m = 2n$ for some $n \in N$.

For example, 2, 6, 10 are even, since $2 = 2 \cdot 1$, $6 = 2 \cdot 3$, and $10 = 2 \cdot 5$. If a natural number is not even, it is said to be *odd*.

multiple In general, if $m \in N$ and $m = pq$, where $p, q \in N$, then *factors* we say that m is a *multiple of p* (it is also a multiple of q). In *(divisors)* that case we also say that p and q are *factors* (*divisors*) of m.

As an example, observe that 6 is a multiple of 2 (and of 3) and that 2 and 3 are factors of 6 since $6 = 2 \cdot 3$.

Note that if $m \in N$, then $m = 1 \cdot m$. Therefore, 1 and m are both divisors of m. We now give the following definition:

prime **Definition.** A natural number p is said to be *prime* if there are exactly two natural numbers (namely, 1 and p) that are divisors of p.

Note that 1 is not prime, since there is only one natural number (namely, 1) that is a divisor of 1. Also 4 is not prime since 1, 2, and 4 are divisors of 4. However, 2, 3, 5, 7, 11 and so on are primes.

Suppose m and n are prime numbers and let $q = mn$. An

interesting question is whether q could have been obtained by multiplying two prime numbers other than m and n.

The answer is no. This follows from a very important theorem called the *fundamental theorem of arithmetic*, which we shall state below. (However, the proof of this theorem is beyond the scope of this book.)

First we must make a remark on notation. We have used letters to represent numbers. If we used the letters $a, b, c, \cdots$ to represent distinct numbers, we could do so for at most twenty-six numbers. To avoid this limitation we use the following device. Among the numbers to be represented we choose one and denote it n_1 (read "n sub one"), then we choose a second one and denote it n_2 (read "n sub two"), then a third one and denote it n_3, and *subscript* so on. In the symbol n_i, the i is called a *subscript*.

Theorem. Let $m \in N - \{1\}$ and suppose m is not prime. Then the following statements are true:

(a) There are prime numbers $p_1, p_2, \cdots, p_n$ (not necessarily distinct) such that $m = p_1 p_2 \cdots p_n$.

(b) $p_1 p_2 \cdots p_n$ is the only way m can be obtained as a product of primes (except for the order of the factors).

For example,

$$60 = 2 \cdot 2 \cdot 3 \cdot 5,$$

and although we can write

$$60 = 3 \cdot 2 \cdot 5 \cdot 2,$$

we cannot find prime numbers other than 2, 2, 3, 5 that will give 60 as their product.

Example 1. Write the numbers $6^3 \cdot 5^2$ and $4^2 \cdot 7^3$ as products of prime numbers and determine whether they are equal.

Solution: We write

$$6^3 \cdot 5^2 = (2 \cdot 3) \cdot (2 \cdot 3) \cdot (2 \cdot 3) \cdot 5 \cdot 5$$
$$= 2 \cdot 2 \cdot 2 \cdot 3 \cdot 3 \cdot 3 \cdot 5 \cdot 5.$$

Also

$$4^2 \cdot 7^3 = (2 \cdot 2) \cdot (2 \cdot 2) \cdot 7 \cdot 7 \cdot 7$$
$$= 2 \cdot 2 \cdot 2 \cdot 2 \cdot 7 \cdot 7 \cdot 7.$$

If the two given numbers were equal, we would have

$$2 \cdot 2 \cdot 2 \cdot 3 \cdot 3 \cdot 3 \cdot 5 \cdot 5 = 2 \cdot 2 \cdot 2 \cdot 2 \cdot 7 \cdot 7 \cdot 7,$$

which is false according to the fundamental theorem of arithmetic.

Example 2. A student (a patient one indeed!) computed the products $5^{157} \cdot 3^{127}$ and $7^{99} \cdot 3^{100}$ and obtained the same result. What can you conclude about his computations?

Solution: From the fundamental theorem of arithmetic these two numbers cannot be equal, since 5 is a prime factor of the first but not of the second. Therefore, the student must have made an error in his calculations.

Example 3. Show that if m is a natural number other than 1, and m^2 is written as a product of primes, then each prime factor of m^2 occurs an even number of times.

Solution: If m is prime, then the unique prime factorization of m^2 is $m \cdot m$, and m occurs two times.

If m is not prime, we write $m = p_1 p_2 \cdots p_k$, where the p_i's are prime numbers. Then

$$m^2 = (p_1 p_2 \cdots p_k) \cdot (p_1 p_2 \cdots p_k)$$

$$= p_1 p_1 p_2 p_2 \cdots p_k p_k.$$

Hence p_1 occurs twice, p_2 twice, and so on. Although the p_i's need not be distinct, it is clear that each prime factor occurs an even number of times.

For example, if $m = 60$, we have

$$60 = 2 \cdot 2 \cdot 3 \cdot 5 \quad \text{and} \quad 60^2 = 2 \cdot 2 \cdot 2 \cdot 2 \cdot 3 \cdot 3 \cdot 5 \cdot 5.$$

Before proceeding to Example 4, the student should read Section 1.4 again. When p, q are propositions, the proposition $p \rightarrow q$ is false when p is true and q is false; it is true in the other three cases. A technique frequently used to prove that a proposition p is true is the following. We assume that $\sim p$ is true and proceed to prove that $\sim p \rightarrow q$ is true where q is known to be false. We are then assured that $\sim p$ is false (otherwise the implication $\sim p \rightarrow q$ would be false). Knowing that $\sim p$ is false, we conclude that p is true.

This type of proof (called proof by contradiction) is illustrated in the next example.

Example 4. Are there two natural numbers m and n such that $m^2 = 2n^2$?

Solution: Suppose there are two natural numbers m and n such that $m^2 = 2n^2$.

First note that the square of any natural number is not a prime number. In particular m^2 is not a prime number. This forces us to conclude that $n \neq 1$. By Example 3, the prime number 2 would either occur an even number of times in the prime factorization of n^2, or else it would not be a factor of n^2. In either case $2n^2$ has a factorization in primes in which the prime number 2 occurs an odd number of times. But then, since $m^2 = 2n^2$, the prime factorization of m^2 would contain the prime number 2 an odd number of times and also an even number of times (see Example 3).

Hence by the fundamental theorem of arithmetic the numbers m^2 and $2n^2$ cannot be equal.

We remark that the last example can be generalized to show that there are no two natural numbers m and n such that $m^2 = pn^2$, where p is a prime.

Exercise 3.6

Read Example 1 before doing Exercises 1–3.

1. Write the numbers $5^4 \cdot 8^7$ and $2^{20} \cdot 5^6$ as products of prime numbers and determine whether they are equal.

2. Repeat Exercise 1 with the numbers $6^4 \cdot 9^5$ and $3^{12} \cdot 2^2 \cdot 6^2$.

3. Repeat Exercise 1 with the numbers $10^5 \cdot 6^4$ and $15^4 \cdot 2^{11}$.

Read Examples 3 and 4 before working Exercises 4–9.

4. Suppose that m is a natural number other than 1 and m^3 is written as a product of primes. If p is a prime factor of m^3 and p occurs k times, then k is divisible by 3. Prove this fact.

5. Are there two natural numbers m and n such that $m^3 = 3n^3$?

6. Are there two natural numbers m and n such that $m^5 = 5n^5$?

7. Are there two natural numbers m and n such that $m^2 = 3n^2$?

8. Are there two natural numbers m and n such that $m^2 = 5n^2$?

9. Are there two natural numbers m and n such that $m^3 = 5n^3$?

10. Some primes are 1 less than a perfect square. For example, $3 = 2^2 - 1$. Can you find other prime numbers of the form $n^2 - 1$? Explain your answer.

11. Some primes are 1 more than a perfect square. For example, $5 = 2^2 + 1$. Find three more primes of the form $n^2 + 1$, where n is a natural number. Is there an end to such primes? (No one knows.)

12. Some primes are of the form $1 + 2^n$, where n is a natural number. For example, $5 = 1 + 2^2$. Find two more primes of this form.

13. Find five primes of the form $n^2 + 1$, where n is a natural number.

14. Let n be a natural number larger than 1 and suppose d is the smallest divisor of n other than 1. Prove that d is a prime.

15. Is it true that if 6 divides the product of two natural numbers it must divide at least one of them?

16. Let $k!$ denote the product of the first k natural numbers. For example, $3! = 1 \cdot 2 \cdot 3 = 6$ and $5! = 1 \cdot 2 \cdot 3 \cdot 4 \cdot 5 = 120$. If $k \neq 1$, is any of the natural numbers $2 + k!, 3 + k!, 4 + k!, \cdots, k + k!$ a prime?

17. Note that

$$4 = 2 + 2,$$
$$6 = 3 + 3,$$
$$8 = 5 + 3,$$
$$10 = 7 + 3.$$

(a) Express the integers 12, 14, 16, 18, and 20 as a sum of two primes.

(b) Use an educated guess to answer the following question: Is it true that every positive even integer other than 2 can be expressed as a sum of two primes?

18. Prove: A natural number is divisible by 2 if its last digit is divisible by 2. [*Hint:* If n is a natural number, write $n = 10a + b$, where b is the last digit.]

19. Prove: A natural number is divisible by 4 if the number formed by the two right-hand digits is divisible by 4.

20. Prove: If p is a prime, then for any two natural numbers m and n, $m^2 \neq pn^2$.

*21. State and prove a sufficient condition for a natural number to be divisible by 8.

3.7 GREATEST COMMON DIVISOR
AND LEAST COMMON MULTIPLE

greatest The *greatest common divisor* (G.C.D.) of several natural num-
common bers $a, b, \cdots, k$ is the largest natural number that divides each
divisor of the numbers $a, b, \cdots, k$. Their *least common multiple*
least (L.C.M.) is the smallest natural number that is a multiple of each
common of these numbers.
multiple   In order to describe a convenient method to find the G.C.D.

and L.C.M. of several natural numbers, we associate with each natural number n a set S_n of symbols in the following way.

The symbol 1 (which represents the number one) is a member of S_n for every natural number n and is the only member of S_1; that is, $S_1 = \{1\}$. If n is prime, then $S_n = \{1, n_1\}$.

If the prime number p occurs k times in the prime factorization of n, then the symbols $p_1, p_2, \cdots, p_k$ are members of S_n. Further, in addition to the symbol 1, S_n will contain only symbols obtained in this manner.

Example 1. Since $72 = 2 \cdot 2 \cdot 2 \cdot 3 \cdot 3$, we have $S_{72} = \{1, 2_1, 2_2, 2_3, 3_1, 3_2\}$. Similarly, $S_{2800} = \{1, 2_1, 2_2, 2_3, 2_4, 5_1, 5_2, 7_1\}$ (since $2800 = 2 \cdot 2 \cdot 2 \cdot 2 \cdot 5 \cdot 5 \cdot 7$), and $S_3 = \{1, 3_1\}$.

Note that although the symbols $2_1, 2_2, 2_3$ are distinct, they represent the same number, namely 2, and $2_1 \cdot 2_2 \cdot 2_3 = 2^3 = 8$. More generally, if p is a prime number, then $p_1 = p_2 = \cdots = p_k = p$ and $p_1 \cdot p_2 \cdot \cdots \cdot p_k = p^k$.

Example 2. If $S_n = \{1, 3_1, 3_2, 5_1, 5_2, 5_3, 11_1\}$, then $n = 3^2 \cdot 5^3 \cdot 11 = 12{,}375$.

For convenience, we also introduce the following notation. If $a, b, \cdots, k$ are natural numbers, the symbols G.C.D.$[a, b, \cdots, k]$ and L.C.M.$[a, b, \cdots, k]$ denote respectively their greatest common divisor and their least common multiple.

We now state the following theorem, which gives a method for finding the G.C.D. and L.C.M. of several natural numbers. The proof of the theorem is not given here, since our primary interest is in its applications.

Theorem. If $a, b, \cdots, k$ are natural numbers, then

$$S_{\text{G.C.D.}[a, b, \cdots, k]} = S_a \cap S_b \cap \cdots \cap S_k,$$

and

$$S_{\text{L.C.M.}[a, b, \cdots, k]} = S_a \cup S_b \cup \cdots \cup S_k.$$

Example 3. Find the G.C.D. and L.C.M. of 60, 14, and 6.

Solution:

$$60 = 2 \cdot 2 \cdot 3 \cdot 5,$$

$$14 = 2 \cdot 7,$$

and

$$6 = 2 \cdot 3.$$

Therefore,

$$S_{60} = \{1, 2_1, 2_2, 3_1, 5_1\},$$
$$S_{14} = \{1, 2_1, 7_1\},$$

and

$$S_6 = \{1, 2_1, 3_1\}.$$

Hence

$$S_{\text{G.C.D.}[60, 14, 6]} = S_{60} \cap S_{14} \cap S_6 = \{1, 2_1\},$$

and

$$S_{\text{L.C.M.}[60, 14, 6]} = S_{60} \cup S_{14} \cup S_6 = \{1, 2_1, 2_2, 3_1, 5_1, 7_1\}.$$

It follows that

$$\text{G.C.D. } [60, 14, 6] = 1 \cdot 2 = 2,$$

and

$$\text{L.C.M. } [60, 14, 6] = 1 \cdot 2 \cdot 2 \cdot 3 \cdot 5 \cdot 7 = 420.$$

Example 4. Find the G.C.D. of 6, 245, and 143.

Solution:

$$6 = 2 \cdot 3,$$
$$245 = 5 \cdot 7 \cdot 7,$$

and

$$143 = 11 \cdot 13.$$

Therefore,

$$S_6 = \{1, 2_1, 3_1\},$$
$$S_{245} = \{1, 5_1, 7_1, 7_2\},$$
$$S_{143} = \{1, 11_1, 13_1\}.$$

The intersection of the foregoing three sets is $\{1\}$. Hence,

$$\text{G.C.D. } [6, 245, 143] = 1.$$

Exercise 3.7

Read Example 1 before working Exercises 1–6. In each of these, a natural number n is given. Find S_n.

    1. $n = 88.$ **2.** $n = 100.$

3. $n = 360.$
4. $n = 5929.$
5. $n = 125 \cdot 169.$
6. $n = 845.$

Read Example 2 before working Exercises 7–10. In each of these S_n is given; find n.

7. $S_n = \{1, 3_1, 3_2, 5_1, 5_2, 5_3, 13_1\}.$
8. $S_n = \{1, 5_1, 5_2, 13_1, 13_2, 17_1\}.$
9. $S_n = \{1, 2_1, 2_2, 2_3, 2_4, 5_1, 7_1, 7_2\}.$
10. $S_n = \{1\}.$

Read Examples 3 and 4 before working Exercises 11–20.

11. Find G.C.D. $[88, 100, 360].$
12. Find L.C.M. $[88, 100, 360].$
13. Find G.C.D. $[5929, 845].$
14. Find L.C.M. $[5929, 845].$
15. Find G.C.D. $[72, 54, 180].$
16. Find L.C.M. $[72, 54, 180].$
17. Find G.C.D. $[169, 75].$
18. Find L.C.M. $[169, 75].$
19. Find G.C.D. $[45, 65, 36, 54, 34].$
20. Find L.C.M. $[45, 65, 36, 54, 34].$

REVIEW EXERCISES

1. Show that ordinary addition is not an operation on the set of odd counting numbers.

For Exercises 2–5 examples from Examples 1–4 in Section 3.1 may be selected.

2. Give an example of an operation that is neither commutative nor associative.

3. Give an example of an operation that is commutative and nonassociative and has identity.

4. Give an example of an operation that is commutative and associative and has identity, and such that every element in the set on which the operation is defined has a unique inverse.

5. Give an example of an operation that is noncommutative and associative and has no identity element.

In Exercises 6–11 write the given expression as a product, using the distributive law.

6. $x^2 + 3x + 2.$
7. $4x^2 - 4x + 1.$
8. $x^2 - 4x + 4.$
9. $5x^2 + 13x + 6.$

10. $5x^2 + 7x - 6$.   **11.** $(x-1)^2 + 4(x-1) + 4$.

12. Write 260 as a product of primes.

13. Explain why the numbers $3^{71} \cdot 5^{101}$, $2^{130} \cdot 5^{99}$ are not equal.

14. Find G.C.D. $[99, 72]$.

15. Find L.C.M. $[99, 72]$.

16. Find G.C.D. $[36, 21, 39]$.

17. Find L.C.M. $[36, 21, 39]$.

18. Determine the solution set of the open sentence

$$(a - b) - x = a - (b - x)$$

if the replacement set is I (the set of integers).

*19. Show that if $\{p, a, b\} \subset N$ (the set of natural numbers) and p is a prime and p divides ab, then p divides a or p divides b. [*Hint:* Use the fundamental theorem of arithmetic.]

*20. Show that if a, b are natural numbers such that L.C.M. $[a, b] = $ G.C.D. $[a, b]$, then $a = b$. [*Hint:* What can you assert about $S_a \cap S_b$ and $S_a \cup S_b$?]

THE
REAL
AND
COMPLEX
NUMBERS

4.1 BASIC PROPERTIES OF RATIONAL NUMBERS

In Section 3.3 we remarked that all integers, except 1 and -1, fail to have multiplicative inverses in I, the set of integers. It follows that no integer is a solution of the open sentence $2x = 1$. The student should be able to cite other open sentences with empty solution sets relative to the replacement set I.

Prompted by the desire to provide multiplicative inverses for as many integers as possible, we now wish to consider an extension of the set I into a larger set, namely the set of rational numbers. We use the symbol Q to denote the set of rational num-

rational bers. The members of Q are called *rational numbers*. The set Q
numbers satisfies the following properties:

    (1) The set of integers I is a proper subset of Q.
    (2) Two operations, addition and multiplication, to be de-

noted by $+$ and $\cdot$, respectively, are defined on Q. Addition and multiplication agree with the corresponding operations on I.

(3) Addition is commutative and associative. The integer 0 is the *additive identity* (that is, $r + 0 = 0 + r = r$ for every rational number r). For each rational number r there is a unique rational number, denoted by $-r$, with the property that $r + (-r) = -r + r = 0$. The rational number $-r$ is called the *additive inverse* of r.

additive identity

additive inverse

(4) Multiplication is commutative and associative. The natural number 1 is the *multiplicative identity* (that is, $1 \cdot r = r \cdot 1 = r$ for every rational number r).

multiplicative identity

(5) Multiplication is distributive over addition — that is, if $a, b, c \in Q$, $a(b + c) = ab + ac$.*

(6) If $s \in Q$ and $s \neq 0$, there is a unique rational number, denoted by s^{-1}, with the property that $s \cdot s^{-1} = s^{-1} \cdot s = 1$. The rational number s^{-1} is called the *multiplicative inverse* of s.

multiplicative inverse

(7) Q is the smallest set satisfying the six properties listed above. That is, if S is a set possessing the six properties, then $Q \subseteq S$.

We note that in the statement of property (3) we bowed to tradition and wrote $-r$ instead of ^-r (see the remark in Section 3.3). Henceforth, we shall always denote the additive inverse of a number c by $-c$.

Returning to the open sentence $2x = 1$, we find that its solution set relative to the replacement set Q is nonempty, because the rational number 2^{-1} is a solution.

We define another operation on Q, symbolically denoted by *subtraction* "$-$" and called *subtraction*, by the rule:

$$\text{If } a, b \in Q, \qquad a - b = a + (-b).$$

Note that this definition conforms to the definition of subtraction for integers as stated in Section 3.5. Observe that if in statements (2), (3), (4), and (5) the term "rational number" is replaced by the term "integer" and the symbol Q is replaced by the symbol I throughout, we obtain true propositions about integers. Any results obtained in Chapter 3 for integers using only those properties which are common to both I and Q will also be valid for rational numbers. For convenience we summarize these results in a theorem.

---

* Often the multiplication symbol is omitted and we write xy in lieu of $x \cdot y$.

Theorem.

(a) If $a \in Q$, then $-(-a) = a$.

(b) If $r \in Q$ and $n \in N$, then $\underbrace{r + r + \cdots + r}_{n \text{ terms}} = nr$.

(c) If $s, r \in Q$, then $s^2 + 2sr + r^2 = (s + r)^2$.

(d) If $s \in Q$, then $s \cdot 0 = 0$.

(e) If $s \in Q$, then $-s = (-1) \cdot s$.

(f) If $a, b \in Q$, then $a \cdot (-b) = -(ab) = (-a) \cdot b$.

(g) If $a, b \in Q$, then $-(a + b) = -a + (-b)$.

(h) If $a, b, c \in Q$, then $a \cdot (b - c) = ab - ac$.

Example 1. Show that 0 has no multiplicative inverse in Q.

Solution: Since $1 \in N$ and $0 \in I - N$, $1 \neq 0$. If a rational number b were the multiplicative inverse of 0, then $b \cdot 0 = 1$. From part (d) of the preceding theorem $b \cdot 0 = 0$. Thus the hypothesis that 0 has a multiplicative inverse in Q leads to the false conclusion that $1 = 0$. It must, therefore, be true that 0 has no multiplicative inverse in Q.

quotient　　If $a, b \in Q$ and $b \neq 0$, we define the *quotient* $\dfrac{a}{b}$ (read: "a divided by b" or "a over b") by the rule $\dfrac{a}{b} = a \cdot b^{-1}$. This rule is

division called *division*. We may write $a \div b$ to denote $\dfrac{a}{b}$. In the quotient $\dfrac{a}{b}$

numerator the numbers a and b are referred to as the *numerator* and *denom-*
denominator *inator*, respectively. Thus, for example,

$$\frac{4}{2} = 4 \cdot 2^{-1} \qquad \text{(by definition of a quotient)}$$

$$= (2 \cdot 2) \cdot 2^{-1}$$

$$= 2 \cdot (2 \cdot 2^{-1}) \qquad \text{(since multiplication is associative)}$$

$$= 2 \cdot 1 \qquad \text{(by definition of } 2^{-1})$$

$$= 2.$$

It must be emphasized that, since 0 has no multiplicative inverse, *division of a rational number by 0 is not defined.*

Example 2. Show that for any nonzero rational number r, $r^{-1} = \dfrac{1}{r}$.

Solution:

$$\frac{1}{r} = 1 \cdot r^{-1} \qquad \text{(by definition of a quotient)}$$

$$= r^{-1} \qquad \text{(since 1 is the multiplicative identity)}.$$

Now we have an alternate way to denote the multiplicative inverse of a nonzero rational number r, namely $\frac{1}{r}$.

Example 3. Express the rational number $\frac{2}{5} + \frac{7}{5}$ in the form $\frac{a}{b}$, where a and b are integers.

Solution:

$\frac{2}{5} + \frac{7}{5} = 2 \cdot \frac{1}{5} + 7 \cdot \frac{1}{5}$ (by definition of a quotient)

$\qquad = (2 + 7) \cdot \frac{1}{5}$ (since multiplication is distributive over addition)

$\qquad = 9 \cdot \frac{1}{5}$

$\qquad = \frac{9}{5}$ (by definition of a quotient).

In Example 3 we succeeded in writing a rational number as a quotient of two integers. One may inquire whether this can always be accomplished. The answer will be provided in the next section after further rules of arithmetic with rational numbers have been developed.

We wish to conclude this section with a theorem of considerable importance. To assist the student, we first make a brief comment on logic.

The student will recall from Chapter 1 that, if p, q are propositions, one may form new propositions $p \rightarrow q, q \rightarrow p$. Further, $p \rightarrow q \wedge q \rightarrow p$ is a proposition that is true precisely when both propositions $p \rightarrow q$, $q \rightarrow p$ are true, and false otherwise. Very often one of the following alternate forms is used to denote the proposition $p \rightarrow q \wedge q \rightarrow p$:

p if and only if q (sometimes abbreviated p iff q, or else $p \leftrightarrow q$).
p is necessary and sufficient for q.
p is equivalent to q.

To prove the truth of the proposition "p if and only if q," one must show that both $p \rightarrow q$ and its converse are true. As an illustration see the proof of the following theorem.

Theorem. Let $c, d \in Q$. Then $c \cdot d = 0$ if and only if $c = 0$ or $d = 0$.

Proof: We want to show (a) if $c \cdot d = 0$, then $c = 0$ or $d = 0$; (b) if $c = 0$ or $d = 0$, then $c \cdot d = 0$. Let us prove each of the stated implications (one at a time!).

(a) Suppose $c \cdot d = 0$. If $d = 0$, then the proposition $c = 0$ or $d = 0$ is true, hence the implication we want to prove is true. If $d \neq 0$, then d^{-1} exists and is a rational number. From the hypothesis that $c \cdot d = 0$, we obtain

$$(c \cdot d) \cdot d^{-1} = 0 \cdot d^{-1}.$$

Hence,

$c \cdot (d \cdot d^{-1}) = d^{-1} \cdot 0$ (since multiplication is associative and commutative)

          $= 0$ (by part (d) of the preceding theorem).

But

$c \cdot (d \cdot d^{-1}) = c \cdot 1$ (by definition of d^{-1})

          $= c$ (since 1 is the multiplicative identity).

Since each of the numbers c and 0 is equal to $c \cdot (d \cdot d^{-1})$, we conclude that $c = 0$.

(b) If $c = 0$ or $d = 0$, then according to part (d) of the preceding theorem, $c \cdot d = 0$.

Example 4. Obtain the solution set of the open sentence $(x - \frac{1}{2})$ $\cdot (x - \frac{1}{3}) = 0$, if the replacement set is Q.

Solution:

$\{x \mid (x - \frac{1}{2}) \cdot (x - \frac{1}{3}) = 0\} = \{x \mid x - \frac{1}{2} = 0 \text{ or } x - \frac{1}{3} = 0\}$
                          (by the preceding theorem)

          $= \{x \mid x - \frac{1}{2} = 0\} \cup \{x \mid x - \frac{1}{3} = 0\}$
                          (by definition of set union)

          $= \{x \mid x = \frac{1}{2}\} \cup \{x \mid x = \frac{1}{3}\}$

          $= \{\frac{1}{2}\} \cup \{\frac{1}{3}\}$

          $= \{\frac{1}{2}, \frac{1}{3}\}.$

Thus the solution set is $\{\frac{1}{2}, \frac{1}{3}\}$.

Example 4 can be readily generalized to show that if a and b are rational numbers, then the solution set of the open sentence $(x - a) \cdot (x - b) = 0$, relative to the replacement set Q, is the set $\{a, b\}$.

Exercise 4.1

Read Example 3 before doing Exercises 1–6. In each of these, express the given rational number in the form $\dfrac{a}{b}$, where a and b are integers.

1. $\frac{4}{5} + \frac{8}{5}$.

2. $\frac{4}{7} + \frac{6}{7}$.

3. $\frac{12}{13} + \frac{6}{13}$.

4. $\frac{8}{17} + \frac{9}{17}$.

5. $\frac{6}{25} - \frac{3}{25}$.

6. $\frac{1}{17} + \frac{23}{17}$.

Read Example 4 before doing Exercises 7–10. In each of these find the solution set of the given open sentence if the replacement set is Q.

7. $(x - 2) \cdot (x - 3) \cdot (x - \frac{1}{4}) = 0$.
8. $(x - 1) \cdot (x + 2) \cdot (x + \frac{1}{4}) = 0$.
9. $(x - 3) \cdot (2x + 1) \cdot (x - 6) \cdot (x + 3) = 0$.
10. $(x - 1)^2 \cdot (x + 1)^2 \cdot (x - \frac{1}{2})^3 = 0$.
11. If $\{a, b\} \subseteq Q - \{1\}$ and $(a - 1) \cdot (a - b) = 0$, what can you conclude about a and b?
*12. Prove: If $\{a, b\} \subseteq Q$, then the solution set of the open sentence $(x - a) \cdot (x - b) = 0$ relative to the set Q is $\{a, b\}$.
13. Write a statement equivalent to: $\{a, b\} \subseteq Q \land ab \neq 0$.
14. Let p and q be propositions such that $p \rightarrow q$ is true and q is false. What can you conclude about the truth of p? Justify your answer. Where was this result used in the solution of Example 1 of this section?
15. Prove that $1^{-1} = 1$. Use Example 2 to show that $\dfrac{1}{1} = 1$.

4.2 ARITHMETIC OF RATIONAL NUMBERS

In Section 4.1 we stated seven basic properties of the set Q, and by using these we derived other useful results. In the following theorem some indispensable rules of arithmetic are stated. Every student of algebra should know them.

Theorem. Let a, b, c, d be rational numbers; then the following statements are true:

(a) If $a \neq 0$, then $(a^{-1})^{-1} = a$; hence $\dfrac{1}{\dfrac{1}{a}} = a$.

(b) If $b \neq 0$, then $\dfrac{a}{b} + \dfrac{c}{b} = \dfrac{a + c}{b}$.

(c) If $bd \neq 0$, then $\dfrac{1}{b} \cdot \dfrac{1}{d} = \dfrac{1}{bd}$.

(d) If $bd \neq 0$, then $\dfrac{a}{b} \cdot \dfrac{c}{d} = \dfrac{ac}{bd}$.

(e) If $d \neq 0$, then $\dfrac{d}{d} = 1$; hence if $\dfrac{a}{b} \in Q$, $\dfrac{ad}{bd} = \dfrac{a}{b}$.

(f) $\dfrac{1}{-1} = \dfrac{-1}{1} = -1$.

(g) If $b \neq 0$, then $-\dfrac{a}{b} = \dfrac{-a}{b} = \dfrac{a}{-b}$.

(h) If $bd \neq 0$, then $\dfrac{a}{b} + \dfrac{c}{d} = \dfrac{ad + bc}{bd}$.

(i) If $bd \neq 0$, then $\dfrac{a}{b} - \dfrac{c}{d} = \dfrac{ad - bc}{bd}$.

(j) If $bcd \neq 0$, then $\dfrac{a}{b} \div \dfrac{c}{d} = \dfrac{ad}{bc}$.

(k) If $\dfrac{a}{b} \in Q$, then $\dfrac{a}{b} = 0$ if and only if $a = 0$.

(l) If $bd \neq 0$, then $\dfrac{a}{b} = \dfrac{c}{d}$ if and only if $ad = bc$.

Proof: (a) If $a \neq 0$, $a \cdot a^{-1} = a^{-1} \cdot a = 1$. This statement tells us that a is the unique multiplicative inverse of a^{-1}. Therefore, $(a^{-1})^{-1} = a$. By Example 2, Section 4.1,

$$(a^{-1})^{-1} = \frac{1}{a^{-1}} = \frac{1}{\dfrac{1}{a}}.$$

If follows that $\dfrac{1}{\dfrac{1}{a}} = a$.

(b) $\dfrac{a}{b} + \dfrac{c}{b} = a \cdot \dfrac{1}{b} + c \cdot \dfrac{1}{b}$ (by definition of a quotient)

$\qquad\qquad = (a + c) \cdot \dfrac{1}{b}$ (since multiplication is distributive over addition)

$\qquad\qquad = \dfrac{a + c}{b}$ (by definition of a quotient).

(c) $\left(\dfrac{1}{b} \cdot \dfrac{1}{d}\right)(bd) = \left(\dfrac{1}{b} \cdot \dfrac{1}{d}\right) \cdot (db)$ (since multiplication is commutative)

$$= \frac{1}{b} \cdot \left[\frac{1}{d} \cdot (db) \right] \qquad \text{(since multiplication is associative)}$$

$$= \frac{1}{b} \cdot \left[\left(\frac{1}{d} \cdot d \right) \cdot b \right] \qquad \text{(since multiplication is associative)}$$

$$= \frac{1}{b} \cdot (1 \cdot b) \qquad \left(\text{since } \frac{1}{d} \cdot d = 1 \right)$$

$$= \frac{1}{b} \cdot b \qquad \text{(since 1 is the multiplicative identity)}$$

$$= 1 \qquad \left(\text{since } \frac{1}{b} \cdot b = 1 \right).$$

Thus $\frac{1}{b} \cdot \frac{1}{d}$ is the unique multiplicative inverse of bd, and therefore

$$\frac{1}{b} \cdot \frac{1}{d} = \frac{1}{bd}.$$

(d) $\dfrac{a}{b} \cdot \dfrac{c}{d} = \left(a \cdot \dfrac{1}{b} \right) \cdot \left(c \cdot \dfrac{1}{d} \right)$

$$= \left(\left(a \cdot \frac{1}{b} \right) \cdot c \right) \cdot \frac{1}{d}$$

$$= \left(a \cdot \left(\frac{1}{b} \cdot c \right) \right) \cdot \frac{1}{d}$$

$$= \left(a \cdot \left(c \cdot \frac{1}{b} \right) \right) \cdot \frac{1}{d}$$

$$= \left((ac) \cdot \frac{1}{b} \right) \cdot \frac{1}{d}$$

$$= (ac) \cdot \left(\frac{1}{b} \cdot \frac{1}{d} \right)$$

$$= (ac) \cdot \frac{1}{bd} \qquad \text{(by part (c) of this theorem)}$$

$$= \frac{ac}{bd} \qquad \text{(by definition of a quotient)}.$$

The student may supply the reason for each of the first six equalities.

(e) $\dfrac{d}{d} = d \cdot \dfrac{1}{d} \qquad \text{(by definition of a quotient)}$

$$= 1 \qquad \left(\text{since } \frac{1}{d} = d^{-1}\right).$$

$$\frac{ad}{bd} = \frac{a}{b} \cdot \frac{d}{d} \qquad \text{(by part (d) of this theorem)}$$

$$= \frac{a}{b} \cdot 1 \qquad \left(\text{since } \frac{d}{d} = 1\right)$$

$$= \frac{a}{b} \qquad \text{(since 1 is the multiplicative identity).}$$

The proofs of the remaining parts of this theorem are left as exercises.

Mastery of arithmetic requires two essential ingredients: knowledge of the basic rules (these have been stated) and practice. The exercises throughout the remainder of this text will provide ample opportunity for practice.

Before considering examples, we wish to comment on the question that was raised following Example 3 of the preceding section: is it possible to express any rational number as a quotient of two integers?

Let $S = \{x \mid x = \frac{p}{q}, \, p, q \in I \text{ and } q \neq 0\}$ — that is, S is the set of all quotients of integers, where the integer in the denominator is not 0. If $p \in I$, $p = \frac{p}{1}$ $\left(\text{since } \frac{p}{1} = p \cdot \frac{1}{1} = p \cdot 1 = p\right)$. Thus $I \subseteq S$. Further, $\frac{1}{2} \in S - I$, therefore $I \subset S$. Thus S satisfies property (1) of Section 4.1. It is not hard to verify that $S \subseteq Q$ and S satisfies properties (2) through (6) of Section 4.1 as well. By property (7) we conclude that $Q \subseteq S$. On the other hand, we have already stated that $S \subseteq Q$. From $Q \subseteq S$ and $S \subseteq Q$ we conclude that $S = Q$. Thus, *every rational number can be expressed as a quotient of two integers*. In fact, the set of all quotients of integers of the form $\frac{p}{q}$, where $q \neq 0$, is the set of rational numbers.

Exercise 4.2

Before working Exercises 1–7, read the proofs of parts (a) through (e) of the theorem of Section 4.2.

1. Prove part (f) of the theorem.
2. Prove part (g) of the theorem.
3. Prove part (h) of the theorem.
4. Prove part (i) of the theorem

5. Prove part (j) of the theorem.

6. Prove part (k) of the theorem.

7. Prove part (l) of the theorem.

8. Use the theorem in this section to simplify the following rational numbers:

(a) $(3^{-1})^{-1}$.   (b) $\frac{1}{3} + \frac{4}{3}$.   (c) $\frac{1}{2} \cdot \frac{1}{3}$.   (d) $\frac{2}{3} \cdot \frac{5}{7}$.

(e) $\frac{21}{6}$.   (f) $\frac{-1}{1}$.   (g) $\frac{-21}{6}$.   (h) $\frac{1}{3} + \frac{2}{5}$.

(i) $\frac{1}{3} - \frac{2}{5}$.   (j) $\frac{2}{7} \div \frac{5}{2}$.   (k) $\frac{0}{3}$.

9. Show that $\frac{11}{13} \neq \frac{22}{25}$.

*10. Let $S = \{x \,|\, x = \frac{p}{q}, p, q \in I \text{ and } q \neq 0\}$. Prove that $S \subseteq Q$ and S has properties (2) through (6) of Section 4.1.

4.3 ARITHMETIC OF RATIONAL NUMBERS – EXAMPLES

relatively prime Two integers are *relatively prime* if their greatest common divisor is 1.

lowest terms A rational number is said to be in *lowest terms* if it is expressed as a quotient of two relatively prime integers. To illustrate, the rational number $\frac{12}{21}$ is not in lowest terms, since G.C.D. $[12, 21] = 3$.

If we write the number $\frac{12}{21}$ in the form $\frac{4}{7}$, this representation is in lowest terms.

Note that $\frac{12}{21} = \frac{4 \cdot 3}{7 \cdot 3} = \frac{4}{7}$ (by part (e) of the theorem in Section 4.2).

Example 1. Express $\frac{19}{24} - \frac{5}{42}$ in lowest terms.

Solution:

$$\frac{19}{24} - \frac{5}{42} = \frac{19 \cdot 42 - 24 \cdot 5}{24 \cdot 42} \qquad \text{(by part (i) of the theorem in Section 4.2)}$$

$$= \frac{678}{(24 \cdot 7) \cdot 6}$$

$$= \frac{113 \cdot 6}{168 \cdot 6}$$

$$= \frac{113}{168} \qquad \text{(by part (e) of the theorem in Section 4.2).}$$

Note that the integers 113 and 168 have no integral factors other than 1 and −1 in common.

Problems such as the one in Example 1 can sometimes be worked in a simpler way if we use the notions of a *least common multiple* (L.C.M.) and *greatest common divisor* (G.C.D.). (See Section 3.7.)

To reduce a given rational number of the form $\frac{a}{b}$, where $a, b \in N$, to lowest terms, one may proceed by first finding the G.C.D. of a and b and then dividing the numerator and denominator of the given quotient by this number. The rational number obtained through this procedure will be in lowest terms. Note that if $m = $ G.C.D. of a and b, then

$$\frac{a}{b} = \frac{a}{b} \cdot \frac{\frac{1}{m}}{\frac{1}{m}} = \frac{a \div m}{b \div m}.$$

The representation on the right is in lowest terms.

Example 2. Express $\frac{7}{60} + \frac{3}{14} - \frac{5}{6}$ in lowest terms.

Solution: By Example 3, Section 3.7, the L.C.M. of 60, 14, and 6 is 420. Note that

$$420 = 60 \cdot 7$$
$$= 14 \cdot 30$$
$$= 6 \cdot 70.$$

Using rules of arithmetic,

$$\frac{7}{60} + \frac{3}{14} - \frac{5}{6} = \frac{7}{60} \cdot \frac{7}{7} + \frac{3}{14} \cdot \frac{30}{30} - \frac{5}{6} \cdot \frac{70}{70}$$

$$= \frac{7 \cdot 7 + 3 \cdot 30 - 5 \cdot 70}{420}$$

$$= -\frac{211}{420}.$$

The answer is in its lowest terms, since 211 is prime and not a divisor of 420.

We conclude the present section with a more complicated example.

Example 3. Simplify the given expression until a rational number in lowest terms is obtained.

$$r = \frac{\frac{1}{3} + 4 - \frac{7}{\frac{2}{3} - 2}}{\frac{5}{2} - \left(1 + \frac{2}{\frac{2}{3} - 2}\right)}.$$

Solution: Since the expression is given in a rather complicated form, we proceed to simplify it piecemeal.

$$\frac{1}{3} + 4 = \frac{1}{3} + \frac{4}{1} \cdot \frac{3}{3}$$

$$= \frac{1}{3} + \frac{4 \cdot 3}{1 \cdot 3}$$

$$= \frac{1}{3} + \frac{12}{3}$$

$$= \frac{13}{3}.$$

$$\frac{7}{\frac{2}{3} - 2} = \frac{7}{\frac{2}{3} - \frac{2}{1} \cdot \frac{3}{3}}$$

$$= \frac{7}{\frac{2}{3} - \frac{2 \cdot 3}{3}}$$

$$= \frac{7}{\frac{2 - 6}{3}}$$

$$= \frac{7}{\frac{-4}{3}}$$

$$= 7 \cdot \frac{3}{-4}$$

$$= \frac{21}{-4}$$

$$= \frac{-21}{4}.$$

$$\frac{1}{3} + 4 - \frac{7}{\frac{2}{3} - 2} = \frac{13}{3} - \left(-\frac{21}{4}\right)$$

$$= \frac{13}{3} + \frac{21}{4}$$

$$= \frac{13 \cdot 4 + 3 \cdot 21}{3 \cdot 4}$$

$$= \frac{115}{12}.$$

By direct computation we also find that

$$1 + \frac{2}{\frac{2}{3} - 2} = -\frac{1}{2}$$

and

$$\frac{5}{2} - \left(1 + \frac{2}{\frac{2}{3} - 2}\right) = 3.$$

Finally,

$$r = \frac{115}{12} \div 3$$

$$= \frac{115}{12} \cdot \frac{1}{3}$$

$$= \frac{115 \cdot 1}{12 \cdot 3}$$

$$= \frac{115}{36}.$$

The number $\frac{115}{36}$ is in lowest terms. Thus $r = \frac{115}{36}$ (in lowest terms).

The student should verify that the G.C.D. of 36 and 115 is 1.

Exercise 4.3

Read Examples 1 and 2 before working Exercises 1–8. In each of these, express the given rational number in lowest terms.

1. $\frac{8}{21} + \frac{13}{35}$.

2. $\frac{5}{24} - \frac{2}{21}$.

3. $\frac{6}{35} + \frac{13}{75}$.

4. $\frac{1}{52} - \frac{5}{39}$.

5. $\frac{7}{60} + \frac{3}{14} - \frac{5}{6}$.

6. $\frac{8}{13} + \frac{4}{17} - \frac{5}{26}$.

7. $\left(\frac{5}{13} + \frac{2}{15}\right) \cdot \left(\frac{3}{4} + \frac{2}{25}\right)$.

8. $\left(\frac{6}{17} + \frac{7}{85}\right) \cdot \left(\frac{5}{4} - \frac{13}{7}\right)$.

Read Example 3 before working Exercises 9–14. In each of these, simplify the given expression until a rational number in lowest terms is obtained.

9. $\dfrac{2 + \frac{3}{2} - \dfrac{1}{5 + \frac{1}{2}}}{3 - \frac{1}{4} + \dfrac{1 + \frac{3}{4}}{2}}$.

10. $\dfrac{\dfrac{1}{4 + \frac{3}{4}} + \dfrac{2}{\frac{1}{3} - \frac{2}{5}}}{1 - \dfrac{3}{4 + \frac{1}{2}}}$.

11. $\dfrac{3 + \frac{1}{4} - \dfrac{1}{\frac{1}{2} + \frac{1}{3}}}{\dfrac{4}{1 + \frac{1}{3}}}$.

12. $\dfrac{1 + \dfrac{1}{1 + \frac{1}{2}}}{2 + \dfrac{3 + \frac{1}{2}}{4 - \frac{3}{4}}}$.

13. $\dfrac{\dfrac{2}{3 + \frac{5}{4}} - \dfrac{1}{\frac{5}{7} - 2}}{8 - \dfrac{6}{1 + \dfrac{5}{7 - \frac{1}{3}}}}$.

14. $\dfrac{5 - \dfrac{2 + \frac{5}{6}}{\frac{7}{3} - 1}}{5 - \dfrac{4 + \frac{1}{3}}{\frac{5}{7} - \frac{5}{9}}}$.

4.4 DECIMAL FORM OF RATIONAL NUMBERS (OPTIONAL)

The student undoubtedly has encountered numbers in decimal form, such as -211.5, 0.275, and so on. As we shall presently see, this form is a short way to denote a sum involving powers of 10.

We define $10^0 = 1$ and, if $n \in N$, $10^{-n} = \dfrac{1}{10^n}$. (This is actually a particular case of a more general definition, which will be stated in Chapter 5). Thus, for example, $10^{-3} = \dfrac{1}{10^3}$.

By definition,

$$-211.5 = -(2 \cdot 10^2 + 1 \cdot 10^1 + 1 \cdot 10^0 + 5 \cdot 10^{-1})$$

and

$$0.275 = 2 \cdot 10^{-1} + 7 \cdot 10^{-2} + 5 \cdot 10^{-3}.$$

More generally, if $n_0, n_1, \cdots, n_k$ is a finite sequence and $m_1, m_2, \cdots$ is any sequence (finite or infinite) where each n_i and each m_j belongs to $\{0, 1, 2, \cdots, 9\}$, then

$$n_k \cdots n_1 n_0 \cdot m_1 m_2 \cdots$$

$$= n_k \cdot 10^k + \cdots + n_0 \cdot 10^0 + m_1 \cdot 10^{-1} + m_2 \cdot 10^{-2} + \cdots.$$

decimal   If a number is expressed in this form, it is said to be in
form   *decimal form*. Numbers that can be represented in decimal form
decimal are called *decimal numbers* or, simply, *decimals*. If a digit or
numbers block of digits is repeated indefinitely after some point, the
periodic decimal is said to be *periodic*.

We indicate a periodic decimal by underscoring the digit or block of digits that is repeated indefinitely.

To illustrate, $0.\underline{37}$ is a periodic decimal. It can also be written $0.37\underline{37}$ or $0.3737\underline{37}$, and so on.

The number of digits in the smallest block that is indefinitely repeated in a periodic decimal is called the *length of the period.* If only one digit is repeated indefinitely, the length of the period is 1.

For example, the length of the period of 3.123$\overline{735}$ is 3.

Example 1. Express $\dfrac{7}{4}$ in decimal form.

Solution:

$$\frac{7}{4} = \frac{7}{4} \cdot \frac{25}{25}$$

$$= \frac{175}{100}$$

$$= \frac{1 \cdot 10^2 + 7 \cdot 10 + 5}{10^2} = \frac{1 \cdot 10^2}{10^2} + \frac{7 \cdot 10}{10^2} + \frac{5}{10^2}$$

$$= 1 + 7 \cdot \frac{1}{10} + 5 \cdot \frac{1}{10^2}$$

$$= 1 + 7 \cdot 10^{-1} + 5 \cdot 10^{-2}$$

$$= 1.75.$$

Of course, the same result could be obtained by performing long division.

Example 2. Express $\dfrac{11}{7}$ in decimal form.

Solution: We use the familiar long-division technique.

```
       1.571428
   7 ⟌ 11
        7
       ‾‾
       40  ←
       35
       ‾‾
       50
       49
       ‾‾
       10
        7              (note that the remainder here is 4)
       ‾‾
       30
       28
       ‾‾
       20
       14
       ‾‾
       60
       56
       ‾‾
        4   ↵
```

It is clear now that the division process will not terminate, since we have obtained the same remainder, namely 4, twice. Moreover, we can predict the succession of digits without further calculations.

For example, the next six digits will be 5, 7, 1, 4, 2, 8 in that order. Thus $\frac{11}{7} = 1.\underline{571428}$. Note that the length of the period is 6.

Example 3. Show that $2.4\underline{31}$ is a rational number.

Solution: Let $r = 2.4\underline{31}$. Then,

$$10^3 \cdot r = 2431.\underline{31} = 2431 + .\underline{31}.$$

Also,

$$10 \cdot r = 24.\underline{31} = 24 + .\underline{31}.$$

Therefore,

$$10^3 \cdot r - 10 \cdot r = 2431 - 24 \qquad (\text{since } .\underline{31} - .\underline{31} = 0).$$

Hence,

$$990 \cdot r = 2407,$$

from which we obtain

$$r = \frac{2407}{990}.$$

The last equality shows that r is a rational number.

Note the technique displayed in the last example. We choose distinct integers k and m such that $10^k r - 10^m r$ is an integer.

Example 4. Show that $1.75\underline{0}$ is a rational number.

Solution: Let $s = 1.75\underline{0}$. Then

$$10^3 \cdot s = 1750.\underline{0}$$

and

$$10^2 \cdot s = 175.\underline{0}.$$

Therefore

$$10^3 \cdot s - 10^2 \cdot s = 1750 - 175.$$

Hence,

$$900 \cdot s = 1575.$$

Consequently

$$s = \frac{1575}{900}$$

and $1.75\underline{0}$ has been shown to be a rational number. Further,

$$\frac{1575}{900} = \frac{7 \cdot 225}{4 \cdot 225}$$

$$= \frac{7}{4}.$$

Comparing the results of Examples 1 and 4, we find that $1.75 = 1.75\underline{0}$. Thus we can consider 1.75 to be a periodic decimal. It should be intuitively clear now why any terminating decimal (one with only a finite number of digits after the decimal point) can be regarded as a periodic decimal.

In Examples 1 and 2 we succeeded in expressing given rational numbers in decimal form. Although we shall not give a proof here, it is true that *every rational number can be expressed as a periodic decimal.* Moreover, if $\frac{a}{b}$ is a rational number and $a, b \in N$, then the length of the period of its periodic decimal form is at most b.

Using the idea displayed in Example 3, one can readily show that *every periodic decimal is a rational number.* The following theorem follows from the foregoing remarks.

Theorem. The set of rational numbers is equal to the set of periodic decimals.

Example 5. Show that $2.\underline{9}$ is an integer.

Solution: Let $t = 2.\underline{9}$. Then

$$10t = 29.\underline{9}$$

and

$$10t - t = 27.$$

It follows that $t = 3$. Thus

$$2.\underline{9} = 3.$$

In this example we showed that $3 = 2.\underline{9}$. Since also $3 = 3.\underline{0}$, we see that the digits appearing after the decimal point in a periodic decimal form of a rational number are not necessarily unique.

Exercise 4.4

Read Examples 1 and 2 before working Exercises 1–6. In each of these, express the given rational number in decimal form.

1. $\frac{2}{7}$.   2. $\frac{5}{8}$.
3. $\frac{12}{13}$.   4. $\frac{1}{3}$.
5. $\frac{91}{7}$.   6. $\frac{135}{17}$.

Read Examples 3 and 4 before working Exercises 7–14. In each of these, show that the given number is rational.

7. $2.3\underline{3}$.   8. $51.2\underline{65}$.
9. $63.\underline{412}$.   10. $0.00\underline{13}$.
11. $151.0\underline{13}$.   12. $5.4\underline{2}$.
13. $71.513\underline{256}$.   14. $0.000\underline{101}$.

Read Example 5 before doing Exercises 15–18. In each of these show that the given rational number is an integer.

15. $3.\underline{9}$.   16. $15.\underline{9}$.
17. $16.\underline{9}$.   18. $110.\underline{9}$.

19. Give an example of a rational number whose decimal representation is not unique.

∗20. Prove that the sum and the product of two periodic decimals are periodic decimals. [*Hint:* Use the theorem in this section.]

4.5 REAL NUMBER LINE

At this point we would like to indicate how to obtain a useful geometric representation of the rational numbers.

origin Choose a point on a horizontal line, denote the point by $P(0)$, and call it the *origin*. Next select a point to the right of $P(0)$ and denote it by $P(1)$. The length of the line segment joining $P(0)$ and $P(1)$ is to be considered one unit. The point one unit to the right of $P(1)$ is denoted by $P(2)$. In general, if $k \in N$, the point $P(k+1)$ is one unit to the right of $P(k)$. Now reflect the point $P(1)$ across the origin to obtain the point $P(-1)$. In general, if $k \in N$, $P(-k)$ is the point obtained by reflecting $P(k)$ across the origin.

This construction associates with each integer a unique point in the manner illustrated in Figure 4.1.

image

coordinate

If $n \in I$, the point $P(n)$ is called the *image of the integer n*, and n is called the *coordinate of* $P(n)$.

It is clear that a point m units to the right of the origin, where $m \in N$, is the image of the natural number m, and a point m units to the left of the origin is the image of the integer $-m$. Next we show how to construct images of rational numbers on this line.

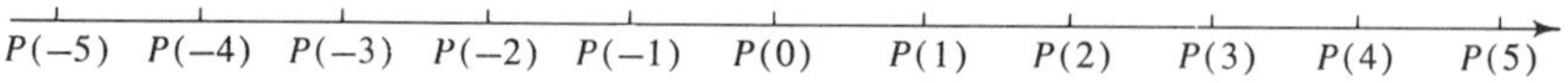

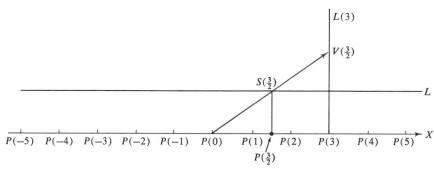

P(−5) P(−4) P(−3) P(−2) P(−1) P(0) P(1) P(2) P(3) P(4) P(5)

Figure 4.1

Let X be a replica of the line in Figure 4.1 and construct a line L one unit above X and parallel to X (see Figure 4.2). Let $r = \dfrac{a}{b} \in Q$, where a and b are natural numbers and r is in lowest terms. To obtain $P(r)$, the image of r, construct through $P(a)$ a line $L(a)$ perpendicular to X. Let $V(r)$ be the point on $L(a)$, b units above $P(a)$. The line through $P(0)$ and $V(r)$ intersects L at the point $S(r)$. The line passing through $S(r)$ and parallel to $L(a)$ intersects X at a point that we denote by $P(r)$, the image of r. Figure 4.2 shows the construction to obtain $P(\frac{3}{2})$.

L(3)
V(3/2)
S(3/2)
L
X
P(−5) P(−4) P(−3) P(−2) P(−1) P(0) P(1) P(2) P(3) P(4) P(5)
P(3/2)

Figure 4.2

If a rational number s cannot be expressed as a quotient of two natural numbers, then either $s = 0$ (in this instance the image of s is the origin) or $s = -\dfrac{c}{d}$, where $c, d \in N$. In the latter case we construct the point $P\left(\dfrac{c}{d}\right)$ and then obtain $P(s)$ by reflecting $P\left(\dfrac{c}{d}\right)$ across the origin.

A moment of thought should reveal that via the indicated construction, one and only one point is associated with each rational number, and if $r, s, \in Q$ with $r \neq s$, then $P(r)$ and $P(s)$ are distinct points. The point $P(r)$ is either r units to the right or $-r$ units to the left of the origin.

Is it true that every point on X is the image of some rational number? The Greek mathematician Pythagoras (circa 500 B.C.) startled his contemporaries when he provided a convincing answer to this question.

The next example deals with a related problem. The answer to the foregoing question is given in Example 2.

Example 1. What is the solution set of the open sentence $x^2 = 3$, if the replacement set is Q?

Solution: Let K be the solution set of $x^2 = 3$, relative to the replacement set Q. Suppose a rational number r is a member of K. Then $r^2 = 3$. Since $(-r)^2 = r^2$, we conclude that $-r$ is also a member of K. It follows that one of the numbers $r, -r$ can be expressed as a quotient of two natural numbers. Hence, there are natural numbers m and n such that $\left(\dfrac{m}{n}\right)^2 = 3$. Consequently, $m^2 = 3n^2$, where $m, n \in N$. This last statement, according to the remark following Example 4 in Section 3.6, is false. Therefore, no rational number is a member of K (that is, no rational number is a solution of $x^2 = 3$), and $K = \varnothing$.

Example 2. Show that there is a point on the number line X in Figure 4.2 that is not the image of a rational number.

Solution: Consider Figure 4.3. We have constructed a right triangle with vertices $P(0), P(1)$, and V. Each leg of this triangle is of unit length. The length of the hypothenuse is denoted by c. We also constructed a right triangle with vertices $P(0), V$, and S whose hypotenuse has length d. The point T is d units to the right of $P(0)$. We proceed to show that T is not the image of a rational number.

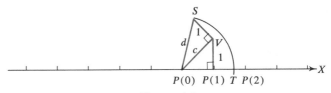

Figure 4.3

From geometry, using Pythagoras' theorem, $c^2 = 1^2 + 1^2$ and $d^2 = c^2 + 1^2$, hence $d^2 = 3$. If the point T were the image of a rational number r, then $r = d$. Since $d^2 = 3$, $r^2 = 3$. Thus r is a rational solution of the open sentence $x^2 = 3$, contrary to our conclusion in Example 1 that no rational number is a solution of this open sentence.

We conclude that T is not the image of a rational number.

In Example 1 we showed that the open sentence $x^2 = 3$ has no rational solutions. This result can be readily generalized to a theorem, stating that every open sentence of the form $x^2 = p$, where p is a prime number, has no solutions that are members of Q. Analogously, the result of Example 2 can be generalized to the assertion that there are infinitely many points on the line X that are not images of rational numbers.

Now we introduce a new set of numbers, namely the set of real numbers. This set is denoted by R. Its members are called *real numbers*. R is defined to be the set of all decimals. The properties of the real numbers can be rigorously developed using rational numbers. This task, however, is beyond the scope of this text. For our purposes it will suffice to state the properties of real numbers without proof.

The following statements are true:

(a) Q is a proper subset of R.

completeness property (b) The set R has the *completeness property*.

Since a thorough discussion of the completeness property is beyond the scope of this text, we give here a brief intuitive description instead. Imagine that a particle rolls or bounces up and down infinitely many times on the number line X, moving from left to right in such a fashion that it never goes beyond a fixed point S. Suppose also that the motion never stops and that we know the points of contact—that is, those points which have been or will be in contact with the particle. The completeness property of the real numbers guarantees that there is a point V on X such that:

(1) no point of contact is to the right of V,
(2) if W is a point to the left of V, then there is a point of contact to the right of W.

Look at Figure 4.4. The points of contact are heavily shaded. The point V in this case is not a point of contact; V is the image of some real number.

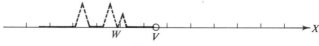

Figure 4.4

The set Q of rational numbers is "incomplete" in some sense, since the images of the rational numbers fail to "complete" the number line, leaving it with many gaps (see the comments following Example 2). On the other hand, the completeness property guarantees that the set R is sufficiently large, so that the images of members of R cover the entire number line, making it "complete."

one-to-one (c) As a consequence of the completeness property, there is a *one-to-one correspondence* between the points on the number line X and the set R (that is, each real number has one and only one image and each point on X is the image of a unique real number). Moreover, this correspondence is such that if a and b are real numbers and a is less than b, then $P(b)$, the image of b, is $b - a$ units to the right of $P(a)$, the image of a.

real number If a line X is in one-to-one correspondence with the set R, and if this correspondence satisfies the foregoing property, then *line (axis)* we call X a *real number line* (also a *real number axis*).

We often identify points on a real number line with real numbers. Thus, when context permits, we may use the number u to denote its image point $P(u)$. We may also use the phrase "the point u," meaning, of course, $P(u)$.

(d) Any open sentence of the form $x^2 = p$, where p is a prime number, has a nonempty solution set relative to the replacement set R.

Further properties of the real numbers will be discussed in Section 4.6.

Exercise 4.5

Read Example 1 and also Exercises 7, 8, and 20 of set 3.6 before working Exercises 1–12. In each of these, find the solution set of the given open sentence if the replacement set is Q.

1. $x^2 = 5$.	**2.** $x^3 = 5$.
3. $x^2 = 7$.	**4.** $x^3 = 7$.
5. $x^2 = 11$.	**6.** $x^3 = 11$.
7. $x^2 = 2$.	**8.** $x^5 = 5$.
9. $(x^2 - 5)(x - 1) = 0$.	**10.** $(x^3 - 7)(x + 2) = 0$.
11. $(x^3 - 25)(x^2 - 7) = 0$.	**12.** $(x^2 - 13)(x + 4) = 0$.

Read Example 2 before working Exercises 13–16. In each of these construct (with a ruler and compass) the point on the number line that is the image of the positive solution to the given open sentence.

13. $x^2 = 2$.   **14.** $x^2 = 7$.
15. $x^2 = 5$.   **16.** $x^2 = 10$.

Study the construction illustrated in Figure 4.2 before working Exercises 17–22. In each of these locate on the number line (with a ruler and compass) the given point.

17. $P(\frac{3}{4})$.   **18.** $P(\frac{2}{3})$.
19. $P(\frac{13}{7})$.   **20.** $P(-\frac{3}{4})$.
21. $P(\frac{7}{5})$.   **22.** $P(\frac{9}{7})$.

23. Suppose a particle bounces up and down infinitely many times on the number line X. Let $\left\{ P(x) \,|\, x = \dfrac{2n-1}{n}, \, n \in N \right\}$ be the set of points of contact.

(a) Draw a real number line and plot on it the images of the points $P\!\left(\dfrac{2n-1}{n}\right)$, where $n \in \{1, 2, 3\}$.

(b) Find a point $P(c)$ on the real number line such that every point of contact is to the left of $P(c)$.

(c) What is the coordinate of the point V, where V is such that (1) no point of contact is to the right of V, (2) if W is a point of contact to the left of V, then there is a point of contact to the right of W?

*24. Refer to Figure 4.4. Explain why the point V must be unique. [*Hint:* Let T be a point for which the properties (1) and (2) hold. Show that T cannot be to the right of V, nor can it be to the left of V.]

4.6 FURTHER PROPERTIES OF THE REAL NUMBERS

In Section 4.5 we defined R, the set of real numbers, to be the collection of all decimals. We also stated four properties of R. Further, two operations, addition and multiplication, to be denoted by $+$ and $\cdot$, respectively, are defined on R. Addition and multiplication agree with the corresponding operations on Q.

The real numbers have properties (3) through (6), which were stated in Section 4.1 for the rational numbers.

An order relation is defined on R. (This will be discussed in Chapter 6.)

We define another operation, symbolically denoted by "$-$" *subtraction* and called *subtraction*, by the rule:

$$\text{If } a, b \in R, \qquad a - b = a + (-b).$$

quotient If $a, b \in R$ and $b \neq 0$, we define the *quotient* $\dfrac{a}{b}$ (read: "a divided by b" or "a over b") by the rule

$$\frac{a}{b} = a \cdot b^{-1}.$$

division This rule is called *division*. We may write $a \div b$ to denote $\dfrac{a}{b}$.*

In the quotient $\dfrac{a}{b}$ the numbers a and b are referred to as the

numerator
denominator *numerator* and *denominator*, respectively. The quotient $\dfrac{a}{b}$ is

fraction also sometimes called a *fraction*. Note that subtraction and division agree with the corresponding rules on Q.

As a consequence of the properties listed above, the *arithmetic of the real numbers is the same as the arithmetic of the rational numbers*. In particular, the theorems we stated in Sections 4.1 and 4.2 for the rational numbers are valid for the real numbers as well. Further, 0 has no multiplicative inverse and if c is a nonzero real number, then $c^{-1} = \dfrac{1}{c}$. Also if a and b are real numbers, then the solution set of the open sentence $(x - a) \cdot (x - b) = 0$ with the replacement set R is the set $\{a, b\}$. (See Example 4 in Section 4.1.)

We stated in Section 4.5 that Q is a proper subset of R. The *irrational* members of the set $R - Q$ are called *irrational numbers*. Here *numbers* are some examples of irrational numbers:

(1) All solutions of the open sentence $x^2 = 3$ with replacement set R.

(2) The number π (recall that π is the length of the circumference of a circle with radius $\frac{1}{2}$).†

(3) The decimal $0 \cdot a_1 a_2 a_3 \cdots$, where

$$a_n = \begin{cases} 0 & \text{if } n = 2^k \text{ for some natural number } k, \\ 1 & \text{otherwise.} \end{cases}$$

This decimal is nonterminating and nonperiodic, hence it is an irrational number. Note that, indicating the first 9 digits after the decimal point,

$$0.a_1 a_2 a_3 \cdots = 0.101011101 \cdots.$$

* Frequently, a/b is used in lieu of $\dfrac{a}{b}$.

† In 1961 the value of π was approximated to 100,265 decimal places by a computer.

We conclude this section with two examples to illustrate some arithmetic with real numbers.

Example 1. Assuming that c is a real number and

$$(c+2) \cdot (c^2 + 5c + 6) \cdot (c^2 + 2c + 1) \neq 0,$$

simplify the fraction

$$\frac{3c^2 + 4c + 1}{c+2} \div \frac{c^2 + 2c + 1}{c^2 + 5c + 6}.$$

Solution:

$$\frac{3c^2 + 4c + 1}{c+2} \div \frac{c^2 + 2c + 1}{c^2 + 5c + 6} = \frac{(3c+1)(c+1)}{c+2} \cdot \frac{c^2 + 5c + 6}{c^2 + 2c + 1}$$

$$= \frac{(3c+1)(c+1)}{c+2} \cdot \frac{(c+2)(c+3)}{(c+1)^2}$$

$$= \frac{(3c+1)(c+3)}{c+1} \cdot \frac{(c+1)(c+2)}{(c+2)(c+1)}$$

$$= \frac{(3c+1)(c+3)}{c+1}.$$

Example 2. If $c, d \in R$ and $c + d \neq 0$,

$$c + d - \frac{2cd}{c+d} = \frac{c+d}{1} \cdot \frac{c+d}{c+d} - \frac{2cd}{c+d}$$

$$= \frac{(c+d)(c+d)}{c+d} - \frac{2cd}{c+d}$$

$$= \frac{(c+d)^2 - 2cd}{c+d}$$

$$= \frac{c^2 + 2cd + d^2 - 2cd}{c+d}$$

$$= \frac{c^2 + d^2}{c+d}.$$

Exercise 4.6

Before doing Exercises 1–4, review properties (1) through (7) stated in Section 4.1.

∗**1.** Prove that a sum of a rational number and an irrational number is an irrational number.

∗**2.** Prove or disprove the statement: The product of a

rational number and an irrational number is an irrational number.

*3. Prove or disprove the statement: The product of a non-zero rational number and an irrational number is an irrational number.

4. Prove or disprove the statement: The sum of two irrational numbers is irrational.

Read Examples 1 and 2 before working Exercises 5–13. Also read Section 3.4 and review the exercises of Section 3.4. In each of the following, assume that the letters involved represent real numbers such that all given expressions are defined (that is, no division by 0 is involved), and simplify.

5. $\dfrac{c^2 + 6c + 8}{c + 3} \cdot \dfrac{c^2 + 4c + 3}{c^2 + 7c + 10}$.

6. $\dfrac{c^2 - 9}{c + 4} - c + 7$.

7. $\left(\dfrac{x + y}{1 - xy} - y\right) \div \left(1 + \dfrac{xy + y^2}{1 - xy}\right)$.

8. $\left(a - 2 - \dfrac{4}{a + 1}\right)\left(a - 1 - \dfrac{8}{a - 3}\right)$.

9. $\left(\dfrac{a + x}{a - x} - \dfrac{a - x}{a + x}\right) \div \left(\dfrac{a}{a + x} + \dfrac{a}{x - a}\right)$.

10. $\dfrac{a^2 - 2ab - 8b^2}{a^2 - 16b^2} \cdot \dfrac{2ab}{2a^2 + 4ab}$.

11. $\left(a - 6 + \dfrac{20}{a + 3}\right) \div \left(a + \dfrac{2}{a - 3}\right)$.

12. $\dfrac{a^2 + ab - 2b^2}{a^2 - 4ab - 5b^2} \div \dfrac{a^2 - 3ab + 2b^2}{a^2 - ab - 2b^2}$.

13. $\left(a - 2 - \dfrac{4}{a + 1}\right) \div \left(a - 1 - \dfrac{3}{a + 1}\right)$.

14. (a) Prove: If c is a real number and c^2 is irrational, then c is irrational.

(b) Why is the following statement true?

$$\{x \,|\, x \in R \wedge x^2 = \pi\} \subseteq R - Q.$$

4.7 ARITHMETIC OF COMPLEX NUMBERS

There are many open sentences with no solutions relative to the replacement set R. For example, no real number is a solution of

$x^2 = -1$ (this fact will be demonstrated in Chapter 6). Solving of problems in both pure and applied mathematics often requires a number system that is more extensive than the system of real numbers. Hence, it has been found useful to introduce a new set of numbers, called the set of complex numbers.

complex **Definition.** A *complex number* is an expression of the form
number $a + bi$, where a and b are real numbers. The set of all complex numbers is denoted by C. Equality, addition (symbolically $+$), and multiplication (symbolically $\cdot$) are defined on C as follows:

$$a + bi = c + di \qquad \text{if and only if } a = c \text{ and } b = d.$$

$$(a + bi) + (c + di) = (a + c) + (b + d)i.$$

$$(a + bi) \cdot (c + di) = (ac - bd) + (ad + bc)i.$$

Further,

$$a + 0 \cdot i = a \qquad \text{and} \qquad 0 + 1 \cdot i = i.$$

Using this definition, one can show that $i^2 = -1$ (see Exercise 21 in the present section). The set C also has the following properties:

(1) $R \subset C$.
(2) Addition is commutative and associative. The integer 0 is the additive identity. Each complex number z has a unique additive inverse in C (we denote this number by $-z$).
(3) Multiplication is commutative and associative. The natural number 1 is the multiplicative identity. If $z \in C$ and $z \neq 0$, then z has a unique multiplicative inverse in C (we denote this number by z^{-1}).
(4) Multiplication is distributive over addition.

These properties follow from the definition of C. Since our main interest in this section is focused on arithmetic, we omit the proof.

Subtraction and division are defined as follows: If $z_1, z_2 \in C$, then

$$z_1 - z_2 = z_1 + (-z_2)$$

and

$$\frac{z_1}{z_2} = z_1 \cdot z_2^{-1}, \qquad \text{provided } z_2 \neq 0.$$

These definitions conform to the corresponding definitions for real numbers (see Section 4.6).

When performing arithmetic with complex numbers the student should keep in mind the following facts:

(a) The rules of arithmetic that are valid for the real numbers are also valid for the complex numbers.

(b) $i^2 = -1$.

If we consider i^n, where $n \in \{1, 2, 3, 4, 5, 6, 7, 8\}$, we find that

$$i^1 = i \quad \text{(by definition of } i^1\text{)}, \qquad i^5 = i,$$

$$i^2 = -1, \qquad\qquad\qquad\qquad\qquad i^6 = -1,$$

$$i^3 = -i, \qquad\qquad\qquad\qquad\qquad i^7 = -i,$$

$$i^4 = 1, \qquad\qquad\qquad\qquad\qquad\, i^8 = 1.$$

More generally, if $m, n \in N$,

$$i^{4m+n} = \underbrace{(i \cdot i \cdots \cdots i)}_{4m \text{ factors}} \underbrace{(i \cdot i \cdots \cdots i)}_{n \text{ factors}}$$

$$= \underbrace{(i^4 \cdot i^4 \cdots \cdots i^4)}_{m \text{ factors}} \cdot i^n$$

$$= \underbrace{(1 \cdot 1 \cdots \cdots 1)}_{m \text{ factors}} \cdot i^n$$

$$= 1 \cdot i^n$$

$$= i^n.$$

Thus we have the useful reduction formula

$$i^{4m+n} = i^n.$$

For example,

$$i^{97} = i^{4 \cdot 24 + 1}$$

$$= i^1$$

$$= i.$$

standard *form* If a complex number z is written in the form $z = a + bi$, where $a, b \in R$, we say that z is in *standard form*. The real *real part* numbers a and b are called the *real part* and the *imaginary part* *imaginary* of the complex number z, respectively. To illustrate, π is the *part* real part and $\frac{7}{25}$ is the imaginary part of the complex number $\pi + \frac{7}{25} \cdot i$.

Example 1. Express $\left(\frac{2}{3}+\frac{1}{5}i\right) \div \left(\frac{4-i}{3}\right)$ in standard form.

Solution:

$$\left(\frac{2}{3}+\frac{1}{5}i\right) \div \left(\frac{4-i}{3}\right) = \left(\frac{2}{3}+\frac{1}{5}i\right) \cdot \frac{3}{4-i}$$

$$= \left(\frac{2}{3}+\frac{1}{5}i\right) \cdot \frac{3 \cdot (4+i)}{(4-i)(4+i)}$$

$$= \left(\frac{2}{3}+\frac{1}{5}i\right) \frac{12+3i}{17}$$

$$= \left[\left(\frac{2}{3}+\frac{1}{5}i\right) \cdot (12+3i)\right] \div 17$$

$$= \left(\frac{37}{5}+\frac{22i}{5}\right) \div 17$$

$$= \frac{37}{85}+\frac{22}{85}i.$$

To determine the standard form of the multiplicative inverse of a nonzero complex number $a + bi$, we write

$$(a+bi)^{-1} = \frac{1}{a+bi} \cdot \frac{a-bi}{a-bi}$$

$$= \frac{a-bi}{(a+bi)(a-bi)}$$

$$= \frac{a-bi}{a^2+b^2}$$

$$= \frac{a}{a^2+b^2} - \frac{b}{a^2+b^2}i.$$

For example,

$$(3+4i)^{-1} = \frac{3}{3^2+4^2} - \frac{4}{3^2+4^2}i$$

$$= \frac{3}{25} - \frac{4}{25}i.$$

The formula

$$(a+bi)^{-1} = \frac{a}{a^2+b^2} - \frac{b}{a^2+b^2}i$$

is sometimes a time-saver in calculations involving complex numbers.

We conclude this section with another example.

Example 2. Express $i^{50} \left(\dfrac{2}{2+i} - \dfrac{i}{3-i} \right)$ in standard form.

Solution:

$$i^{50} = i^{4 \cdot 12 + 2}$$

$$= i^2$$

$$= -1.$$

$$\frac{2}{2+i} - \frac{i}{3-i} = \frac{2(3-i) - i(2+i)}{(2+i)(3-i)}$$

$$= \frac{7 - 4i}{7 + i}$$

$$= \left(\frac{7-4i}{7+i} \right) \left(\frac{7-i}{7-i} \right)$$

$$= \frac{45 - 35i}{50}$$

$$= 0.9 - 0.7i.$$

Therefore

$$i^{50} \left(\frac{2}{2+1} - \frac{i}{3-1} \right) = (-1)(0.9 - 0.7i)$$

$$= -0.9 + 0.7i.$$

Each set of numbers that the student is familiar with is a subset of complex numbers. The set of complex numbers is the largest set of numbers that we have described. We illustrate this with the following diagram:

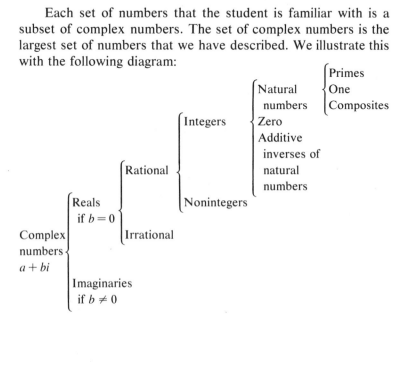

Exercise 4.7

Study the definition at the beginning of Section 4.7 before working Exercises 1–8. In each of these, perform the indicated operations.

1. $(3 + 4i) + (2 - 5i)$.
2. $(5 - 2i) + (-4 + 5i)$.
3. $(2 - 6i) + (-2 + 6i)$.
4. $(4 + 5i) + [(6 - 2i) + 4 - 3i]$.
5. $(3 + i)(5 - 2i)$.
6. $(2 - i)(2 + i)$.
7. $(3 + 2i)[(5 + 3i) + (2 - 5i)]$.
8. $(1 - 6i)[(2 + i) + (-3 - 4i)]$.

Read the discussion following the definition and study the reduction formula given before Example 1. Then simplify the expressions given in Exercises 9–14.

9. i^{27}. **10.** i^{51}.
11. i^{13}. **12.** $i^{53}(2 + i)^2$.
13. $i^{17}(3 - i)^2$. **14.** $(3 + 2i)(5 - 4i)i^{54}$.

Read Examples 1 and 2 before doing Exercises 15–20. In each of these, express the given complex number in standard form.

15. $\left(\dfrac{3}{4} + \dfrac{1}{5}i\right) + \left(\dfrac{4 - 3i}{2}\right)$.

16. $\left(\dfrac{2}{3} + \dfrac{5}{4}i\right) + \left(\dfrac{5 + 6i}{2}\right)$.

17. $\left(\dfrac{17}{13} + \dfrac{5}{26}i\right) + \left(\dfrac{2 - 3i}{5}\right)$.

18. $i^{35}\left(\dfrac{3}{2 - i} - \dfrac{i}{3 + i}\right)$.

19. $(2 + 3i)\left[\dfrac{4}{2 + 3i} - \dfrac{3i}{2 - 3i}\right]$.

20. $(5 - 6i) + \left[\dfrac{2i}{3 - i} + \dfrac{3}{3 - i}\right]$.

21. Prove that $i^2 = -1$. [*Hint:* $i = 0 + 1i$, therefore $i^2 = (0 + 1i)(0 + 1i)$. Use the definition of product.]

In Exercises 22–25 express the given complex number in standard form.

22. $\dfrac{1}{3+i}.$ **23.** $\dfrac{1}{\frac{1}{2}-\frac{1}{3}i}.$

24. $\left(3+\dfrac{1}{i}\right)^{-1}.$ **25.** $\dfrac{1+2i}{2-3i}.$

26. Give an example of (a) an imaginary number, (b) an irrational number, (c) a rational number that is not an integer, (d) an integer that is not a natural number, (e) a natural number that is not a prime.

27. Let $z = a + bi$ be a complex number. Show that $a = \dfrac{z+\bar{z}}{2}$, where $\bar{z} = a - bi$, and $b = \dfrac{z-\bar{z}}{2i}$.

28. Prove: A complex number z is real if and only if $z = \bar{z}$. [*Hint:* Use the definition of equality of complex numbers and consult Exercise 27 in this section.]

REVIEW EXERCISES

1. Find the solution set of the open sentence $(x^2 - 3) \cdot (x + \frac{1}{2}) = 0$, if the replacement set is Q (the set of rational numbers).

2. Let $S = \{x \mid x = \dfrac{p}{q}, \ \{p, q\} \subset I \text{ and } q \neq 0\}$. State which of the following statements are true and which are false:
(a) $S \subset Q$,   (b) $S \cap Q = \varnothing$,
(c) $Q \subset S$,   (d) $S = Q$.

In Exercises 3–6 express the given rational number in lowest terms.

3. $\frac{3}{5} + \frac{9}{11}.$ **4.** $\frac{5}{3} \cdot \left(\frac{4}{5} - \frac{7}{6}\right).$

5. $\left(\dfrac{1 + \frac{1}{4}}{1 - \frac{1}{4}}\right) \div \left(3 + \frac{7}{3}\right).$ **6.** $\dfrac{3 + \frac{11}{5}}{5 + \frac{11}{3}} - \left(\frac{1}{2} - \frac{1}{3}\right).$

7. Express $\frac{11}{4}$ in decimal form.

8. Show that $1.7\underline{3}$ is a rational number.

9. Show that the number $\pi + 1$ is irrational. (Recall that π is irrational.)

10. Show that if a and b are nonzero real numbers, then

$$\left[\left(\frac{a}{b}\right)^2 + \left(\frac{b}{a}\right)^2\right] \cdot \frac{ab^2}{a^4 + b^4} = \frac{1}{a}.$$

11. Simplify the expression $\dfrac{x^3 y - 4xy^3}{(x + 2y)(x - 2y)}.$ You may assume that x and y are real numbers and $(x + 2y)(x - 2y) \neq 0$.

In Exercises 12–17 express the given complex numbers in standard form.

12. $(5 + i)(5 - i)$.   **13.** $i^{14}(i^2 + i^3)$.

14. $(3 + 2i) \div (1 + 5i)$.   **15.** $3 + 7i - (\frac{1}{2} + \frac{1}{5}i)$.

16. $\dfrac{1 + i}{1 - i} + \dfrac{2 + i}{1 + i}$.   **17.** $\dfrac{4}{i} + \dfrac{3}{i + 1}$.

18. Find the multiplicative inverse of $1 + 3i$.

19. Find the additive inverse of $1 + 3i$.

In Exercises 20–25 label each statement as true or false. If a statement is false, illustrate this fact with an example.

20. Every integer is a rational number.

21. Some rational number is not an integer.

22. Every real number is a rational number.

23. Every real number is a complex number.

24. The sum of any two imaginary numbers is an imaginary number.

25. The product of any two imaginary numbers is an imaginary number.

CHAPTER 5

ALGEBRA
OF
POLYNOMIALS
AND
FACTORING

⬡

5.1 LAWS OF INTEGRAL EXPONENTS

If a is a complex number and $n \in N$, then we use the symbol a^n to denote the product

$$\underbrace{a \cdot a \cdot \ldots \cdot a.}_{n \text{ factors}}$$

Thus,

$$a^1 = a,$$

$$a^2 = a \cdot a,$$

$$a^3 = a \cdot a \cdot a, \qquad \text{and so on.}$$

nth power We call a^n the *nth power* of a. In the expression a^n, n is
exponent said to be the *exponent*. To illustrate, in the expression π^5, 5
is the exponent, and π^5 is called the fifth power of π.

If $n, m \in N$ and $a \in C$, then

$$a^n \cdot a^m = \underbrace{a \cdot a \cdot \ldots \cdot a}_{n \text{ factors}} \cdot \underbrace{a \cdot a \cdot \ldots \cdot a}_{m \text{ factors}}$$

$$= \underbrace{a \cdot a \cdot \ldots \cdot a}_{(n+m) \text{ factors}}$$

$$= a^{n+m}.$$

The foregoing argument proves the first statement in the
following theorem.

Theorem. If $a, b \in C$ and $n, m \in N$, then

    (1) $a^n \cdot a^m = a^{n+m}$ *(first basic law of exponents)*,
    (2) $(a^n)^m = a^{n \cdot m}$ *(second basic law of exponents)*,
    (3) $(a \cdot b)^n = a^n \cdot b^n$ *(third basic law of exponents)*.

The proofs of the last two statements are left as exercises.

Example 1.

$10^3 \cdot 10^6 = 10^{3+6} = 10^9$ (by the first basic law of exponents).

$(2^{15})^7 = 2^{15 \cdot 7} = 2^{105}$ (by the second basic law of exponents).

$(3i)^5 = 3^5 \cdot i^5 = 243i$ (by the third basic law of exponents).

We now wish to investigate the possibility of extending the
scope of validity of the basic laws of exponents by replacing the
restriction "$n, m \in N$" with the restriction "$n, m \in I$." Hence,
we must assign a suitable precise meaning to expressions such
as $2^0, 3^{-2}, \pi^{-5}, (1 + i)^{-3}$.

Let us for a moment assume that a is a nonzero complex
number and $a^m \cdot a^n = a^{m+n}$ is true for all integers m and n. Then,
if $n \in N$

$$a^n \cdot a^0 = a^{n+0} = a^n.$$

From the equality $a^n \cdot a^0 = a^n$ we conclude that $a^0 = 1$. Further,

$$a^n \cdot a^{-n} = a^{n-n}$$

$$= a^0$$

$$= 1.$$

From the equality $a^n \cdot a^{-n} = 1$, we conclude that a^{-n} is the
multiplicative inverse of a^n, that is $a^{-n} = (a^n)^{-1}$. Hence, if

$a \neq 0$ and $a^m \cdot a^n = a^{m+n}$ for all integers m and n, then $a^0 = 1$, and if $n \in N$, $a^{-n} = (a^n)^{-1}$. We thus give the following:

Definition. If $a \in C$ and $a \neq 0$, then $a^0 = 1$ and, whenever $n \in N$, $a^{-n} = (a^n)^{-1}$.

Note that the discussion preceding the definition is not a proof but simply an attempt to show the reason for this definition. That is, the definition has been chosen in such a way that the basic laws of exponents will hold for all integer exponents.

Remarks. (1) We do not assign a meaning to 0^n if $n \in I - N$. In particular, 0^0 is undefined.

(2) If $a \neq 0$ and $n \in I$, there are three possibilities: $n \in N$, $n = 0$, or $-n \in N$. If $n \in N$, then

$$a^{-n} = (a^n)^{-1} \qquad \text{(by the foregoing definition)}$$

$$= \frac{1}{a^n}.$$

If $n = 0$, then

$$a^{-0} = a^0 \qquad \text{(since } -0 = 0)$$

$$= 1 \qquad \text{(since } a^0 = 1)$$

$$= \frac{1}{a^0}.$$

If $-n \in N$, then

$$a^{-n} = [(a^{-n})^{-1}]^{-1} \qquad \text{(why?)}$$

$$= (a^{-(-n)})^{-1} \qquad \text{(by the foregoing definition)}$$

$$= (a^n)^{-1}$$

$$= \frac{1}{a^n}.$$

We can conclude now that for any integer n,

$$a^{-n} = \frac{1}{a^n}, \qquad \text{hence} \qquad a^n \cdot a^{-n} = 1 \qquad (a \neq 0).$$

(3) If $a \neq 0$, $n, -m \in N$, and $n + m = 0$, then $m = -n$. Hence,

$$a^n \cdot a^m = a^n \cdot a^{-n}$$

$$= 1 \qquad \text{(by Remark (2))}$$

$$= a^0 \qquad \text{(since } a^0 = 1)$$

$$= a^{n+m} \qquad \text{(since } n + m = 0).$$

(4) If $a \neq 0$, $n, -m \in N$, and $n + m \in N$, then

$$a^n \cdot a^m = (a^{n+m} \cdot a^{-m}) a^m \qquad \text{(by the first basic law}$$
$$\text{of exponents)}$$

$$= a^{n+m} \cdot (a^{-m} \cdot a^m) \qquad \text{(why?)}$$

$$= a^{n+m} \cdot 1 \qquad \text{(by Remark (2))}$$

$$= a^{n+m}.$$

(5) If $a \neq 0$, $n, -m \in N$, and $-(n + m) \in N$, then

$$a^{-(n+m)} \cdot (a^n \cdot a^m) = (a^{-(n+m)} a^n) a^m$$

$$= a^{-(n+m)+n} \cdot a^m \qquad \text{(by the first basic law}$$
$$\text{of exponents)}$$

$$= a^{-m} \cdot a^m$$

$$= 1. \qquad \text{(by Remark (2))}$$

Hence,

$$a^n \cdot a^m = \frac{1}{a^{-(n+m)}}$$

$$= a^{-[-(n+m)]} \qquad \text{(by Remark (2))}$$

$$= a^{n+m}.$$

(6) If $a \neq 0$ and $-n, -m \in N$, then

$$a^n \cdot a^m = a^{-(-n)} \cdot a^{-(-m)}$$

$$= \frac{1}{a^{-n}} \cdot \frac{1}{a^{-m}} \qquad \text{(by Remark (2))}$$

$$= \frac{1}{a^{-n} \cdot a^{-m}}$$

$$= \frac{1}{a^{-n-m}} \qquad \text{(by the first basic law of exponents)}$$

$$= a^{n+m}. \qquad \text{(by Remark (2))}$$

(7) If $a \neq 0$, $n, m \in I$, and $n = 0$, then

$$a^n \cdot a^m = a^0 \cdot a^m$$

$$= 1 \cdot a^m \qquad \text{(since } a^0 = 1)$$

$$= a^m$$

$$= a^{0+m}$$

$$= a^{n+m} \qquad \text{(since } n = 0).$$

Theorem. If $a, b \in C, n, m \in I$ and $ab \neq 0$, then

(1) $a^n a^m = a^{n+m}$ (*first law of integral exponents*),

(2) $(a^n)^m = a^{n \cdot m}$ (*second law of integral exponents*),

(3) $(ab)^n = a^n \cdot b^n$ (*third law of integral exponents*),

(4) $\dfrac{a^n}{a^m} = a^{n-m}$ (*fourth law of integral exponents*),

(5) $\dfrac{a^n}{b^n} = \left(\dfrac{a}{b}\right)^n$ (*fifth law of integral exponents*).

Proof: The first law of integral exponents follows from the first basic law of exponents and Remarks (3) through (7).

The second and third laws of integral exponents can be established by considering the cases $n \in N, n = 0, -n \in N$. The details are left as exercises.

(4) $\dfrac{a^n}{a^m} = \dfrac{a^{n-m}a^m}{a^m}$ (by the first law of integral exponents)

$= a^{n-m}.$

(5) $\dfrac{a^n}{b^n} = a^n \cdot \dfrac{1}{b^n}$

$= a^n \cdot b^{-n}$ (by Remark (2))

$= a^n \cdot (b^{-1})^n$ (by the second law of integral exponents)

$= (a \cdot b^{-1})^n$ (by the third law of integral exponents)

$= \left(\dfrac{a}{b}\right)^n.$

Remark. Recall that we did not define 0^0. Of course we had the privilege of defining it if we so desired — but there is no point in making a definition unless it is useful.

The next example illustrates some applications of the foregoing theorem.

Example 2. Simplify the following expressions. Assume that $a, b \in C, ab \neq 0$ and $n \in I$.

(1) $a^{n+2} \cdot a^{-5} \cdot b^{n-3}$,

(2) $\dfrac{(a^2)^5}{a^8 \cdot b^2}$,

(3) $(a^{-1} - b^{-2})^{-3}$ (assume also that $b^2 - a \neq 0$).

Solution:

(1) $a^{n+2} \cdot a^{-5} \cdot b^{n-3} = a^{(n+2)+(-5)} \cdot b^{n-3}$ (by the first law of integral exponents)

$= a^{n-3} \cdot b^{n-3}$

$= (ab)^{n-3}$ (by the third law of integral exponents).

(2) $\dfrac{(a^2)^5}{a^8 \cdot b^2} = \dfrac{a^{2 \cdot 5}}{a^8 \cdot b^2}$ (by the second law of integral exponents)

$= \dfrac{a^{10}}{a^8} \cdot \dfrac{1}{b^2}$

$= a^{10-8} \cdot \dfrac{1}{b^2}$ (by the fourth law of integral exponents)

$= \dfrac{a^2}{b^2}$

$= \left(\dfrac{a}{b}\right)^2$ (by the fifth law of integral exponents).

(3) $(a^{-1} - b^{-2})^{-3} = \left(\dfrac{1}{a} - \dfrac{1}{b^2}\right)^{-3}$ $\left(\text{since } a^{-1} = \dfrac{1}{a} \text{ and } \right.$

$\left. b^{-2} = \dfrac{1}{b^2}\right)$

$= \left[\left(\dfrac{b^2 - a}{ab^2}\right)^{-1}\right]^3$ (by the second law of integral exponents)

$= \left(\dfrac{ab^2}{b^2 - a}\right)^3$ $\left(\text{since the multiplicative inverse of } \right.$

$\left. \dfrac{b^2 - a}{ab^2} \text{ is } \dfrac{ab^2}{b^2 - a}\right).$

Exercise 5.1

Review Example 1 before working Exercises 1–6. Apply the basic laws of exponents to the following expressions and state the law used in each case.

1. $3^8 \cdot 3^2$.   **2.** $(7a)^5$.

3. $(25^6)^9$.

4. $i^7 \cdot i^4$.

5. $[(a+b)^2]^3$.

6. $[u(v-w)]^9$.

Study Example 2 before doing Exercises 7–18. Simplify the following expressions. Assume that $\{a, b, c, d\} \subset C$, $\{m, n\} \subset I$, and that a, b, c, d are such that none of the given expressions involves division by zero.

7. $(a^5)^2(-a^{-2})^3$.

8. $(2ab^3)^3(-4a^2b^{-1})^2$.

9. $\dfrac{(c^5)^2}{(cd)^5}$.

10. $\dfrac{(-a)^3(a^2b)^3}{(ab^{-1})^3}$.

11. $(a^{-1}-b^{-1})^{-3}$.

12. $\dfrac{(c^{-2}-d^{-2})^{-5}}{(c^{-2}+d^{-2})^{-5}}$.

13. $a^{n-m}(a^{n+m})^2b^{3n+m}$.

14. $(a+b)^{4m+n}(a+b)^{-3m+n}(a+b)^m$.

15. $\dfrac{(c^n)^{-4}d^{3m-n}}{c^{-3m+n}(d^4)^n}$.

16. $(c^{-n}+c^m)^{-m}$.

17. $(a^{-n}-b^{-n})^{-n}(a^{-n}+b^{-n})^n$.

18. $(c^{-n}-c^2)^{m+n}(c^{-m}-c^2)^{-m-n}$.

*19. Prove the second basic law of exponents.

*20. Prove the third basic law of exponents.

*21. Prove the second law of integral exponents.

*22. Prove the third law of integral exponents.

23. Show that the proposition "$(a+b)^n = a^n + b^n$ for all numbers a, b and positive integers n" is false by replacing a, b, n by specific numbers. Under what conditions does the equality hold?

5.2 ADDITION AND SUBTRACTION OF POLYNOMIALS

Let x be a variable with replacement set C (recall that C denotes the set of complex numbers). Expressions of the form cx^n or c, where c is a specific complex number and $n \in N$, are called *monomials* in the single variable x over C.

For example, the expressions $2x^5$, x, $1 + i$, 0 are monomials over C. Note that whenever we replace x in the monomial $2x^5$ by a complex number, we obtain a complex number. To illustrate, $2i^5$, $2 \cdot 3^5$ are complex numbers.

Question. Why is $2x^5$ not an open sentence?

polynomial A *polynomial* in a single variable x over C is a monomial in a single variable x over C or a finite sum of monomials in a single variable x over C.

Although it is possible to consider both monomials and polynomials in more than one variable (such as $xy^2, 3xy + x$), to simplify matters we shall not discuss them here. Within the context of our discussion, in this chapter the term "polynomial" will always refer to a polynomial in a single variable over C.

Example 1. The following expressions are polynomials: $3x^3 + ix, 3x, 2$. On the other hand, $x + x^2 + x^3 + \cdots$ (the three dots indicate that the sum is continued indefinitely) is not a polynomial (why?). Neither is the expression x^{-2} a polynomial, since the exponent of x, namely -2, is not a natural number, and x^{-2} is not a specific complex number.

It is a very common practice to write polynomials in the order of descending or ascending powers of the variable. Thus we often write $3x^2 + x + 1$ or $1 + x + 3x^2$ instead of $1 + 3x^2 + x$. If a polynomial $T(x)$ (read "T of x") is expressed in the form

standard
form

$$a_n x^n + a_{n-1} x^{n-1} + \cdots + a_1 x + a_0,$$

leading
coefficient
term
coefficients
degree

linear
polynomial
quadratic
polynomial

zero
polynomial

where each a_j is a complex constant and $a_n \neq 0$, then we say that $T(x)$ is in *standard form*. In this representation a_n is called the *leading coefficient*. Each of the monomials $a_0, a_1 x, \cdots$, $a_{n-1} x^{n-1}, a_n x^n$ is said to be a *term* of $T(x)$, and the complex numbers $a_0, a_1, \cdots, a_{n-1}, a_n$ are called *coefficients* of $T(x)$. If $n \neq 0$ or $a_0 \neq 0$, then the *degree* of $T(x)$ is n. If the degree of $T(x)$ is 1, then $T(x)$ is said to be a *linear polynomial*. A *quadratic polynomial* is a polynomial of degree 2. The term *zero polynomial* specifies the number 0.

Example 2. (1) The standard form of the polynomial $T(x)$, where $T(x) = 1 + 2x + 3x^2$, is $3x^2 + 2x + 1$. The leading coefficient is 3. The terms of $T(x)$ are: $1, 2x, 3x^2$. Each of the numbers $1, 2, 3$ is a coefficient of $T(x)$ and $T(x)$ is a quadratic polynomial.

  (2) The polynomial $x + 1$ is linear.

  (3) Every nonzero complex number may be regarded as a polynomial of degree 0.

If $T(x)$ and $S(x)$ are polynomials, they are considered equal if and only if the solution set of the open sentence $T(x) = S(x)$ is C. Symbolically, $T(x) = S(x)$ if and only if $\{x \mid T(x) = S(x)\} = C$. To emphasize that $\{x \mid T(x) = S(x)\} = C$, we may use the notation $T(x) \equiv S(x)$ (read "$T(x)$ is identically equal to $S(x)$").

It can be shown that $T(x)$ is identically equal to $S(x)$ if and only if the corresponding coefficients of these two polynomials are equal. That is,

$a_0 x^n + a_1 x^{n-1} + \cdots + a_{n-1} x + a_n$

$$\equiv b_0 x^n + b_1 x^{n-1} + \cdots + b_{n-1} x + b_n$$

if and only if

$$a_0 = b_0 \wedge a_1 = b_1 \wedge \cdots \wedge a_{n-1} = b_{n-1} \wedge a_n = b_n.$$

(See Exercises 26, 27, 28 in Section 5.3 for some special cases of this theorem.)

To illustrate, if $T(x) = 2x^2 + 1$ and $S(x) = x^2 + 1$, then $T(x) \neq S(x)$, since the complex number 1 is not a solution of the open sentence $2x^2 + 1 = x^2 + 1$. (To see this, note that the proposition $2 \cdot 1^2 + 1 = 1^2 + 1$ is false.)

We perform addition and subtraction of polynomials in a natural way. That is, to add two polynomials or to subtract a polynomial from a polynomial, we regard the variable as an unspecified complex number and use the properties of complex numbers. In particular, the commutative and associative properties of addition, the distributive law, and the properties of additive inverses and the number 0 are used extensively.

Example 3. Add the polynomials $3x^2 + 7x + 1$ and $5x^2 - 7x - 2$.

Solution:

$(3x^2 + 7x + 1) + (5x^2 - 7x - 2)$

$$= (3x^2 + 5x^2) + (7x - 7x) + (1 - 2)$$

$$= (3 + 5)x^2 + (7 - 7)x + (-1)$$

$$= 8x^2 + 0 \cdot x - 1$$

$$= 8x^2 - 1.$$

Example 4. Express the difference

$(15x^7 + 16x^6 + 3x^3 - 2x^2 + 1) - (32x^8 + 15x^6 - 7x^5 + x^3 - x^2 + 2)$

in standard form.

Solution:

$(15x^7 + 16x^6 + 3x^3 - 2x^2 + 1) - (32x^8 + 15x^6 - 7x^5 + x^3 - x^2 + 2)$

$= 15x^7 + 16x^6 + 3x^3 - 2x^2 + 1$

$\quad - 32x^8 - 15x^6 + 7x^5 - x^3 + x^2 - 2$

$$= -32x^8 + 15x^7 + (16 - 15)x^6 + 7x^5 + (3 - 1)x^3$$
$$+ (-2 + 1)x^2 + (1 - 2)$$
$$= -32x^8 + 15x^7 + x^6 + 7x^5 + 2x^3 - x^2 - 1.$$

Remark. If the given polynomials are lengthy expressions and we wish to perform addition or subtraction, we may reduce both the amount of writing and the possibilities of errors caused by oversight by using the following schematic device:

$$15x^7 + 16x^6 + 0x^5 + 0x^4 + 3x^3 - 2x^2 + 0x + 1$$
$$\frac{-(32x^8 + \ \ 0x^7 + 15x^6 - 7x^5 + 0x^4 + \ \ x^3 - \ \ x^2 + 0x + 2)}{-32x^8 + 15x^7 + \ \ \ \ x^6 + 7x^5 + 0x^4 + 2x^3 - \ \ x^2 + 0x - 1} .$$

The student should compare the foregoing with the solution in Example 4.

Example 5. If a is a complex number and $a \neq 1$, express the following in standard form:

$$[ax - (a + x)] - [2x - [3x - (x - a)] + a].$$

Solution:

$$ax - (a + x) = ax - a - x = (a - 1)x - a,$$
$$3x - (x - a) = 3x - x + a = 2x + a,$$
$$2x - [3x - (x - a)] + a = 2x - (2x + a) + a$$
$$= 2x - 2x - a + a$$
$$= 0.$$

Therefore,

$$[ax - (a + x)] - [2x - [3x - (x - a)] + a] = (a - 1)x - a - 0$$
$$= (a - 1)x - a.$$

We use the symbol $C[x]$ to denote the set of all polynomials in a single variable x over C. Although we do not give a proof here, it is true that if $Q(x), S(x) \in C[x]$, then $Q(x) + S(x)$ and $Q(x) - S(x)$ are unique members of $C[x]$. Thus addition and subtraction are indeed operations on the set $C[x]$.

Remark. It is customary to use one of the letters from the end of the alphabet to denote a variable, while letters at the beginning of the alphabet usually represent fixed numbers. For example, unless otherwise specified, the expression $au^2 + bu + c$ would be considered a polynomial in the variable u, degree two, with leading coefficient a, assuming $a \neq 0$.

Exercise 5.2

Read Example 1 before working Exercises 1 and 2. Which of the following expressions are polynomials?

1. (a) $x + i$, (b) 3, (c) $4u^{-1}$, (d) $\frac{1}{2} + \frac{1}{3}x + \frac{1}{4}x^2 + \cdots$.
2. (a) $2ix$, (b) $2V^2 - 2$, (c) $y^2 + y + y^{-2}$,
   (d) $x + 2x^2 + 3x^3 + \cdots$.

Study Example 2 before working Exercises 3–7. Write each of the following polynomials in standard form. Also in each case write down (a) the degree, (b) the leading coefficient, (c) the terms, (d) the coefficients.

3. $T(x) = 3 - 2x$.
4. $S(t) = -it + 3t^4$.
5. $T(x) = 5$.
6. $T(y) = a + by + cy^2$, where $\{a, b, c\} \subset C$ and $c \neq 0$.
7. $(2 + i)x^2 - 5x^4 + 3x - (2 + 4i)$.

Review Example 3 before working Exercises 8–19. Add the given polynomials as illustrated in Example 3.

8. $2x + 1$ and $5x - 2$.
9. $4 - 7y$ and $10y - 6$.
10. $4y^2 - 2y$ and $8y^2 + y - 1$.
11. $7x - 5x^2 + 2$ and $1 + x$.
12. $-12x^2 + 4$ and $15x^2 - 3x$.
13. $6u^2 - (i - 1)u - (-3 + 4i)$ and
    $2u^2 + (1 + i)u + (4i + 3)$.
14. $8x^6 - 7x^5 + 6x^4 - x + 1$ and $3x - 4x^3 + 5x^5 - 6x^6$.
15. $33V^{12} - 13V^{10} + 10V^8 - V^7 + V^2$ and
    $9V^{11} + 6V^9 - 6V^6 + 7V^5 - V^2$.
16. $17m^7 + 18im^6 - 12m^4 + im^3 - 16m$ and
    $18m^8 + im^7 - (6 + i)m^4 + 6m^3 - 2im$.
17. $6w^8 - 4iw^7 + 6w^6 - (2 - i)w^5 + w^4 - (i - 3)w^3 + 2w^2$
    and $-5w^8 + 4iw^7 - 5w^6 - (i - 2)w^5 - (3 - i)w^3 - w^2$.
18. $(a^2 - b)x^5 + (a^2 + c)x^4 - (-a^2 - d)x^3 - (-a^2 + e)x^2$
    $+ a^2$ and $bx^5 - cx^4 - dx^3 + ex^2 + a^2x$.
19. $(a + bi^3)z^4 + ci^5z^3 + (d + ei)z^2 - (a + bi)$ and

    $(-bi - a)z^4 + aiz^3 - (ai^3 + d)z^2 + bi^5z + ai^4$.

Study Example 4 before working Exercises 20–27. Subtract the second polynomial from the first and express the result in standard form.

20. $5x^2 + 3x$, $4x^2 + 2x - 1$.

21. $8x^3 - 7x^2 + x + 4$, $-2x^3 + 3x^2 - 9x - 6$.

22. $4 - 3x + 5x^2 + 6x^5$, $2 + 3x - 3x^3 + 3x^5 + x^4 - 2x^6$.

23. $25u^8 - 12u^7 + 4u^5 - 8u^4 + 9u^3$,
$$15u^8 - 8u^7 + u^6 + 3u - u^2 + 3u^5 - 1.$$

24. $13y^6 + 3y^3 - 21y + 39$, $41 - 20y + y^2 - 4y^3 - 17y^4$
$$+ 5y^6 + 41y^9.$$

25. $(6 + i)u^5 - 4iu^4 + 3u^3 + 6u^2 - 2iu - 3$,
$$3u^4 - iu^3 + 2iu^2 + 7u + i.$$

26. $ai - ai^2V + bi^3V^2 - ci^5V^3$, $-ci^3V^3 + aiV^4 + bi^3V^5$.

27. $(a + bi) - (b + ci)y + (c - di)y^2 - (d + ei)y^3$,
$$a - ciy + cy^2 - eiy^3.$$

Study Example 5 before working Exercises 28–33. Write each expression in standard form. Assume that $\{a, b, c, d\} \subset C$.

28. $[2x - (3 + 4x)] - \{5x - [7x - (1 - 2x)] + 8x\}$.

29. $-[y + 2 - (2y - 1)] - \{4y - 2[y + 3(2y - 1)] - 2y\}$.

30. $2[ax - 2(x - a)] + \{7x - 3[ax + (a - x)]\}$.

31. $[3u - a(u + b) + 2au] - \{-4au + a[-u -$
$$(ab - 5u)] - 6au\}.$$

32. $-c\{[az - b] - a - (d - bz)]\}$
$$- \{-dbz - [cdz - ac] + adz\}.$$

33. $-\{[bx + c] - a + cx - (c + dx)]\}$
$$- \{ax - [dx - a] - bx - d\}.$$

5.3 MULTIPLICATION AND DIVISION OF POLYNOMIALS

We perform multiplication of polynomials in a natural way. That is, to multiply two polynomials, we regard the variable as an unspecified complex number and use the properties of the complex numbers and the basic laws of exponents. Since the process of multiplication employs the distributive law quite extensively, we make a brief comment on this property.

If $a, c_1, c_2 \in C$, then according to the distributive law $a(c_1 + c_2) = ac_1 + ac_2$. Now suppose $a, c_1, c_2, c_3 \in C$. Then

$$a(c_1 + c_2 + c_3) = a((c_1 + c_2) + c_3)$$

$$= a(c_1 + c_2) + ac_3 \qquad \text{(by the distributive law)}$$

$$= (ac_1 + ac_2) + ac_3 \qquad \text{(by the distributive law)}$$

$$= ac_1 + ac_2 + ac_3.$$

More generally, it can be shown that, if $a, c_1, c_2, \cdots, c_n \in C$, then

$$a(c_1 + c_2 + \cdots + c_n) = ac_1 + ac_2 + \cdots + ac_n.$$

generalized The foregoing is called the *generalized distributive law*.
distributive The next example illustrates multiplication of two poly-
law nomials, using the generalized distributive law.

Example 1. Multiply $3x^2 + x$ and $x^2 + x + 1$.

Solution:

$$(3x^2 + x)(x^2 + x + 1) = 3x^2(x^2 + x + 1) + x(x^2 + x + 1)$$
$$= 3x^2 \cdot x^2 + 3x^2 \cdot x + 3x^2 \cdot 1 + x \cdot x^2$$
$$+ x \cdot x + x \cdot 1$$
$$= 3x^4 + 3x^3 + 3x^2 + x^3 + x^2 + x$$
$$= 3x^4 + (3 + 1)x^3 + (3 + 1)x^2 + x$$
$$= 3x^4 + 4x^3 + 4x^2 + x.$$

Therefore,

$$(3x^2 + x)(x^2 + x + 1) = 3x^4 + 4x^3 + 4x^2 + x.$$

Remark. The foregoing multiplication could be also shown as
follows:

$$3x^2 + x$$
$$\underline{x^2 + x + 1}$$
$$\overline{x^3 + x^2 + x} \qquad \text{(note that}$$
$$\qquad\qquad x^3 + x^2 + x = x(x^2 + x + 1))$$

$$\underline{+ \quad 3x^4 + 3x^3 + 3x^2} \qquad \text{(note that}$$
$$3x^4 + 4x^3 + 4x^2 + x \qquad 3x^4 + 3x^3 + 3x^2 = 3x^2(x^2 + x + 1)).$$

Expressions of the form $\dfrac{F(x)}{G(x)}$, where $F(x)$ and $G(x)$ are
rational polynomials and $G(x)$ is not the zero polynomial, are called
expressions *rational expressions*.

To divide a polynomial $F(x)$ by a nonzero polynomial
$G(x)$ will always mean to express $\dfrac{F(x)}{G(x)}$ in the form

$$\frac{F(x)}{G(x)} = T(x) \qquad \text{or} \qquad \frac{F(x)}{G(x)} = T(x) + \frac{S(x)}{G(x)},$$

where $T(x)$ and $S(x)$ are polynomials and $S(x)$ is a nonzero
polynomial whose degree is less than the degree of $G(x)$. Note
that the replacement set of $\dfrac{F(x)}{G(x)}$ may be a proper subset of C,

since we cannot replace x by a complex number b for which $G(b) = 0$.

Example 2. Divide $x^2 + x$ by x.

Solution:

$$\frac{x^2 + x}{x} = \frac{x^2}{x} + \frac{x}{x}$$

$$= x + 1.$$

Thus

$$\frac{x^2 + x}{x} = x + 1.$$

Example 3. Verify that

$$(x^3 + 1) \div (x^2 - 2x + 1) = x + 2 + \frac{3x - 1}{x^2 - 2x + 1}.$$

Solution:

$$\left[(x + 2) + \frac{3x - 1}{x^2 - 2x + 1} \right] (x^2 - 2x + 1)$$

$$= (x + 2)(x^2 - 2x + 1) + \left(\frac{3x - 1}{x^2 - 2x + 1} \right)(x^2 - 2x + 1)$$

$$= (x^3 - 3x + 2) + (3x - 1)$$

$$= x^3 + 1.$$

Example 4. Divide $2x^3 + x + 1$ by $x^2 + 2$.

Solution: We wish to find polynomials $T(x)$ and $S(x)$ such that

$$(2x^3 + x + 1) \div (x^2 + 2) = T(x) + \frac{S(x)}{x^2 + 2},$$

where the degree of $S(x)$ is less than 2 or $S(x)$ is the zero polynomial. If such polynomials $T(x)$ and $S(x)$ exist, then

$$2x^3 + x + 1 = T(x) \cdot (x^2 + 2) + S(x).$$

It is clear that the degree of $T(x)$ must be 1. Let $T(x) = a_1 x + a_0$. Then we must have

$$2x^3 + x + 1 = a_1 x(x^2 + 2) + a_0(x^2 + 2) + S(x).$$

The leading coefficients of the right-hand and left-hand polynomials are a_1 and 2, respectively. Thus we are induced to choose $a_1 = 2$. Now we have

$$2x^3 + x + 1 = 2x(x^2 + 2) + a_0(x^2 + 2) + S(x).$$

Hence,

$$(2x^3 + x + 1) - 2x(x^2 + 2) = a_0(x^2 + 2) + S(x).$$

That is,

$$-3x + 1 = a_0(x^2 + 2) + S(x).$$

Since $S(x)$ is the zero polynomial or the degree of $S(x)$ is less than two, we are induced to choose $a_0 = 0$. It then follows that $S(x) = -3x + 1$ and $T(x) = 2x$. It is easy to verify that

$$2x^3 + x + 1 = 2x(x^2 + 2) + (-3x + 1).$$

From the foregoing equality we conclude that

$$(2x^3 + x + 1) \div (x^2 + 2) = 2x + \frac{-3x + 1}{x^2 + 2}.$$

The preceding example justifies the steps in the following schematic device, called the *long-division technique of poly-nomials*.

$$
\begin{array}{r}
2x \\
x^2 + 2 \overline{\smash{\big)}\, 2x^3 + \ x + 1} \\
\underline{2x^3 + 4x} \\
-3x + 1
\end{array}
$$
(note that $2x^3 + 4x = 2x(x^2 + 2)$)
(note that
$-3x + 1 = (2x^3 + x + 1) - (2x^3 + 4x)$).

Example 5. Divide $3x^3 - 2x^2 + x + 1$ by $x^2 + 5$.

Solution: We use the long-division technique of polynomials.

$$
\begin{array}{r}
3x \ - 2 \\
x^2 + 5 \overline{\smash{\big)}\, 3x^3 - 2x^2 + \ \ x + \ 1} \\
\underline{3x^3 \qquad\quad + 15x} \\
-2x^2 - 14x + \ 1 \\
\underline{-2x^2 \qquad\quad - 10} \\
- 14x + 11 \ .
\end{array}
$$

Therefore

$$(3x^3 - 2x^2 + x + 1) \div (x^2 + 5) = 3x - 2 + \frac{-14x + 11}{x^2 + 5}.$$

In Section 5.2 we introduced the symbol $C[x]$ to represent the set of all polynomials in a single variable x over C. It can be shown that if $F(x), G(x) \in C[x]$, then the *product* $F(x) \cdot G(x)$ is a unique member of $C[x]$ (the proof of this statement will not be given here). Thus multiplication of polynomials is an operation on $C[x]$. On the other hand, the polynomials x and x^2 are members of $C[x]$; however, $\dfrac{x}{x^2} \notin C[x]$. Hence division of polynomials is not an operation on $C[x]$.

Exercise 5.3

Study Example 1 before working Exercises 1–6. In each of these, multiply the given polynomials.

1. $2x^3 + x^2 + 1$	and	$3x^2 - 2.$
2. $3x^4 - x^2$	and	$4x^3 - x + 1.$
3. $5x^2 - x + 1$	and	$5x^2 + x - 1.$
4. $4u^4 + u^3 - 1$	and	$5u^5 + u^2 - u.$
5. $8y^7 - 3y^5 + 9y^3 - 2y$	and	$y^6 - 2y^4.$
6. $-7z^5 + z^4 - z^3$	and	$2z^3 + 3z^2.$

Review Example 2 before working Exercises 7–10. In each of these, divide the first polynomial by the second.

7. $y^4 + 3y^3 - y^2;\ y^2.$

8. $(x - 1)^3 - 3(x - 1)^2 + (x - 1);\ x - 1.$

9. $(u + 3)^5 - 2(u + 3)^3 + (u + 3);\ u + 3.$

10. $2(x^2 + x - 1)^3 - 3(x^2 + x - 1)^2;\ x^2 + x - 1.$

Before doing Exercises 11–14, study Example 3. Verify each of the following.

11. $(x^4 - 1) \div (x^2 + 2x + 1) = x^2 - 2x + 3 - \dfrac{4x + 4}{x^2 + 2x + 1}.$

12. $(x^3 + x^2 + 1) \div (x^2 + 1) = x + 1 + \dfrac{-x}{x^2 + 1}.$

13. $(x^4 + x^3 + x^2 + x + 1) \div (x^2 + x + 1) = x^2 + \dfrac{x + 1}{x^2 + x + 1}.$

14. $(4y^5 - 3y^4 + 3y + 1) \div (y^3 + 3y - 1) = 4y^2 - 3y - 12 + \dfrac{13y^2 + 36y - 11}{y^3 + 3y - 1}.$

Study Example 4 before working Exercises 15–18.

15. Divide $4x^3 - 3x + 2$ by $2x^2 - 1.$

*16. Divide $2y^5 + 3y^3 + 1$ by $y^2 + y + 1.$

*17. Divide $u^6 - 1$ by $u^2 + 2.$

*18. Divide $3V^6 + 4V^4 - V + 6$ by $V^3 - V + 2.$

Review Example 5 before working Exercises 19–25. In these exercises, use the long-division technique of polynomials.

19. Divide $6x^3 + x - 2$ by $3x^2 + 2$.
20. Divide $4u^6 - 2u^3 + u + 3$ by $2u^3 + 1$.
21. Divide $x^4 - 1$ by $x - 1$.
22. Divide $x^3 + 3x - 5$ by $x + 2$.
23. Divide $3x^4 - 2x^3 + 2x - 7$ by $3x^2 + 5x - 1$.
24. Divide $y^3 + 27$ by $y + 3$.
25. Divide $6x^5 - x^2 + 2$ by $x^3 + 3x^2 - 5x + 1$.
26. Prove: If $3x + 1 \equiv a_1 x + a_0$, then $a_1 = 3$ and $a_0 = 1$.
[*Hint:* If $3x + 1 \equiv a_1 x + a_0$, then $\{x \mid 3x + 1 = a_1 x + a_0\} = C$, hence

$$\{0, 1\} \subset \{x \mid 3x + 1 = a_1 x + a_0\}.]$$

27. Prove: If $ax + b \equiv c$, then $a = 0$ and $b = c$. [*Hint:* If you have difficulties, do Exercise 26 first.]
28. Prove: If $a_1 x + a_0 \equiv b_1 x + b_0$, then $a_1 = b_1 \land a_0 = b_0$. [*Hint:* Use Exercise 27.]

5.4 THE REMAINDER THEOREM AND SOME OF ITS APPLICATIONS

In Example 4 of Section 5.3 we determined polynomials $T(x)$ and $S(x)$, each of degree 1 such that $2x^3 + x + 1 = T(x) \cdot (x^2 + 2) + S(x)$. This enabled us to divide $2x^3 + x + 1$ by $x^2 + 2$. The following theorem allows us to conclude that we can always divide a polynomial $F(x)$ by a nonzero polynomial $G(x)$.

Theorem (Division Algorithm for Polynomials). If $F(x)$, $G(x) \in C[x]$ and $G(x)$ is not the zero polynomial, then there are polynomials $T(x)$ and $S(x)$ in $C[x]$ such that $F(x) = T(x) \cdot G(x) + S(x)$, where the degree of $S(x)$ is less than the degree of $G(x)$ or $S(x)$ is the zero polynomial. (The polynomial $S(x)$ is called *remainder* the *remainder.*) Moreover, $T(x)$ and $S(x)$ are uniquely determined by the given polynomials $F(x)$ and $G(x)$.

A proof of this theorem is beyond the scope of this text. We wish, however, to point out some consequences of the division algorithm for polynomials.

Suppose $F(x) \in C[x]$ and c is a complex number. Then according to the foregoing theorem there are polynomials $T(x)$ and $S(x)$ in $C[x]$ such that $F(x) = T(x)(x - c) + S(x)$, where the degree of $S(x)$ is less than the degree of the polynomial $x - c$ or $S(x)$ is the zero polynomial (that is, the degree of $S(x)$

is the number 0 or $S(x) \equiv 0$). We conclude now that there is a complex number K such that $F(x) = T(x)(x - c) + K$. An interesting question arises, if we inquire whether the number K is related to $F(x)$. Before investigating the answer, we want to make a brief comment on notation.

If $F(x)$ is a polynomial and c is a complex number, then $F(c)$ is the complex number obtained by substituting c for x in the polynomial expression represented by $F(x)$.

Example 1. If $F(x) = 2x^4 + x + 1$, then

$$F(1) = 2 \cdot 1^4 + 1 + 1 = 4,$$

$$F(\tfrac{1}{2}) = 2 \cdot (\tfrac{1}{2})^4 + \tfrac{1}{2} + 1 = \tfrac{13}{8},$$

$$F(i) = 2 \cdot i^4 + i + 1 = 3 + i,$$

and if $c \in C$,

$$F(c) = 2c^4 + c + 1.$$

Theorem (The Remainder Theorem). If $F(x) \in C[x]$ and $c \in C$, then there is a polynomial $T(x)$ in $C[x]$ such that $F(x) = T(x)(x - c) + F(c)$.

Proof: We showed earlier in this section that there is a polynomial $T(x)$ and a complex number K such that $F(x) = T(x) \cdot (x - c) + K$. Hence,

$$F(c) = T(c)(c - c) + K$$

$$= T(c)0 + K$$

$$= K.$$

Therefore, $F(x) = T(x)(x - c) + F(c)$.

Example 2. Show that a natural number is divisible by 9 if and only if the sum of its digits is divisible by 9.

Solution: Let q be a natural number and write

$$q = a_n 10^n + a_{n-1} 10^{n-1} + \cdots + a_1 10 + a_0,$$

where each $a_j \in \{0, 1, 2, \cdots, 9\}$. Let

$$F(x) = a_n x^n + a_{n-1} x^{n-1} + \cdots + a_1 x + a_0.$$

Since $F(x) \in C[x]$, by the remainder theorem there is a polynomial $T(x)$ such that $F(x) \equiv T(x)(x - 1) + F(1)$. Using the fact that the coefficients of $F(x)$ are integers, we can show that the coefficients of $T(x)$ are integers. In particular, $F(10) = T(10)(10 - 1) + F(1)$. Since $F(10) = q$ and $T(10)$ is an integer, we conclude that $q = 9m + F(1)$ for some integer m. From the equality $q = 9m + F(1)$ we reason that q is divisible by 9 if and only if $F(1)$ is divisible by 9. Finally, observing that

$$F(1) = a_n + a_{n-1} + \cdots + a_1 + a_0$$

(that is, $F(1)$ is the sum of the digits of q), we have that q is divisible by 9 if and only if the sum of its digits is divisible by 9.

Example 3. Are the numbers 179,654 and 213,111,153 divisible by 9?

Solution: The sum of the digits of 179,654 is $1 + 7 + 9 + 6 + 5 + 4 = 32$. Similarly the sum of the digits of 213,111,153 is 18. Since 32 is not divisible by 9, while 18 is divisible by 9, from Example 2 we conclude that 179,654 is not divisible by 9 and 213,111,153 is divisible by 9.

The student will recall from Section 3.6 that if $\{m, p, q\} \subset N$ and $m = pq$, then the numbers p and q are factors (divisors) of m and m is a multiple of p (it is also a multiple of q). Similarly, if $\{F(x), G(x), H(x)\} \subset C[x]$ and $F(x) = G(x) \cdot H(x)$, then the polynomials $G(x)$ and $H(x)$ are called *factors* (*divisors*) of *factors* $F(x)$ and $F(x)$ is said to be a *multiple of* $G(x)$ (it is also a multi- *(divisors)* ple of $H(x)$).

Example 4. Since $x^2 - 1 = (x + 1)(x - 1)$, each of the polynomials $x + 1$, $x - 1$ is a factor (divisor) of $x^2 - 1$, and $x^2 - 1$ is a multiple of $x + 1$ and of $x - 1$. On the other hand, $x + i$ is not a factor of $x^{98} + x + 1$. The student has two choices to verify the last statement. He may divide $x^{98} + x + 1$ by $x + i$ to find that the remainder is not 0, or he may use the following theorem, called the *factor theorem*.

Theorem. If $F(x) \in C[x]$ and $c \in C$, then $x - c$ is a factor of $F(x)$ if and only if $F(c) = 0$.

Proof: If $x - c$ is a factor of $F(x)$, then there is a polynomial $G(x)$ in $C[x]$ such that $F(x) = (x - c)G(x)$. Hence $F(c) = (c - c)G(c) = 0 \cdot G(c) = 0$.

If $F(c) = 0$, then according to the remainder theorem there is a polynomial $T(x)$ in $C[x]$ such that $F(x) = T(x)(x - c)$, hence $x - c$ is a factor of $F(x)$.

Example 5. Is $x - 1$ a factor of $5x^{12} + 7x^7 - 12x^2 + x - 1$?

Solution: $5 \cdot 1^{12} + 7 \cdot 1^7 - 12 \cdot 1^2 + 1 - 1 = 0$. By the factor theorem we conclude that $x - 1$ is a factor of $5x^{12} + 7x^7 - 12x^2 + x - 1$.

Example 6. Show that if $n \in N$ and $b \in C$, then $x - b$ is a factor of $x^n - b^n$.

Solution: Let $F(x) = x^n - b^n$. Then $F(b) = b^n - b^n = 0$. By the factor theorem, $x - b$ is a factor of $F(x)$. Since $F(x) = x^n - b^n$, $x - b$ is a factor of $x^n - b^n$.

Exercise 5.4

Study Example 1 before working Exercises 1–4.

1. If $F(x) = 3x^3 - x - 1$, find   (a) $F(1)$,   (b) $F(0)$,   (c) $F(i)$,   (d) $F(a)$.
2. If $G(u) = 4u^4 + u^2 + 3$, find   (a) $G(-1)$,   (b) $G(2)$,   (c) $G(-2i)$,   (d) $G(b)$.
3. If $T(y) = y^4 - 1$, find   (a) $T(1)$,   (b) $T(-1)$,   (c) $T(0)$,   (d) $T(-c)$.
4. If $F(V) = V^2 + V + 1$, find   (a) $F(-3)$,   (b) $F(2 + i)$,   (c) $F(a - b)$,   (d) $F(a - bi)$.

Review Examples 2 and 3 before working Exercises 5–8. In these, which of the given numbers are divisible by 9?

5. (a) 117, (b) 198.
6. (a) 8972, (b) 7938.
7. (a) 67,284, (b) 79,241.
8. 4,301,669, (b) 777,777,777.

Study Examples 4 and 5 before working Exercises 9–14.

9. Is $u - 1$ a factor of $u^{95} - u^{94} - u^{38} + u^{37} + 3u - 3$?
10. Is $x - 2$ a factor of $2x^{10} - 4x^9 - 3x - 6$?
11. Is $y - i$ a factor of $y^{97} + 3y^{50} + y^{27} + 3$?
12. Is $x + 1$ a factor of $x^{99} - 3x^{47} - 5x^{31} + 2x^{15} - 4$?
13. Is $V - i$ a factor of $4V^{121} - 8V^{80} + 4V^{63} + 2V^{32} + 2V^{27} + 8$?
14. Is $y - 5$ a factor of $y^{20} + y^{10} + 1$?

Study Example 6 before working Exercises 15 and 16.

15. Show that if $n \in N$, n is odd, and $b \in C$, then $x + b$ is a factor of $x^n + b^n$.
16. Use the factor theorem to show that $2x - 1$ is a factor of the polynomial $4x^3 - 31x + 15$.
*17. To test whether a given natural number is divisible by 11, one may proceed as follows: let n_1 be the sum of the digits in the odd positions, where the positions are counted from right

to left, and let n_2 be the sum of the digits in the even positions. The given number is divisible by 11 if and only if $n_1 - n_2$ is divisible by 11. (a) Prove the last statement. (b) Test the numbers in Exercises 5–8 for divisibility by 11.

18. Let $q = 378,950 + K$ where $K \in \{0, 1, 2, \cdots, 9\}$. Determine K if (a) q is divisible by 9, (b) q is divisible by 11. (c) Is it possible that q is divisible by 99? Justify your answer.

5.5 FACTORING OF POLYNOMIALS

It is an asset to know how to factor certain frequently encountered polynomials. For convenience we present here some useful *factorization* *factorization formulas*.

formulas   If $b \in C$, then

(1) $x^2 - b^2 = (x - b)(x + b)$,
(2) $x^2 + 2bx + b^2 = (x + b)^2$,
(3) $x^2 - 2bx + b^2 = (x - b)^2$,
(4) $x^3 - b^3 = (x - b)(x^2 + bx + b^2)$,
(5) $x^3 + b^3 = (x + b)(x^2 - bx + b^2)$,
(6) $x^n - b^n = (x - b)(x^{n-1} + bx^{n-2} + \cdots + b^{n-2}x + b^{n-1})$,
     if $n \in N - \{1\}$.

Every one of these formulas can be readily established by multiplying the corresponding right-hand members of each asserted equality. Note that formulas (1), (4), and (5) are special cases of formula (6).

Example 1.

(1) $$x^2 - 16 = x^2 - 4^2$$
$$= (x - 4)(x + 4).$$

(2) $$4x^2 + 12x + 9 = (2x)^2 + 12x + 3^2$$
$$= (2x + 3)^2.$$

(3) $$16x^2 - 24x + 9 = (4x)^2 - 24x + 3^2$$
$$= (4x - 3)^2.$$

(4) $$x^3 - 27 = x^3 - 3^3$$
$$= (x - 3)(x^2 + 3x + 9).$$

(5) $$8x^3 + 27 = (2x)^3 + 3^3$$
$$= (2x + 3)(4x^2 - 6x + 9).$$

$$(6) \qquad x^5 - 1 = x^5 - 1^5$$

$$= (x - 1)(x^4 + x^3 + x^2 + x + 1).$$

completely We say that a given polynomial is *completely factored* if it is
factored expressed as a product of polynomials, each of which is linear.
In more advanced texts it is shown that every polynomial in
$C[x]$ of degree one or higher can be completely factored. To
keep matters simple, we shall restrict our attention to elementary
factorization techniques, using examples as illustrations.

Example 2. Factor completely $4b^2x^3 - 6bx^2 + 8bcx^2$ $(b \neq 0)$.

Solution: We use the generalized distributive law.

$$4b^2x^3 - 6bx^2 + 8bcx^2 = 2bx^2(2bx - 3 + 4c)$$

$$= (2bx)x(2bx - 3 + 4c).$$

Note that the given polynomial is now completely factored, since
it is expressed as a product of three linear polynomials.

The clue in the foregoing problem was the circumstance that
every term in the given polynomial had a common factor, namely,
$2bx^2$. The student should practice to develop his ability to recognize such common factors at a glance.

Example 3. Factor completely $x^2 - 7x + ax - 4a + 12$.

Solution:

$$x^2 - 7x + ax - 4a + 12 = x^2 - 7x + 12 + a(x - 4)$$

$$= x^2 - 4x - 3x + 12 + a(x - 4)$$

$$= x(x - 4) - 3(x - 4) + a(x - 4)$$

$$= (x - 3 + a)(x - 4)$$

$$= (x + a - 3)(x - 4).$$

Remark. We recognized immediately that the terms ax and $-4a$
have a common factor, namely a. Since $ax - 4a = a(x - 4)$, we
were induced to investigate whether $x - 4$ is a factor of $x^2 - 7x$
$+ 12$. By writing $-7x$ in the form $-4x - 3x$, we found that $x - 4$
is indeed a factor of $x^2 - 7x + 12$.

The technique illustrated in Example 2 is sometimes referred
to as *factoring out a common factor*. Example 3 illustrates the
technique sometimes called *factoring by grouping terms* or
factoring by associating terms.

Next we illustrate a technique that sometimes enables us
to factor completely a quadratic polynomial with integral coeffi-

cients. After the student has studied Chapter 7, he should be able to factor completely *any* quadratic polynomial.

Example 4. Factor completely $x^2 + 2x - 35$.

Solution: If there is an integer a such that $x - a$ is a factor of $x^2 + 2x - 35$, then a is a factor of -35. Since the only integral factors of -35 are $-35, -7, -5, -1, 1, 5, 7, 35$, we may check these eight "candidates." Recalling from the factor theorem that $x - c$ is a factor of $x^2 + 2x - 35$ if and only if $c^2 + 2c - 35 = 0$, we proceed as follows:

$$(-35)^2 + 2(-35) - 35 \neq 0$$

and

$$(-7)^2 + 2(-7) - 35 = 0.$$

Hence $x + 7$ is a factor of $x^2 + 2x - 35$. Division of $x^2 + 2x - 35$ by $x + 7$ yields the other factor, namely $x - 5$. Thus

$$x^2 + 2x - 35 = (x + 7)(x - 5).$$

Example 5. Factor completely $2x^2 + 7x + 6$.

Solution: If there are integers a, b, c, d such that

$$2x^2 + 7x + 6 = (ax + b)(cx + d),$$

then $ac = 2$ and $bd = 6$. If we make the choice $a = 1, c = 2$, then we are faced with the simpler problem of determining integers b and d such that

$$2x^2 + 7x + 6 = (x + b)(2x + d).$$

Keeping the restriction $bd = 6$ in mind, we may try the eight possible combinations of integral values for b and d until we discover that the choice $b = 2, d = 3$ does the job. That is,

$$2x^2 + 7x + 6 = (x + 2)(2x + 3).$$

Remark. We could have used the factor theorem in Example 5 as well, but we wanted to illustrate another technique. The student should keep in mind that his skill in elementary factoring will depend on the experience gained through practice. After some practice the student should be able to do problems such as those in Examples 4 and 5 by inspection of the given quadratic polynomials. This is why we choose to name techniques displayed in Examples 4 and 5 *factoring of quadratic polynomials by inspection.* We caution the student that there are quadratic polynomials that cannot be factored by inspection. In fact, if a polynomial has integer coefficients and we wish to factor it as a product of linear factors with integer coefficients only, then the task may be impossible. For example, $x^2 + 1$ cannot be factored as a product of linear factors with integer coefficients, since $x^2 + 1 = (x - i)(x + i)$, and i is not an integer.

We conclude this section with a more complicated example, displaying several factorization techniques.

Example 6. Factor completely $4x^2 + 4ax + a^2 + 2x + a - 2$.

Solution:

$$4x^2 + 4ax + a^2 + 2x + a - 2 = ((2x)^2 + 2(2x)a + a^2)$$
$$+ (2x + a) - 2$$
$$= (2x + a)^2 + (2x + a) - 2$$

(using Formula (2)).

Let $y = 2x + a$. Then

$$(2x + a)^2 + (2x + a) - 2 = y^2 + y - 2.$$

We factor $y^2 + y - 2$ by inspection as follows:

$$y^2 + y - 2 = (y + 2)(y - 1).$$

Remembering that $y = 2x + a$, we have now that

$$(2x + a)^2 + (2x + a) - 2 = (2x + a + 2)(2x + a - 1).$$

Thus,

$$4x^2 + 4ax + a^2 + 2x + a - 2 = (2x + a + 2)(2x + a - 1).$$

Exercise 5.5

Study Example 1 before working Exercises 1–12. In each of these, factor the given polynomial.

1. $x^2 - 49$.
2. $9y^2 - 12y + 4$.
3. $4x^2 + 20x + 25$.
4. $V^3 - 64$.
5. $z^3 + 8$.
6. $x^6 - 1$.
7. $x^5 + 1$.
8. $4V^2 - 9u^2$.
9. $16u^2 + 24bu + 9b^2$.
10. $16z^2 - 24az + 9a^2$.
11. $8y^3 - 27c^3$.
12. $27x^3 + 64a^6$.

Before working Exercises 13–18, study Example 2. Factor completely each of the following polynomials by factoring out a common factor.

13. $8b^2x^2 - 12bx + 6abx$.

14. $-16bdy^2 + 12dy^2 - 4bdy$.

15. $a^2bx^3 - ab^2x^3 + abcx^2 - abdx^2$.

16. $9a^2x^3 + 6ax^2 - 12acx^2$.

17. $(a - b)^4u^4 - 3(a - b)^3u^3 + 2(a - b)^2u^2$.

18. $(a^2 + b^2)^3x^4 + 2(a^2 + b^2)^2x^3 + (a^2 + b^2)x^2$.

Study Example 3 before working Exercises 19–24. Factor by grouping terms.

19. $3x + a - 6x^2 - 2ax$.

20. $2u^2 + 3u - 2au - 3a$.

21. $b^2y^3 + y^3 - b^2 - 1$.

22. $x^2y^3 - 8x^2 - 4y^3 + 32$.

23. $V^2 + 7V + 2aV + 4a + 10$.

24. $w^2 + 2bw - 5w - 8b + 4$.

Study Example 4 before working Exercises 25–30. Factor completely each of the given polynomials by inspection.

25. $x^2 + 11x + 30$.   **26.** $y^2 + y - 12$.

27. $y^2 - y - 20$.   **28.** $z^2 - 11z + 24$.

29. $x^2 - 2x - 35$.   **30.** $t^2 + 4t - 32$.

Before working Exercises 31–36, study Example 5. Factor completely each of the following polynomials by inspection.

31. $3x^2 + 5x + 2$.   **32.** $4t^2 - 11t + 6$.

33. $3r^2 - 5r - 12$.   **34.** $y^2 + 3y - 18$.

35. $6x^2 + x - 2$.   **36.** $6u^2 + 9u - 6$.

Study Example 6 before working Exercises 37–44. Factor each of the following polynomials.

37. $4x^2 - 4ax + a^2 + 6x - 3a + 2$.

38. $9r^2 + 6br + b^2 + 3r + b - 2$.

39. $16y^2 - 24ky + 9k^2 - 8y + 6k + 1$.

40. $25t^2 + 20\,at + 4a^2 - 4$.

41. $9cx^3 + 6abcx^2 + a^2b^2cx - 9cx^2 - 3abcx - 10cx$.

42. $a^2cx^3 - 2abcx^2 + b^2cx + 4acx^2 - 4bcx + 4cx$.

*43. $a^4 + a^2y^2 + y^4$.

44. $a^4 - 6a^2y^2 + y^4$.

45. Show that $x^2 + x - 6$ is a factor of $2x^4 + 5x^3 - 11x^2 - 20x + 12$ and then express this polynomial of fourth degree as a product of four polynomials of the first degree.

46. Suppose that u is a number for which $u^3 + u^2 + u + 1 = 0$. Express the following products in terms of powers of u

lower than the third: (a) $(u^2 + u + 1)(u^3 - 1)$, (b) $(u^4 - 1)$ $(u^2 + 2u + 3)$.

REVIEW EXERCISES

1. State the five laws of integral exponents.

In Exercises 2–7 simplify the given expressions.

2. $[(x^4)^3 + x^5 \cdot x] + x^5 \cdot x^7 + 3(x^3)^2$.

3. $(a^2 + b^3)^2 + a(a^3 + 2ab^3)$.

4. $\dfrac{(a + b)^{-1}}{(a + b)^{-3}}$, $(a + b \neq 0)$.

5. $\dfrac{(abc^{-3}) \cdot (a^{-1}b^{-2}c^2)}{a^2b^{-1}c^4}$, $(abc \neq 0)$.

6. $\dfrac{(x^{-3}y^{-5})^{-2}}{(x^{-1}y)^{-2}}$, $(xy \neq 0)$.

7. $\left(\dfrac{2a}{3y}\right)^2\left(\dfrac{3y}{4a}\right)^3$, $(ay \neq 0)$.

8. Express the following polynomials in standard form and state their degree:
  (a) $7 + x^2 + x + 5x^7 + x^3$,
  (b) $3x + x^2 + 1$,
  (c) 3.

In Exercises 9–14 perform the indicated operations; whenever possible, express your answer in standard form.

  9. $(3x^5 + 7x + 3) - (x^5 + 2x^2 - 3x)$.
  10. $(x^3 + x + 1)(2x^2 + i)$.
  11. $(x^3 + x + 1) \div (3x - 1)$.
  12. $(2x^3 - 1) \div (3x + 5)$.
  13. $(4x^4 + x^3 - 1) \div (x^2 + x + 1)$.
  14. $(3x^3 - 1) \div (2x^2 + x)$.

15. Give an example of a quadratic polynomial.
16. Give an example of a linear polynomial.
17. If $F(x) = 3x^3 + x^2 - 1$, find $F(0)$, $F(1)$, and $F(a)$.
18. Which of the following numbers are divisible by 9:
(a) 13,457,114, (b) 73,154,898, (c) 888,888,888?
19. (a) State the factor theorem. (b) Using the factor theorem, show that $x + i$ is a factor of $x^{99} + x$.

In Exercises 20–25 factor the given polynomials completely.

20. $4x^2 + 28x + 49$.

21. $2x^3 - 8x$.

22. $x^2 - 4c + 3 - 2cx + 4x + c^2$.

23. $10x^2 + 11x - 6$.

24. $x^4 - 5x^2 + 4$.

25. $9s^2 + 6bs + b^2 + 3s + b - 2$.

26. Find a polynomial $F(x)$ of degree three such that $F(4) = 12$ and $(x - 1)$, $(x - 2)$, $(x - 3)$ are factors of $F(x)$.

27. Factor $x^2 + 9$. [*Hint:* $9 = -(3i)^2$.]

28. Find the solution set of the open sentence "$7 \cdot 10^6 + 3 \cdot 10^5 + 2 \cdot 10^4 + 4 \cdot 10^3 + x$ is divisible by 9," if the replacement set is $\{0, 1, 2, 3, 4, 5, 6, 7, 8, 9\}$.

*29. In the solution of Example 2, Section 5.4, we stated that the coefficients of $T(x)$ are integers. Prove this fact. [*Hint:* Two polynomials are equal if and only if their corresponding coefficients are equal.]

INEQUALITIES

6.1 THE ORDER PROPERTIES OF THE REAL NUMBERS

Let X be a real number line (see Figure 6.1). Recall from Section *image* 4.5 that if $c \in R$, the point $P(c)$ on X is called the *image* of c, *coordinate* c is the *coordinate* of $P(c)$, and $P(-c)$ is the point obtained by reflecting $P(c)$ across the origin. Every point on X is the image of one and only one real number.

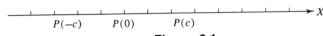

$$P(-c) \qquad P(0) \qquad P(c)$$

Figure 6.1

The collection of all those points which are located to the *positive* right of the origin is called the *positive X-axis*. It is plausible to *X-axis* expect the following statements to be true:

(1) If $P(c) \in X$, then precisely one of the following propositions is true: $P(c)$ is on the positive X-axis, $P(-c)$ is on the positive X-axis, $P(c)$ coincides with the origin.
(2) If $P(b)$ and $P(c)$ are points on the positive X-axis, then $P(b+c)$ and $P(b \cdot c)$ are points on the positive X-axis.

For example, the points $P(2), P(3), P(2+3)$, and $P(2 \cdot 3)$ are on the positive X-axis.

The two foregoing statements form a geometric interpretation of the algebraic order properties of the real numbers, which we now state as follows:

[122]

positive
real
numbers

There is a subset $\mathscr{P}$ of R, whose members are called *positive real numbers*. The statement "$c \in \mathscr{P}$" may be expressed in the equivalent form "c is positive." The sets $\mathscr{P}$ and R have the following properties:

0-1. If $c \in R$, then precisely one of the following statements is true: $c \in \mathscr{P}, -c \in \mathscr{P}, c = 0$.

0-2. If $b, c \in \mathscr{P}$, then $b + c \in \mathscr{P}$ and $b \cdot c \in \mathscr{P}$.

Note that the positive X-axis is the collection of all those points which are the images of members of $\mathscr{P}$.

order
relations

Next we define two *order relations* $<$ and $\leqslant$.

Definition. If $a, b \in R$, $a < b$ if and only if $b - a \in \mathscr{P}$, and $a \leqslant b$ if and only if $a < b$ or $a = b$. The statement $a < b$ is read "*a is less than b*" or, equivalently, "*b is greater than a.*" The statement $a \leqslant b$ is read "*a is less than or equal to b*" or, equivalently, "*b is greater than or equal to a.*"

To illustrate, $1 < 2$ if and only if $2 - 1$ is in $\mathscr{P}$, and $3 \leqslant 3$ if and only if $3 < 3$ or $3 = 3$.

Question: (a) Why is the statement $3 \leqslant 3$ true? (b) What geometric interpretation can you give to the statement $a < b$?

Example 1. Show that $0 < a$ if and only if a is positive.

Solution: By the definition of the relation $<$, $0 < a$ if and only if $a - 0 \in \mathscr{P}$. Since $a - 0 = a$, we have now:

$$0 < a \qquad \text{if and only if } a \in \mathscr{P}.$$

Finally, recalling that the statements "$a \in \mathscr{P}$," "a is positive" are equivalent, we have that $0 < a$ if and only if a is positive.

negative
non-positive
non-negative

If $c < 0$, we say that c is *negative*. If $c \leqslant 0$, we say that c is *non-positive*. If $0 \leqslant c$, we say that c is *non-negative*.

We may use the notation $b > a$ in lieu of $a < b$. By the notation $b \geqslant a$ we mean, of course, $a \leqslant b$. To illustrate,

$$\pi > 3 \qquad \text{if and only if } 3 < \pi.$$

$$2 \geqslant 1 \qquad \text{if and only if } 1 \leqslant 2.$$

We are now ready to state further properties of the order relations $<$ and $\leqslant$.

law of
trichotomy

Theorem. (a) If $a, b \in R$, then precisely one of the following statements is true: $a < b, b < a, a = b$. (This property is often called the *law of trichotomy*.)

(b) $a < b$ if and only if $a + c < b + c$ for any real number c. Also, $a \leqslant b$ if and only if $a + c \leqslant b + c$ for any real number c.

(c) If $0 < c$, then $a < b$ if and only if $ac < bc$. Also, if $0 < c$, then $a \leqslant b$ if and only if $ac \leqslant bc$.

(d) If $c < 0$, then $a < b$ if and only if $bc < ac$. Also, if $c < 0$, then $a \leqslant b$ if and only if $bc \leqslant ac$.

(e) If $a < b$ and $b < c$, then $a < c$. Also, if $a \leqslant b$ and $b \leqslant c$, then $a \leqslant c$. (These properties are sometimes referred to as the *transitive laws*.)

Proof: This theorem is a consequence of the order properties of the real numbers. We prove here part (a) only.

(a) If $a, b \in R$, $b - a \in R$ (since subtraction is an operation on R). By 0-1 precisely one of the following statements is true:

$$b - a \in \mathscr{P}$$

$$-(b - a) \in \mathscr{P}$$

$$b - a = 0.$$

We know that

$b - a \in \mathscr{P}$ if and only if $a < b$,

$-(b - a) \in \mathscr{P}$ if and only if $b < a$ (since $-(b - a) = a - b$),

$b - a = 0$ if and only if $a = b$.

Hence precisely one of the following statements is true: $a < b$, $b < a$, $a = b$.

Example 2. Show that if $a \in R - \{0\}$, then $a^2 > 0$.

Solution: If $a \in R - \{0\}$, then by 0-1 either $a \in \mathscr{P}$ or $-a \in \mathscr{P}$. If $a \in \mathscr{P}$, then by Example 1, $a > 0$, hence by part (c) of the preceding theorem $a \cdot a > 0 \cdot a$, that is $a^2 > 0$. If $-a \in \mathscr{P}$, then by Example 1, $-a > 0$, hence by part (c) of the preceding theorem $(-a)(-a) > 0 \cdot (-a)$, that is, $a^2 > 0$.

Remarks. (1) Since $1 \in R - \{0\}$, from Example 2, we obtain that $1^2 > 0$. Thus, we conclude that $1 > 0$. Repeated application of part (b) of the preceding theorem yields that $1 + 1 > 0 + 1$, that is, $2 > 1$, from which $2 + 1 > 1 + 1$, that is, $3 > 2$. Now we have that $3 > 2$ and $2 > 1$ and $1 > 0$. From $2 > 1$ and $1 > 0$, by transitivity of the relation $>$, we conclude that $2 > 0$. Similarly, $3 > 0$. More generally, it can be shown that every natural number is positive. The fact that this is convincingly displayed by the real number line should reaffirm our faith in the real number line as a useful geometric model of the real numbers.

(2) Since $0^2 = 0$ is true, and since $0 < 0$ is false according to the law of trichotomy, we conclude that $0^2 < 0$ is false. In view of Example 2, we now know that $a^2 < 0$ is false for every real number a. Thus,

$$\{x \mid x \in R \text{ and } x^2 < 0\} = \varnothing.$$

From $1 > 0$, we obtain that $1 + (-1) > 0 + (-1)$, hence $0 > -1$, that is, $-1 < 0$. It is clear that

$$\{x \mid x \in R \text{ and } x^2 = -1\} \subseteq \{x \mid x \in R \text{ and } x^2 < 0\}.$$

Since the set on the left is the solution set of the open sentence $x^2 = -1$ with the replacement set R and the set on the right is the empty set, we conclude that the open sentence $x^2 = -1$ has empty solution set relative to the replacement set R (recall that this fact was stated without proof in Section 4.7).

Example 3. Show that a real number a is positive if and only if a^{-1} is positive.

Solution: Suppose that a is a positive real number. Then $0 < a$ (see Example 1). We know that $0 < 1$. Since $0 \cdot a = 0$ and $a^{-1} \cdot a = 1$, we have now that $0 \cdot a < a^{-1} \cdot a$, hence by part (c) of the theorem in this section, $0 < a^{-1}$.

  If a^{-1} is a positive real number, then the above argument allows us to conclude that $(a^{-1})^{-1}$ is positive. Since $a = (a^{-1})^{-1}$, a is positive.

Question: Why is every rational number that can be expressed as a quotient of two natural numbers positive?

Example 4. Determine the solution set of the open sentence $\dfrac{1}{x-2} > 0$, if the replacement set is $R - \{2\}$.

Solution:

$$\left\{x \,\middle|\, \frac{1}{x-2} > 0\right\} = \{x \mid x - 2 > 0\} \qquad \text{by Example 3,}$$

$$= \{x \mid x - 2 + 2 > 0 + 2\} \qquad \text{by part (b) of the theorem in this section,}$$

$$= \{x \mid x > 2\}.$$

Thus the solution set is $\{x \mid x > 2\}$.

Exercise 6.1

Study Example 1 before working Exercises 1–6. In each of these, determine whether the given proposition is true or false.

1. If $a \in R$, $0 < -a$ if and only if $-a$ is positive.
2. If $c \in R$, $c < 0$ if and only if $-c$ is positive.
3. If $a, b \in R$ and $a < b$, then $a + b \in \mathscr{P}$.
4. If $a, b \in R$ and $a < b$, then $a - b \in \mathscr{P}$.
5. If $b, c, d \in R$ and $c < d$, then $bc < bd$.
6. If $x, y \in R$, then $(x - y)(y - x) \leqslant 0$.

Study Examples 2 and 3 before doing Exercises 7 and 8.

7. Show that if $a \in R$, then $a^2 \geqslant 0$.
8. Is it true that a real number b is negative if and only if b^{-1} is negative?

Before working Exercises 9–16, study Example 4. Find the "largest" suitable replacement set $S \subseteq R$, and the solution set of each of the following sentences.

9. $\dfrac{3}{x - 1} > 0.$

10. $\dfrac{91}{u - 7} \geqslant 0.$

11. $\dfrac{-6}{4x + 1} \leqslant 0.$

12. $\dfrac{-2}{3 - V} < 0.$

13. $0 < \dfrac{1}{x}.$

14. $\dfrac{a}{b - x} \leqslant 0$, where $a < 0.$

15. $\dfrac{m}{bn - cy} > 0$, where $m < 0$, $c < 0.$

16. Find the solution set of the open sentence $0 < \dfrac{1}{x}$ with replacement set $R - \mathscr{P}$.

*17. Prove that if a and b are real numbers and $a < b$, then $a < \dfrac{a + b}{2} < b$.

*18. Prove that if a and b are positive real numbers and $a < b$, then $a^2 < ab < b^2$.

*19. Prove that a real number a is negative if and only if a^{-1} is negative. [*Hint:* Study Example 3.]

*20. Prove that $\dfrac{a}{b} + \dfrac{b}{a} > 2$ if a and b are unequal positive real numbers.

6.2 CHAINS OF OPEN SENTENCES

It is clear that statements such as $0 < 1$, or $3 \leqslant 2$ are propositions. Also, sentences such as $\dfrac{1}{x - 2} > 0$ (see Example 4 in

Section 6.1) or $x^2 \leqslant -1$ can be regarded as open sentences relative to any subset K of R, provided K does not contain numbers that, when substituted for the variable, produce meaningless expressions. To illustrate, if we substitute for x in the open sentence $\dfrac{1}{x-2} > 0$ any real number other than 2, we obtain a proposition. For example, $\dfrac{1}{3-2} > 0$ is a true proposition, while $\dfrac{1}{1-2} > 0$ is a false proposition. Since 2 is a real number that produces a meaningless expression, it should not be a member of any replacement set for the open sentence $\dfrac{1}{x-2} > 0$. We say that the open sentence $\dfrac{1}{x-2} > 0$ has *implicit restrictions relative to the set R*.

Statements such as $0 < \pi$, $3 \leqslant 2$, $\dfrac{1}{x-2} > 0$ are examples *inequality* of inequalities. An *inequality* is a symbolic statement relating two expressions by means of one of the symbols $<, \leqslant, >, \geqslant$. An open sentence in one variable that is an inequality is called an *inequality in one variable*. To *solve a given inequality $p(x)$ over a specified set K* means to express the set

$$\{x \mid x \in K \wedge p(x) \text{ is true}\}$$

in a form sufficiently simple so that one can readily check whether any given member of K is in this set.

We wish to state a technique that is very useful to solve inequalities. Before doing this, we digress in order to introduce some new concepts concerning open sentences.

Let U be the universal set (that is, U is the replacement set for every open sentence presently under discussion). If $p(x)$ and $q(x)$ are open sentences such that $p(a) \to q(a)$ is a true proposition for every a in U (that is, if the solution set of $p(x)$ is a subset of the solution set of $q(x)$), then we say that $p(x)$ *implies* $q(x)$. Note that this statement is a proposition (why?). The proposition "$p(x)$ implies $q(x)$" may be stated in any one of the following alternate forms:

$p(x) \to q(x)$.

If $p(x)$, then $q(x)$.

$p(x)$ is sufficient for $q(x)$.

$q(x)$ is necessary for $p(x)$.

For example, if $U = R$, then $x = 1 \rightarrow x^2 = 1$, since

$$\{1\} = \{x \,|\, x = 1\} \subseteq \{x \,|\, x^2 = 1\} = \{-1, 1\}.$$

If $p_1(x), p_2(x), \cdots, p_k(x)$ are several open sentences, such that $p_1(x) \rightarrow p_2(x), p_2(x) \rightarrow p_3(x), \cdots, p_{k-1}(x) \rightarrow p_k(x)$, *chain of open* then the sequence $p_1(x), p_2(x), \cdots, p_k(x)$ is called *a chain of* *sentences* *open sentences*. The statement "$p_1(x), p_2(x), \cdots, p_k(x)$ is a chain of open sentences" may be denoted by either of the following symbolic forms:

$$
U \left\downarrow
\begin{array}{l}
p_1(x) \\
p_2(x) \\
\cdot \\
\cdot \\
\cdot \\
p_k(x)
\end{array}
\right.
\qquad
U \left\uparrow
\begin{array}{l}
p_k(x) \\
\cdot \\
\cdot \\
\cdot \\
p_2(x) \\
p_1(x)
\end{array}
\right.
$$

Note that we simply list the open sentences in order and then place a single-headed vertical arrow to the left to indicate the flow of implications. If it is clear from the context that the replacement set is U, we may omit the symbol U in the foregoing notation.

To illustrate, if $U = R$, the following propositions are true: $x = 1 \rightarrow x^2 = 1$, $x^2 = 1 \rightarrow x^4 = 1$. Thus we may write

$$
R \left\downarrow
\begin{array}{l}
x = 1 \\
x^2 = 1 \\
x^4 = 1
\end{array}
\right.
\qquad \text{or, alternately,} \qquad
R \left\uparrow
\begin{array}{l}
x^4 = 1 \\
x^2 = 1 \\
x = 1
\end{array}
\right.
$$

We wish to emphasize that the order in which we list the open sentences may be of crucial importance. Thus, while the above statements are true,

$$
R \left\downarrow
\begin{array}{l}
x^2 = 1 \\
x = 1 \\
x^4 = 1
\end{array}
\right.
$$

is false, since $x^2 = 1 \rightarrow x = 1$ is false (note that $-1 \in \{x \,|\, x^2 = 1\} - \{x \,|\, x = 1\}$).

If two open sentences $p(x)$ and $q(x)$ have the same solution set, then we say that $p(x)$ *is equivalent to* $q(x)$. The proposition "$p(x)$ is equivalent to $q(x)$" may be stated in any one of the following alternate forms:

$p(x) \leftrightarrow q(x)$.

$p(x)$ if and only if $q(x)$.

$p(x)$ iff $q(x)$.

$p(x)$ is necessary and sufficient for $q(x)$.

$p(x) \rightarrow q(x)$ and $q(x) \rightarrow p(x)$.

For example, if $U = R$,

$$x + 3 = 0 \leftrightarrow x = -3,$$

since $\{x \mid x + 3 = 0\} = \{x \mid x = -3\}$.

If $q_1(x), q_2(x), \cdots, q_n(x)$ are several open sentences such that $q_1(x) \leftrightarrow q_2(x)$, $q_2(x) \leftrightarrow q_3(x)$, $\cdots$, $q_{n-1}(x) \leftrightarrow q_n(x)$, then the sequence $q_1(x), q_2(x), \cdots, q_n(x)$ is called a *chain of equivalent open sentences*. The statement "$q_1(x), q_2(x), \cdots, q_n(x)$ is a chain of equivalent open sentences" may be symbolically denoted as follows:

$$
U \left\updownarrow
\begin{array}{l}
q_1(x) \\
q_2(x) \\
\cdot \\
\cdot \\
\cdot \\
q_n(x)
\end{array}
\right.
$$

Note that we simply list the open sentences and then place a double-headed vertical arrow to the left. Although the order of the listing is no longer crucially important, common sense suggests that the listing be such that the equivalence of any two adjacent open sentences can be readily justified. The symbol U may be omitted in the foregoing notation if no ambiguity results.

Example 1. Determine the solution set of the open sentence $\dfrac{1}{x-2} > 0$, if the replacement set is $R - \{2\}$.

Solution:

$$
\left\updownarrow
\begin{array}{l}
\dfrac{1}{x-2} > 0. \\[2mm]
x - 2 > 0. \\[2mm]
x - 2 + 2 > 0 + 2. \\[2mm]
x > 2.
\end{array}
\right.
$$

Thus, the solution set is $\{x \mid x > 2\}$. The student should consult Example 4 in Section 6.1 and compare the two solutions. He should not fail to notice that the notation in the foregoing solution is more compact.

Now we are ready to state a basic technique for solving inequalities in one or more variables.

Whenever possible, we generate a chain of equivalent open sentences until we obtain an open sentence sufficiently simple so that the members of its solution set can be readily determined. The following sections will provide enough illustrations of this technique. The criterion for equivalence of any two given open sentences will always have to be solidly founded on known true facts or on one's ability to show that their solution sets are equal. In particular, the theorem in Section 6.1 will be one of our main tools in solving inequalities.

In our later work it will be sometimes convenient to denote the statement $s_1 \rightarrow s_2 \land s_2 \rightarrow s_3 \land \cdots \land s_{k-1} \rightarrow s_k$, where each s_j is a proposition, as follows:

$$
\begin{array}{l}
s_1 \\
s_2 \\
\cdot \\
\cdot \\
\cdot \\
s_k
\end{array}
$$

The notation

$$
\begin{array}{l}
s_1 \\
s_2 \\
\cdot \\
\cdot \\
\cdot \\
s_k
\end{array}
$$

will occasionally be used to state that the propositions $s_1, s_2, \cdots, s_k$ are equivalent.

Exercise 6.2

Review Example 1 before doing Exercises 1–13. In each of these, find the "largest" replacement set $K \subseteq R$ and determine the solution set by generating a chain of equivalent open sentences.

1. $\dfrac{3}{x-9} > 0.$ 　　　　　**2.** $\dfrac{4}{w+2} > 0.$

3. $\dfrac{-2}{7y-1} > 0.$ 　　　　　**4.** $0 < \dfrac{11}{3x-a}$, where $a \in R.$

5. $\dfrac{-r}{2y+7} < 0$, where $r > 0.$

6. $\dfrac{-b}{5-4x} \leqslant 0$, where $b > 0.$

7. $\dfrac{a+b}{-u-6} > 0$, where $a+b > 0.$

8. $\dfrac{a}{bx+c} \leqslant 0$, where $a < 0, b > 0.$

9. $\dfrac{k}{cv-d} \geqslant 0$, where $k > 0, c > 0.$

10. $\dfrac{m}{ay+f} < 0$, where $m < 0, a < 0.$

11. $\dfrac{p+q}{mx-a+nx-b} < 0$, where $p+q < 0, m+n < 0.$

12. $\dfrac{c+d}{c(u-b)+d(u-b)} > 0$, where $c+d < 0.$

13. $2 - \dfrac{a}{x+a} \leqslant \dfrac{x}{x+a}.$

14. A student scores 60, 70, and 80 on three tests. How well must he do on two more tests in order to average 80 or better on all five?

15. A student tried to solve the inequality $x + 1 < 4x + 4$ as follows:

$$x + 1 < 4x + 4,$$

$$x + 1 < 4(x + 1),$$

$$1 < 4.$$

He concluded that the solution set is R. Is this correct? Explain.

6.3 LINEAR INEQUALITIES IN ONE VARIABLE. LINE GRAPHS

Throughout the remainder of this chapter we assume that the replacement set for all open sentences under discussion is the set of real numbers R or the largest subset of R that is suitable to be the replacement set for the given open sentence.

First we present a solution of an inequality in one variable using the technique mentioned at the conclusion of Section 6.2.

Example 1. Solve $2x + 1 < 0$.

Solution: We write a chain of equivalent inequalities as follows:

$$2x + 1 < 0,$$
$$2x + 1 + (-1) < 0 + (-1),$$
$$2x < -1,$$
$$\tfrac{1}{2} \cdot 2x < \tfrac{1}{2} \cdot (-1),$$
$$x < -\tfrac{1}{2}.$$

R

The reason for the equivalence of any two adjacent open sentences in the above chain can be found in the theorem of Section 6.1.

The solution set of $2x + 1 < 0$ is $\{x \mid x < -\tfrac{1}{2}\}$.

Sometimes we encounter propositions such as: $0 < 3 \wedge 3 < 4$. This particular proposition may be denoted by $0 < 3 < 4$. More generally, if $a, b, c \in R$, then the proposition "$a < b \wedge b < c$" may be denoted by "$a < b < c$". Extensions of this notation to more than three numbers are obvious. Thus $a < b < c < d$ denotes the proposition $a < b \wedge b < c \wedge c < d$. We may also write $k \leq m \leq n$ instead of $k \leq m \wedge m \leq n$. The meaning of a statement such as $a < b \leq c$ should now be entirely clear.

If $p(x)$ and $q(x)$ are inequalities with replacement set S, then $p(x) \wedge q(x)$ is an open sentence with the same replacement set. The next example illustrates a solution of an open sentence of this form.

Example 2. Solve $3 < x + 1 \wedge x + 1 < 4$.

Solution: We solve this inequality using two different methods of notation.

Solution 1:

$\{x \mid 3 < x + 1 \wedge x + 1 < 4\}$

$\qquad = \{x \mid 3 < x + 1\} \cap \{x \mid x + 1 < 4\}$

$\qquad = \{x \mid 3 - 1 < x + 1 - 1\} \cap \{x \mid x + 1 - 1 < 4 - 1\}$

$\qquad = \{x \mid 2 < x\} \cap \{x \mid x < 3\}$

$$= \{x \mid 2 < x \land x < 3\}$$
$$= \{x \mid 2 < x < 3\}.$$

Solution 2:

R $\quad$
$\uparrow$ $\quad 3 < x + 1 \land x + 1 < 4.$

$\quad 3 - 1 < x + 1 - 1 \land x + 1 - 1 < 4 - 1.$

$\quad 2 < x \land x < 3.$

$\downarrow$ $\quad 2 < x < 3.$

The two solutions yield the same solution set, namely, $\{x \mid 2 < x < 3\}$.

Example 3. Solve $x \leqslant -3x + 1 < 2x + 5$.

Solution:

$\uparrow$ $\quad x \leqslant -3x + 1 < 2x + 5.$

$\quad x \leqslant -3x + 1 \land -3x + 1 < 2x + 5.$

$\quad x + 3x \leqslant -3x + 1 + 3x \land -3x + 1 + 3x - 5 < 2x + 5 + 3x - 5.$

$\quad 4x \leqslant 1 \land -4 < 5x.$

$\quad x \leqslant \frac{1}{4} \land -\frac{4}{5} < x.$

$\downarrow$ $\quad -\frac{4}{5} < x \leqslant \frac{1}{4}.$

Thus the desired solution set is $\{x \mid -\frac{4}{5} < x \leqslant \frac{1}{4}\}$.

Note that the symbol R has not been written to the left of the arrow since it is clear from context that the replacement set is the set of real numbers.

Very often it is helpful to have a graphical representation of the solution set of a given open sentence. If X is a real number line and $S \subseteq R$, the *line graph* of S is the collection of all those points on X which are images of numbers in S. For example, the line graph of the set $\{\frac{1}{2}, 1\}$ may be shown as in Figure 6.2.

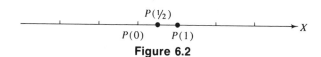

Figure 6.2

The line graph of $\{\frac{1}{2}, 1\}$ is the set of points $\{P(\frac{1}{2}), P(1)\}$. We have indicated this on X by placing heavy dots on the points $P(\frac{1}{2})$ and $P(1)$.

As another example, the line graph of the solution set of the open sentence $2x + 1 < 0$ (see Example 1 in this section) may be displayed as shown in Figure 6.3.

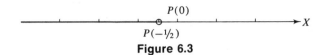

Figure 6.3

The heavy line shows the points of the graph. We have placed a small circle around the endpoint $P(-\frac{1}{2})$ to indicate that it is not on the graph.

Example 4. Solve $2 \leqslant -x < 4$ and graph the solution set.

Solution:

$$2 \leqslant -x < 4.$$
$$2 \leqslant -x \wedge -x < 4.$$
$$x \leqslant -2 \wedge -4 < x.$$
$$-4 < x \leqslant -2.$$

Note that $2 \leqslant -x \leftrightarrow x \leqslant -2$, also $-x < 4 \leftrightarrow -4 < x$. Thus the solution set is $\{x \mid -4 < x \leqslant -2\}$. Its line graph is shown in Figure 6.4. Note that this time we have identified points on the real number line with real numbers. As mentioned in Section 4.5, this is a very common practice.

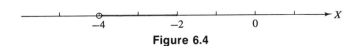

Figure 6.4

In this section we have considered only the simplest types of *linear* inequalities, among them *linear inequalities* (inequalities in *inequalities* which the variable appears only in polynomial expressions of degree one). Other types of inequalities will be discussed in the subsequent sections of this chapter.

Exercise 6.3

Review Example 1 before working Exercises 1–6. Solve each of the following inequalities.

1. $2 - 8x \leqslant 0$.

2. $\dfrac{-7y + 11}{-4} > 0$.

3. $\dfrac{-6x + 1}{-3} \geqslant 0$.

4. $at + b < 0$, where $a < 0$.

5. $\dfrac{ax + b}{c} \leqslant 0$, where $a < 0$, $c < 0$.

6. $\dfrac{(m + 1)^2 - (m^2 - 1)x}{m + 1} > 0$, where $m > 1$.

Study Example 2 before working Exercises 7–12. For each of these, give two solutions (as was done in Example 2).

7. $-1 < x + 3 \wedge x + 3 < 0$.
8. $7 \leqslant y - 8 \wedge y - 8 < 11$.
9. $5 < 2u + 3 \wedge 2u + 3 < 4$.
10. $a < 2x + b \wedge 2x + b \leqslant c$, where $a < c$.
11. $0 \leqslant ax + n - m \wedge ax + n - p < 0$, where $a > 0$, $m < p$.
12. $-c < ax + b \wedge ax + b < c$, where $c > 0$, $a < 0$.

Before working Exercises 13–18, study Example 3. Solve the following inequalities.

13. $-x < -4x + 1 \leqslant x + 5$.
14. $x - 4 < 5x - 1 < -x$.
15. $-2y < -y - 5 \leqslant -3y + 1$.
16. $3t + 4 \leqslant -5t + 2 \leqslant t - 1$.
17. $av \leqslant 3v - 1 < v - b$, where $b < 0$, $a < 1$.
18. $ay - b < -my - n < cy - d$,
      where $a + m < 0$, $c + m < 0$.

Study Example 4 before working Exercises 19–22. Solve each inequality and graph its solution set.

19. $-1 < -x < 3$.
20. $-2 \leqslant 1 - y < 2$.
21. $-5 \leqslant -4 - z \leqslant -3$.
22. $6x - 8 < 1 - 3x < x + 1$.

6.4 DIRECTED DISTANCES. ABSOLUTE VALUE

Consider the real number line X in Figure 6.5.

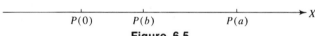

$P(0)$ $P(b)$ $P(a)$

Figure 6.5

directed   If $P(a)$ and $P(b)$ are points on X, the *directed distance*
distance from $P(a)$ to $P(b)$ is defined to be the number $b - a$. We write
$\overline{P(a)P(b)}$ to denote the directed distance from $P(a)$ to $P(b)$.
To illustrate,

$$\overline{P(1)P(\tfrac{3}{2})} = \tfrac{3}{2} - 1 = \tfrac{1}{2}$$

and

$$\overline{P(\tfrac{3}{2})P(1)} = 1 - \tfrac{3}{2} = -\tfrac{1}{2}.$$

It is clear now that in general $\overline{P(a)P(b)} \neq \overline{P(b)P(a)}$.

Questions. (a) What must be true about the points $P(a)$, $P(b)$
if $\overline{P(a)P(b)} = \overline{P(b)P(a)}$?
(b) What is the numerical value of $\overline{P(a)P(b)}$ if $\overline{P(b)P(a)}$
$= -3$?
The statement $\overline{P(1)P(\tfrac{3}{2})} = \tfrac{1}{2}$ tells us that $P(\tfrac{3}{2})$ is $\tfrac{1}{2}$ unit to
the right of $P(1)$. Similarly, from the statement $\overline{P(\tfrac{3}{2})P(1)} = -\tfrac{1}{2}$
we conclude that $P(1)$ is $\tfrac{1}{2}$ unit to the left of $P(\tfrac{3}{2})$. More generally, if $\overline{P(a)P(b)} > 0$, then the point $P(b)$ is $b - a$ units to
the right of $P(a)$, and if $\overline{P(a)P(b)} < 0$, then the point $P(b)$
is $-(b - a)$ units to the left of $P(a)$. Finally, if $\overline{P(a)P(b)} = 0$,
then the points $P(a)$ and $P(b)$ coincide.

Example 1. Show that $\overline{P(a)P(b)} = \overline{P(a)P(c)} + \overline{P(c)P(b)}$
where $P(a)$, $P(b)$, $P(c)$ are any points on X.

Solution:

$$\overline{P(a)P(b)} = b - a$$
$$= b + (-c + c) - a$$
$$= (c - a) + (b - c)$$
$$= \overline{P(a)P(c)} + \overline{P(c)P(b)}.$$

For example,

$$\overline{P(2)P(99)} + \overline{P(99)P(7)} = \overline{P(2)P(7)} = 7 - 2 = 5.$$

Note that $\overline{P(0)P(a)} = a - 0 = a$. Thus the *coordinate of
a point is the directed distance from the origin to the point.* By
the *distance between* two given points we understand, of course,
the length of the line segment joining the points. What then is the
distance between the points $P(0)$ and $P(a)$? We reason as follows: if $\overline{P(0)P(a)} > 0$, then the point $P(a)$ is a units to the right
of the origin $P(0)$, hence the distance is a units. If $\overline{P(0)P(a)}$
$= 0$, then the point $P(a)$ coincides with the origin, hence this
distance is 0. If $\overline{P(0)P(a)} < 0$, then the point $P(a)$ is $-a$ units

to the left of the origin, hence the distance is $-a$ units. Summarizing these observations, we have that the distance between $P(0)$ and $P(a)$ is a units (if $a \geqslant 0$) or $-a$ units (if $a < 0$).

The next definition introduces a concept that is useful when we are dealing with distances of points on a real number line. This concept also enables us to discover and state some further properties of the real numbers.

absolute **Definition.** If a is a real number, then the *absolute value* of a is *value* denoted by $|a|$ and defined by: $|a| = a$ if $a \geqslant 0$, $|a| = -a$ if $a < 0$.

distance The *distance* between the points $P(a)$ and $P(b)$ is defined to be the number $|b - a|$. This definition is in full agreement with our earlier statements about distances. Note that $|a|$ is the distance between the points $P(0)$ and $P(a)$.

Example 2. (a) $|\frac{2}{3}| = \frac{2}{3}$, since $\frac{2}{3} > 0$.

(b) $|-1| = -(-1) = 1$, since $-1 < 0$.

(c) The distance between the points $P(5)$ and $P(3)$ is $|3 - 5| = 2$.

It follows immediately from the definition of absolute value that the absolute value of a real number is always a non-negative real number. The next two examples establish further properties of absolute values.

Example 3. Show that for any real number a, $|a|^2 = a^2$.

Solution: If $a \geqslant 0$, then $|a| = a$, hence $|a|^2 = a^2$. If $a < 0$, then $|a| = -a$, hence $|a|^2 = (-a)^2$. Since $(-a)^2 = a^2$, we conclude that $|a|^2 = a^2$.

Example 4. Show that for any real numbers a and b the following statements are true:

(1) $|a| < |b|$ if and only if $a^2 < b^2$,

(2) $|ab| = |a| \cdot |b|$.

Proof: (1)

$$|a| < |b|.$$
$$0 < |b| - |a|.$$
$$0 < (|b| - |a|)(|b| + |a|).$$
$$0 < |b|^2 - |a|^2.$$
$$|a|^2 < |b|^2.$$
$$a^2 < b^2.$$

(2) Note that $|ab|^2 = (ab)^2$, $||a| \cdot |b|| = |a| \cdot |b|$ and

$(|a| \cdot |b|)^2 = (ab)^2$. If $|ab| < |a| \cdot |b|$ or $|a| \cdot |b| < |ab|$ then by part (1) we have the false statement $(ab)^2 < (ab)^2$. Thus $|ab| < |a| \cdot |b|$, $|a| \cdot |b| < |ab|$ are false statements. By the law of trichotomy, $|ab| = |a| \cdot |b|$ is a true statement.

We remark that as a consequence of the preceding example,

$$|-a| = |(-1)a| = |-1||a| = |a|.$$

Thus $|-a| = |a|$. To illustrate, $|-3| = |3|$.

Question: What geometric interpretation can you give to the algebraic statement $|-3| = |3|$?

Example 5. Solve the inequality $|x - 3| < |x - 5|$ and graph the solution set.

Solution: By the preceding example,

$$|x - 3| < |x - 5| \leftrightarrow (x - 3)^2 < (x - 5)^2.$$

Thus we have the following chain of equivalent open sentences.

$|x - 3| < |x - 5|$.

$(x - 3)^2 < (x - 5)^2$.

$x^2 - 6x + 9 < x^2 - 10x + 25$.

$(x^2 - 6x + 9) + (-x^2 + 10x - 9) < (x^2 - 10x + 25) + (-x^2 + 10x - 9)$.

$4x < 16$.

$x < 4$.

The desired solution set is $\{x \mid x < 4\}$. Its graph is shown in Figure 6.6.

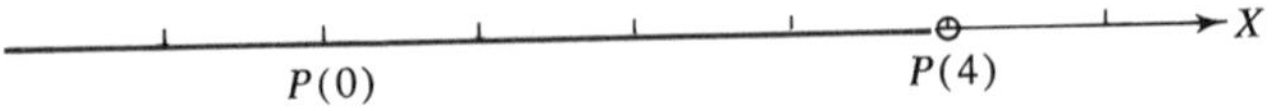

Figure 6.6

Exercise 6.4

Study Example 1 before working Exercises 1–4.

1. Determine the directed distance $\overline{P(a)P(b)}$, if (a) $a = \frac{1}{2}$, $b = \frac{3}{5}$; (b) $a = -3$, $b = 2$; (c) $a = -7$, $b = -10$.
2. Determine $\overline{P(a)P(c)} + \overline{P(c)P(b)}$, if (a) $a = -2$, $b = 4$, $c = 33$; (b) $a = 5$, $b = -1$, $c = 0$; (c) $a = (s + t)^2$, $b = (s - t)^2$, $c = s^2 - t^2$.

3. Show that $\overline{P(a)\,P(b)} = -\overline{P(b)\,P(a)}$ for any $a, b \in R$.
4. Show that $\overline{P(c)\,P(a)} + \overline{P(b)\,P(c)} = \overline{P(b)\,P(a)}$ for any $a, b, c, \in R$.

Review the definition of distance before working Exercises 5–10. In each of these, find the distance between the given points.

5. $P(2)$ and $P(4)$.
6. $P(-7)$ and $P(-1)$.
7. $P(-1)$ and $P(-7)$.
8. $P(m^2 + 2mn + n^2)$ and $P(m^2 - 2mn + n^2)$, where $mn > 0$.
9. $P(k - a - b)$ and $P(1 - a - b)$.
10. $P(-p - q)$ and $P(-p - q - r)$.

Study Examples 3 and 4 before doing Exercises 11–14.

11. Show that for any real number a, $|-a|^2 = a^2$.
12. Show that for any real numbers a, b, $|a| \leqslant |b|$ if and only if $a^2 \leqslant b^2$.
13. Express each absolute value as a product of two absolute values: (a) $|x^4 - 1|$, (b) $|x^4 y^4 - 3x^2 y^2 + 2|$, (c) $|u^3 v^3 + a^3 b^3|$.
14. Express each absolute value as a product of two absolute values: (a) $|x^2 - 1|$, (b) $|x^4 y^4 + 3x^2 y^2 + 2|$, (c) $|u^3 v^3 - a^3 b^3|$.

Study Example 5 before doing Exercises 15–20. In each of these, solve the given inequality. Also sketch the graph of its solution set (except for Exercises 19 and 20).

15. $|x - 1| \leqslant |x + 1|$.   16. $|y - 2| > |3 - y|$.
17. $|3u + 2| < |7 - 3u|$.   18. $2|1 - 2y| \geqslant |8 - 4y|$.
19. $|cx + b| \leqslant |d - cx|$, where $b + d > 0$, $c > 0$.
20. $|(a + b)x - (a + b)^2| < |(a + b)x - (ak + bk)|$, where $a \neq -b$, $a + b < k$.
*21. (a) Show that $x + y \leqslant |x| + |y|$ and $-x - y \leqslant |x| + |y|$.
(b) Using part (a), show that $|x + y| \leqslant |x| + |y|$ (triangle inequality).

6.5 SOME OTHER TYPES OF INEQUALITIES

From the theorem in Section 6.1 it follows that *the product of two real numbers is positive if and only if both numbers are positive or both numbers are negative.* Further, *the product of two real numbers is negative if and only if one of the numbers is positive and the other negative.*

The foregoing statements are very useful in solving certain types of inequalities.

Example 1. Solve the inequality $x^2 - 3x + 2 < 0$.

Solution: Note that $x^2 - 3x + 2 = (x-1)(x-2)$. Moreover, $(x-1)(x-2) < 0$ if and only if

$$(x - 1 < 0 \wedge x - 2 > 0) \vee (x - 1 > 0 \wedge x - 2 < 0).$$

We have now the following chain of equivalent open sentences:

$$x^2 - 3x + 2 < 0.$$
$$(x - 1)(x - 2) < 0.$$
$$(x - 1 < 0 \wedge x - 2 > 0) \vee (x - 1 > 0 \wedge x - 2 < 0).$$
$$(x < 1 \wedge x > 2) \vee (x > 1 \wedge x < 2).$$
$$(x < 1 \wedge x > 2) \vee (1 < x < 2).$$

Thus the solution set is

$$\{x \mid x < 1 \wedge x > 2\} \cup \{x \mid 1 < x < 2\}.$$

Since $\{x \mid x < 1 \wedge x > 2\} = \emptyset$, we can write the solution set in the simpler form $\{x \mid 1 < x < 2\}$.

Example 2. Solve the inequality $|x - a| < c$, where $a, c \in R$ and $c > 0$.

Solution: Recall that $|x - a| < c$ if and only if $(x - a)^2 < c^2$ (see Example 4 in Section 6.4 and note that $|c| = c$, since $c > 0$). We have now:

$$|x - a| < c.$$
$$(x - a)^2 < c^2.$$
$$(x - a)^2 - c^2 < 0.$$
$$(x - a - c)(x - a + c) < 0.$$
$$(x - a - c < 0 \wedge x - a + c > 0) \vee$$
$$(x - a - c > 0 \wedge x - a + c < 0).$$
$$(x < a + c \wedge x > a - c) \vee (x > c + a \wedge x < a - c).$$

Thus the solution set is

$$\{x \mid x < a + c \wedge x > a - c\} \cup \{x \mid x > c + a \wedge x < a - c\}.$$

Since $c > 0$, $-c < c$ and therefore $a - c < c + a$. It is clear now that $\{x \mid x > c + a \wedge x < a - c\} = \emptyset$. Thus we can write the solution set of $|x - a| < c$ in the simpler form

$$\{x \mid a - c < x < a + c\}.$$

Example 3. Solve the inequality $|x - 3| < 2$ and graph the solution set.

Solution:

$|x - 3| < 2.$

$(x - 3)^2 < 2^2.$

$x^2 - 6x + 9 < 4.$

$x^2 - 6x + 5 < 0.$

$(x - 1)(x - 5) < 0.$

$(x - 1 < 0 \wedge x - 5 > 0) \vee (x - 1 > 0 \wedge x - 5 < 0).$

$(x < 1 \wedge x > 5) \vee (x > 1 \wedge x < 5).$

$(x < 1 \wedge x > 5) \vee (1 < x < 5).$

Noting that $\{x \mid x < 1 \wedge x > 5\} = \emptyset$, we can write the solution set of $|x - 3| < 2$ simply as $\{x \mid 1 < x < 5\}$. Observe that, using the result of Example 2, we could have immediately written

$$\{x \mid |x - 3| < 2\} = \{x \mid 3 - 2 < x < 3 + 2\}$$
$$= \{x \mid 1 < x < 5\}.$$

The graph of the solution set is shown in Figure 6.7.

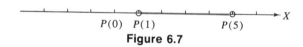

$P(0) \quad P(1) \qquad\qquad P(5)$

Figure 6.7

Example 4. Solve the inequality $|x - a| > b$, if $a, b \in R$.

Solution: We consider two cases: $b < 0$, and $b \geqslant 0$.

CASE 1. $b < 0$. Since the absolute value of any real number is a non-negative number, we see that in this case the solution set is R.

CASE 2. $b \geqslant 0$. Then, by Example 4 in Section 6.4,

$$|x - a| > b \leftrightarrow (x - a)^2 > b^2.$$

Now we have the following chain of equivalent open sentences:

$$|x - a| > b.$$
$$(x - a)^2 > b^2.$$
$$(x - a)^2 - b^2 > 0.$$
$$(x - a - b)(x - a + b) > 0.$$
$$(x - a - b > 0 \land x - a + b > 0) \lor$$
$$(x - a - b < 0 \land x - a + b < 0).$$
$$(x > a + b \land x > a - b) \lor (x < a + b \land x < a - b).$$

Thus the solution set is

$$\{x \mid x > a + b \land x > a - b\} \cup \{x \mid x < a + b \land x < a - b\}.$$

Since $b \geq 0$, $-b \leq 0$, hence $a - b \leq a + b$. It follows that

$$\{x \mid x > a + b \land x > a - b\} = \{x \mid x > a + b\},$$

also

$$\{x \mid x < a + b \land x < a - b\} = \{x \mid x < a - b\}.$$

Hence, the solution set is

$$\{x \mid x > a + b\} \cup \{x \mid x < a - b\}.$$

The student should convince himself that if $b < 0$,

$$\{x \mid x > a + b\} \cup \{x \mid x < a - b\} = R.$$

Thus in any case it is correct to state that the solution set of $|x - a| > b$ is $\{x \mid x > a + b\} \cup \{x \mid x < a - b\}$. For an alternate solution, see Exercise 25 below.

Example 5. Solve the inequality $|x + 2| > 1$ and graph the solution set.

Solution: Observe that the given inequality is of the form $|x - a| > b$. In our problem $a = -2$ and $b = 1$. From Example 4, we conclude that the solution set is $\{x \mid x > -2 + 1\} \cup \{x \mid x < -2 - 1\}$, that is,

$$\{x \mid x > -1\} \cup \{x \mid x < -3\}.$$

The graph is shown in Figure 6.8.

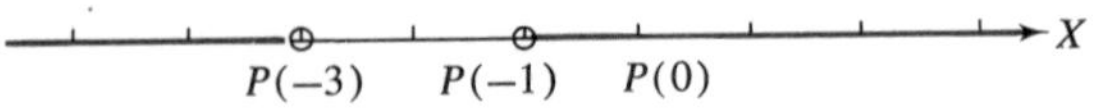

Figure 6.8

Example 6. Solve the inequality $\dfrac{x+2}{x-2} \geqslant 6$ and graph the solution set.

Solution:

$$R - \{2\} \left\{ \begin{array}{l} \dfrac{x+2}{x-2} \geqslant 6. \\[2ex] \dfrac{x+2}{x-2} - 6 \geqslant 0. \\[2ex] \dfrac{x+2-6(x-2)}{x-2} \geqslant 0. \\[2ex] \dfrac{-5x+14}{x-2} \geqslant 0. \\[2ex] \dfrac{5x-14}{x-2} \leqslant 0. \\[2ex] (5x-14 \geqslant 0 \wedge x-2 < 0) \ \vee \\ \quad (5x-14 \leqslant 0 \wedge x-2 > 0). \\[1ex] (x \geqslant \tfrac{14}{5} \wedge x < 2) \vee (x \leqslant \tfrac{14}{5} \wedge x > 2). \\[1ex] (x \geqslant \tfrac{14}{5} \wedge x < 2) \vee (2 < x \leqslant \tfrac{14}{5}). \end{array} \right.$$

Noting that $\{x \,|\, x \geqslant \tfrac{14}{5} \wedge x < 2\} = \varnothing$, we describe the desired solution set as $\{x \,|\, 2 < x \leqslant \tfrac{14}{5}\}$. Its graph is shown in Figure 6.9.

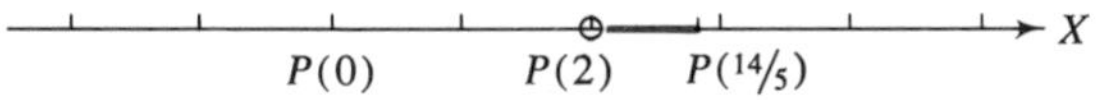

Figure 6.9

Exercise 6.5

Study Example 1 before working Exercises 1–6. In each of these solve the given inequality.

1. $x^2 - 1 < 0.$ 2. $y^2 - y - 6 \geqslant 0.$
3. $2u^2 - 9u - 5 < 0.$ 4. $6x^2 + 7x + 2 > 0.$
5. $0 \leqslant t^2 - (a+b)t + ab$, where $a < b.$
6. $k^2 x^2 + (ak^2 - km)x - akm \leqslant 0$, where $k \neq 0$

    and $\dfrac{m}{k} < -a.$

Before working Exercises 7–12, study carefully Examples 2 and 3. Solve each inequality and for Exercises 7, 8 graph the solution sets.

7. $|x + 1| < 3.$ **8.** $|x + 4| \leq 1.$

9. $0.0001 > |10V - a|.$ **10.** $|u - b| < e.$

11. $b + e > ||a|y - b|,$ where $a \neq 0, b + e > 0.$

12. $|(m + n)x - (m^2 - n^2)| < d(m + n),$ where $m + n > 0, d > 0.$

Review Examples 4 and 5 before working Exercises 13–18. In each of these, solve the given inequality, and in Exercises 13–15 sketch the graph of the solution sets.

13. $|x - 7| > 3.$ **14.** $2 \leq |3x + 1|.$

15. $5 < |7w - 4|.$ **16.** $a < |3y - b|,$ where $a > 0.$

17. $|mx + q| > p + q,$ where $m > 0, p > 0, q > 0.$

18. $a^2c|b| < |a^2bx - a^4b^2 + a^2bc|,$ where $c > 0, a \neq 0, b \neq 0.$

Study Example 6 before working Exercises 19–24. In each of these, solve the given inequality and graph its solution set.

19. $\dfrac{x - 1}{x + 1} \geq 2.$ **20.** $\dfrac{x + 1}{2x - 1} < 4.$

21. $\dfrac{2r + 3}{4r - 1} \leq -3.$ **22.** $\dfrac{5t - 2}{5 - 2t} > 1.$

23. $\dfrac{6 - 21y}{4y + 1} \geq -6.$ **24.** $\dfrac{x - 2}{3x + 2} \leq -2.$

25. To solve $|x - a| > b$ we may proceed as follows. First note that $|x - a| > b$ if and only if $\sim (|x - a| \leq b)$. Thus, we may find the solution set K of $|x - a| \leq b$ and conclude that the solution set of $|x - a| > b$ is $R - K$. Give an alternate solution for Example 4 using this technique.

26. Solve the inequalities:
(a) $(x - 1)^2 \geq 0.$
(b) $(x - 1)^2 > 0.$

27. Determine the solution set of the open sentence

$$|x - 3| < 2 \land |x + 1| < 3.$$

$*$**28.** Solve $x^3 > x^2 + 2x.$

$*$**29.** Without using decimal approximations, determine the proper order relationship $>$, $=$, or $<$ between the numbers $\sqrt{15} - \sqrt{14}$ and $\sqrt{32} - \sqrt{31}$.

$*$**30.** Solve the inequality $\dfrac{4}{x} + 9x > 12.$

31. Solve the inequality $|x - a| < c$ where $c \leq 0.$

REVIEW EXERCISES

1. State the law of trichotomy.

2. State the transitive laws for the relations $<$ and $\leqslant$.

In Exercises 3–16 solve the given inequalities over R and graph their solution sets.

3. $x \leqslant x.$

4. $\dfrac{1}{x} > 0.$

5. $\dfrac{1}{x+1} > 0.$

6. $x^2 + x - 6 > 0.$

7. $x \geqslant |x - 1|.$

8. $\overline{P(x) P(2)} > x - 2.$

9. $\dfrac{y-1}{y} > 1.$

10. $|y| > |y + 3|.$

11. $|x - 3| < 5.$

12. $|x + 1| > 2.$

13. $\dfrac{x+1}{x-2} > 2.$

14. $\dfrac{|x+1|}{x-2} > 2.$

∗15. $\dfrac{x+1}{|x-2|} > 2.$

16. $2x^2 + x - 3 < 0.$

In Exercises 17–22 solve the given open sentences over R.

17. $2x < 4 \wedge x + 1 > 0.$ **18.** $|x - 3| < 2 \wedge |x - 2| \leqslant 1.$
∗19. $x(x - 1)(x + 2) \geqslant 0.$ **20.** $y^4 - 1 < 0.$
21. $|x - 2| \leqslant 3 \wedge |x - 1| \geqslant 4.$

22. $\dfrac{x-5}{x-1} > 1.$

∗23. **(a)** Prove that for any real number c, $c^2 + 2c + 2 > 0$.
   (b) Solve the inequality $c^2 - 1 + cx + x > 0$, $(c > -1)$
      over R.
24. Solve the inequality $(x + 1)(x - 3) < 0$ over I (the set of integers).
25. Solve the inequality $(3x + 1) \cdot |x - 2| > 0$ over R and graph the solution set.

CHAPTER

7

EQUATIONS

7.1 EQUATIONS AS OPEN SENTENCES

equation
left-hand
member
right-hand
member

A symbolic sentence asserting equality of two expressions is called an *equation*. We write an equation by displaying expressions on each side of the equality sign. The expressions on the left and right of the equality sign are referred to as the *left-hand* and *right-hand members* of the equation, respectively.

To illustrate, here are some equations:

$$\tfrac{1}{3} + \tfrac{7}{5} = \tfrac{26}{15},$$

$$A = \pi r^2,$$

$$x^2 - x = 7$$

(assume that x is a variable with replacement set R).

In the example cited last, namely $x^2 - x = 7$, we have an equation involving a variable. Note that this equation is an open sentence. In fact, every equation involving a variable (or variables) may be regarded as an open sentence relative to a specified suitable replacement set.

equation in one
variable

In the present chapter we consider *equations in one variable* (that is, equations involving only one variable). Later, especially in Chapters 10 and 11, we shall encounter equations in more than one variable.

Suppose that an open sentence $p(x)$ containing a single variable x is an equation that is given along with a specified set S. *To solve the equation over S* means to represent the set $\{x \mid x \in S$ and $p(x)$ is true$\}$ in a form sufficiently simple so that one can readily check whether any given member of S is in this set. We accomplish this by regarding the given equation as an open sentence relative to the largest subset of S that is suitable to serve as a replacement set for the equation at hand and then arriving at the solution set through the use of a valid argument.

Example 1. Solve the equation $\frac{1}{x} = 1$ over R.

Solution: We regard the equation $\frac{1}{x} = 1$ as an open sentence relative to the set $R - \{0\}$. Note that this is the largest subset of R that is suitable to be a replacement set for $\frac{1}{x} = 1$. (Why did we exclude 0 from R to obtain the replacement set?)

A member c of $R - \{0\}$ is a solution of $\frac{1}{x} = 1$ if and only if $\frac{1}{c} = 1$. The last statement is true if and only if $c = 1$. Thus the solution set is $\{1\}$.

Many problems in physical and other sciences can be stated mathematically as equations. Thus one's capability to cope with problems in engineering, physics, chemistry, and other sciences depends rather heavily on one's ability to solve equations. For example, to determine the time it takes a freely falling body to travel 81 feet in a vacuum, the scientist may be interested in the solutions of the equation $16t^2 + v_0 t = 81$, where v_0 is the initial velocity.

Before we discuss techniques for solving equations, it will be helpful to consider equations in general.

Since any equation involving variables may be regarded as an open sentence, the notation of Section 6.2 is applicable to equations as well. In particular, if $p_1(x), p_2(x), \cdots, p_k(x)$ are equations with a common replacement set U, such that

$$p_1(x) \rightarrow p_2(x), \quad p_2(x) \rightarrow p_3(x), \quad \cdots, \quad p_{k-1}(x) \rightarrow p_k(x),$$

chain of then $p_1(x), p_2(x), \cdots, p_k(x)$ is called a *chain of equations.*
equations We denote this by writing

$$U \begin{array}{c} \left| \begin{array}{c} p_1(x) \\ p_2(x) \\ \cdot \\ \cdot \\ \cdot \\ p_k(x) \end{array} \right\downarrow \end{array} \qquad \text{or, alternatively,} \qquad U \begin{array}{c} \left\uparrow \begin{array}{c} p_k(x) \\ \cdot \\ \cdot \\ \cdot \\ p_2(x) \\ p_1(x) \end{array} \right| \end{array}$$

If it is clear from the context that the replacement set is U, we may omit the symbol U in the foregoing notation.

Similarly, if $q_1(x)$, $q_2(x)$, $\cdots$, $q_n(x)$ are equations with a common replacement set U, such that

$$q_1(x) \leftrightarrow q_2(x) \leftrightarrow q_3(x), \cdots, \quad q_{n-1}(x) \leftrightarrow q_n(x),$$

equivalent then $q_1(x)$, $q_2(x)$, $\cdots$, $q_n(x)$ is a *chain of equivalent equations,*
equations and we denote this by writing

$$U \left\uparrow\downarrow \begin{array}{l} q_1(x) \\ q_2(x) \\ \vdots \\ q_n(x) \end{array}\right.$$

If context permits, we may omit the symbol U in the foregoing notation.

Example 2. If $x - 1 = 0$ and $(x - 1)(x - 2) = 0$ are equations over C, then

$$\{x \,|\, x - 1 = 0\} = \{1\}$$

and

$$\{x \,|\, (x - 1)(x - 2) = 0\} = \{1, 2\},$$

hence the following statement is true:

$$C \left\downarrow \begin{array}{l} x - 1 = 0, \\ (x - 1)(x - 2) = 0. \end{array}\right.$$

On the other hand, the statement

$$C \left\downarrow \begin{array}{l} (x - 1)(x - 2) = 0, \\ x - 1 = 0, \end{array}\right.$$

is false.

Example 2 displays two equations $p(x)$, $q(x)$ such that

$$\{x \,|\, x \in C \wedge p(x) \text{ is true}\} \subset \{x \,|\, x \in C \wedge q(x) \text{ is true}\}.$$

defective In a situation like this we say that $p(x)$ is *defective* relative to
redundant $q(x)$ (also, $q(x)$ is *redundant* relative to $p(x)$). Thus in Example 2 the equation $x - 1 = 0$ is defective relative to the equation $(x - 1)(x - 2) = 0$, and $(x - 1)(x - 2) = 0$ is redundant relative to $x - 1 = 0$.

Example 3. Solve $\dfrac{1}{x} + \dfrac{1}{x - 1} = \dfrac{2x - 1}{x(x - 1)}$ over R.

Solution: It is clear from the context that the replacement set of the given equation is $R - \{0, 1\}$. If $c \in R - \{0, 1\}$, then

$$\frac{1}{c} + \frac{1}{c-1} = \frac{c-1+c}{c(c-1)}$$

$$= \frac{2c-1}{c(c-1)}.$$

Thus the solution set of the given equation is the replacement set itself, namely, $R - \{0, 1\}$.

Example 3 is a particular case of the following more general situation: if $p(x)$ is an equation with replacement set S and the solution set of $p(x)$ is equal to S, then $p(x)$ is called an *identity* *identity on S*. If the solution set of $p(x)$ is a proper subset of S, *conditional* then we say that $p(x)$ is a *conditional equation on S*.
equation
For example, the equation $\dfrac{1}{x} + \dfrac{1}{x-1} = \dfrac{2x-1}{x(x-1)}$ in Example

3 is an identity on the set $R - \{0, 1\}$; on the other hand the equation

$$\frac{1}{x} + \frac{1}{x-1} = \frac{2x}{x(x-1)}$$

is a conditional equation on the set $R - \{0, 1\}$, since its solution set is a proper subset of $R - \{0, 1\}$. (Note that

$$\frac{1}{2} + \frac{1}{2-1} = \frac{2 \cdot 2}{2(2-1)}$$

is false.)

Exercise 7.1

Read Example 1 before doing Exercises 1–6. In each of these, solve the given equation over R.

1. $\dfrac{1}{y} = 1.$ 2. $\dfrac{1}{2x} = \dfrac{1}{2}.$

3. $\dfrac{1}{au} = 1,\ a \neq 0.$ 4. $\dfrac{1}{x-1} = 1.$

5. $\dfrac{1}{bx+3} = 1,\ b \neq 0.$ 6. $\dfrac{1}{ax+b} = 1,\ a \neq 0.$

Study Example 2 before doing Exercises 7–12. In each of these,

$p(x)$ and $q(x)$ represent equations over C. Write the chain of equations

(a) $\Bigg\downarrow \begin{array}{c} p(x) \\[4pt] q(x) \end{array}$ (b) $\Bigg\downarrow \begin{array}{c} q(x) \\[4pt] p(x) \end{array}$

In each case, indicate whether the implication is true (T) or false (F). [*Note:* If $p(x)$ represents an equation, as, for example, $x + 2 = 0$, we write $p(x): x + 2 = 0$.]

7. $p(x): x - 2 = 0$; $q(x): (x - 2)(x - 1) = 0$.
8. $p(x): x + 5 = 0$; $q(x): (x - 6)(x + 5) = 0$.
9. $p(y): y = 0$; $q(y): y(y + 3) = 0$.
10. $p(x): (x - a)(x - b) = 0$, $(a \neq b)$; $q(x): x - b = 0$.
11. $p(z): (z - a)(z - b) = 0$; $q(z): (b - z)(a - z) = 0$.
12. $p(x): x + a = 0$; $q(x): (x - a)(x + b) = 0$,
$\quad (a \neq 0, a \neq b)$.

Study Example 3 before doing Exercises 13–20. Solve each equation over R.

13. $\dfrac{1}{x} + \dfrac{1}{x - 3} = \dfrac{2x - 3}{x(x - 3)}$.

14. $\dfrac{1}{u} + \dfrac{1}{u - 9} = \dfrac{2u - 9}{u(u - 9)}$.

15. $\dfrac{1}{3y} + \dfrac{3}{2y - 2} = \dfrac{11y - 2}{6y(y - 1)}$.

16. $\dfrac{1}{2v} - \dfrac{1}{2v - 14} = \dfrac{-7}{2v(v - 7)}$.

17. $\dfrac{1}{x - 1} + \dfrac{1}{x + 1} = \dfrac{2x}{x^2 - 1}$.

18. $\dfrac{1}{x - a} + \dfrac{1}{x - b} = \dfrac{2x - (a + b)}{(x - a)(x - b)}$.

19. $\dfrac{x + a}{x + b} = 1, a \neq b$.

20. $\dfrac{y}{y + b} = k - \dfrac{b}{y + b}, b, k \in R \wedge k \neq 1$.

∗21. What must be true about the real number c if the equations

$$x^2 - 6x + 9 = c^2, \qquad |x - 3| = c$$

are equivalent relative to the replacement set R?

*22. Solve the equation

$$\frac{3(x^2 + 4)}{x - 3} = 3 + 2x + \frac{13x}{x - 3} \qquad \text{over } R.$$

*23. Solve the equation

$$\frac{3}{x - 3} + 4 = \frac{x}{x - 3} \qquad \text{over } R.$$

24. Is $|x - 2| = |2 - x|$ an identity on R?

25. Is $|x - 2| = 2 - x$ an identity on R?

26. Is $|x + 2| = x + 2$ an identity on N (the set of natural numbers)?

27. Are the equations $(x - 1)(x + 3) = 0$ and $x + 3 = 0$ equivalent if their common replacement set is $R - \{1\}$?

28. Are the equations $(x - 1)(x + 3) = 0$ and $x + 3 = 0$ equivalent if their common replacement set is R?

29. Is it possible to find a subset K of C such that

$$K \quad \begin{array}{c} \uparrow \\ \\ \downarrow \end{array} \quad \begin{array}{l} x + 1 = x, \\ \\ x^2 + 1 = x^2, \end{array}$$

is false? Justify your answer.

30. For what value(s) of c is $(x + c)^2 = x^2 + c^2$ an identity on C?

7.2 EQUATIONS—BASIC FACTS AND TECHNIQUES

We are now ready to state a general method to solve a given equation over a specified set S. We generate a chain of equations $p_1(x), p_2(x), \cdots, p_k(x)$, where $p_1(x)$ is the given equation and each $p_j(x)$ has the same replacement set as $p_1(x)$. The last equation, $p_k(x)$, should be sufficiently simple so that one can readily determine the members of its solution set. In case $p_1(x), p_2(x), \cdots, p_k(x)$ is a chain of equivalent equations, we know that the solution set of $p_1(x)$ is the same as the solution set of $p_k(x)$. If $p_1(x), p_2(x), \cdots, p_k(x)$ is a chain of equations such that we are uncertain of their equivalence, then we know that the solution set of $p_1(x)$ is a subset of the solution set of $p_k(x)$. This knowledge is usually sufficient to determine the solution set of $p_1(x)$ through a process called *checking the solutions*.

The remainder of the present chapter contains examples illustrating the method described in the foregoing paragraph.

First we wish to state some facts about the equality relation and equations.

We begin by reminding the student of the *fundamental properties of equality:*

If S is a set and a, b, c are members of S, then the following statements are true:

reflexive   (1) $a = a$ *(reflexive property)*.

symmetric   (2) If $a = b$, then $b = a$ *(symmetric property)*.

transitive   (3) If $a = b$ and $b = c$, then $a = c$ *(transitive property)*.

Example 1. $4 = 4$, by the reflexive property. If $9 = 3^2$, then $3^2 = 9$, by the symmetric property. If $a = b$ and $b = 0.12$, then $a = 0.12$, by the transitive property.

Our interest henceforth is focused on solving equations over subsets of C (the set of complex numbers). The truth of the following useful statements is a consequence of the properties of complex numbers.

If $p(x)$ is an equation over some subset of C, then addition or subtraction of the same polynomial to both members of $p(x)$ produces an equation that is equivalent to $p(x)$. Also, multiplication of both members of $p(x)$ by the same nonzero complex number produces an equation that is equivalent to $p(x)$.

Example 2. Solve the equation $ax + b = 0$ over C, if $a, b \in C$ and $a \neq 0$.

Solution: In view of the foregoing discussion we have the following chain of equivalent equations:

$$ax + b = 0.$$

$$ax + b + (-b) = 0 + (-b).$$

$$ax = -b.$$

$$\frac{1}{a}ax = \frac{1}{a}(-b).$$

$$x = -\frac{b}{a}.$$

Hence the solution set is $\left\{ -\dfrac{b}{a} \right\}$.

We remark that an equation of the form $ax + b = 0$, $a \neq 0$, linear is called a *linear equation in one variable*. The foregoing example equation shows that such equations can be readily solved.

Example 3. Solve $2x + 1 = x - 7$ over R.

Solution:

$$2x + 1 = x - 7.$$

$$2x + 1 - x - 1 = x - 7 - x - 1 \qquad \text{(note that } -x - 1 \text{ is a polynomial).}$$

$$x = -8.$$

Hence the solution set is $\{-8\}$.

Example 4. In an attempt to solve $\dfrac{1}{x} + \dfrac{1}{x+1} = \dfrac{2x-1}{x(x-1)}$ over R, a student wrote

$$\frac{1}{x} + \frac{1}{x-1} = \frac{2x-1}{x(x-1)},$$

$$\left(\frac{1}{x} + \frac{1}{x-1}\right) x(x-1) = \frac{2x-1}{x(x-1)} \, x(x-1),$$

$$x - 1 + x = 2x - 1,$$

$$2x - 1 = 2x - 1,$$

and concluded that the solution set is R. Is this correct?

Answer: R is not the solution set, since the numbers 0 and 1 are not solutions. The correct solution set is $R - \{0, 1\}$ (see Example 3, Section 7.1). The equation $2x - 1 = 2x - 1$ is redundant relative to the equation $\dfrac{1}{x} + \dfrac{1}{x-1} = \dfrac{2x-1}{x(x-1)}$.

Example 5. The student mentioned in Example 4, unperturbed by his earlier fiasco, attempted to solve the equation $x^2 = x$ over R as follows:

$$x^2 = x,$$

$$\frac{1}{x}x^2 = \frac{1}{x}x,$$

$$x = 1.$$

He concluded that the solution set is $\{1\}$. Is this correct?

Answer: It will be shown in Example 1, Section 7.3, that the correct solution set is $\{0, 1\}$. It is now clear that the equation $x = 1$ is defective relative to $x^2 = x$.

In Example 4 we saw that multiplication of both members of an equation by an expression involving a variable may produce an equation that is redundant relative to the given equation.

Example 5 shows that division of both members of an equation by an expression involving a variable may produce an equation that is defective relative to the given equation. These two examples should caution the student to avoid the pitfalls of uncritical reasoning. The following theorem, stated here without proof, provides useful tools for solving equations correctly.

Theorem. If $P(x)$, $Q(x)$, and $S(x)$ are expressions in a variable x such that $P(c)$, $Q(c)$, $S(c)$ are complex numbers for every c in a subset K of C, then relative to the replacement set K the following statements are true:

(1) The equations $P(x) = Q(x)$ and $P(x) + S(x) = Q(x) + S(x)$ are equivalent.

(2) $P(x) = Q(x) \rightarrow P(x) \cdot S(x) = Q(x) \cdot S(x)$.

(3) $P(x) = Q(x) \leftrightarrow P(x) \cdot S(x) = Q(x) \cdot S(x)$ if and only if every solution of $S(x) = 0$ over K is also a solution of $P(x) = Q(x)$.

Example 6. (1) The equations $x^2 = -1$ and $x^2 + \dfrac{1}{x} = -1 + \dfrac{1}{x}$ are equivalent relative to the replacement set $C - \{0\}$.

(2) Relative to the replacement set R, $x^2 = 1 \rightarrow x^2(x - 2) = x - 2$.

(3) Relative to the replacement set $R - \{2\}$, $x^2 = 1 \leftrightarrow x^2(x - 2) = x - 2$, although relative to the replacement set R, $x^2 = 1$ is not equivalent to $x^2(x - 2) = x - 2$. (Why?)

Exercise 7.2

Read Example 1 before doing Exercises 1–6. In each statement, which fundamental property of equality is used? Assume that a, b, c, d are real numbers.

1. If $16 = 4^2$, then $4^2 = 16$.
2. $11 = 11$.
3. If $a = d$ and $d = 0.1$, then $a = 0.1$.
4. $3a + b = 3a + b$.
5. If $a + b = c$ and $c = d$, then $a + b = d$.
6. If $a(b + c) = ab + ac$, then $ab + ac = a(b + c)$.

Study Examples 2 and 3 before doing Exercises 7–12. In each of these solve the given linear equation over R.

7. $10x + 3 = x - 6$.
8. $3x + 2 = x + 7$.

9. $2x - a = x - b$.

10. $\frac{2}{3}ax - \frac{3}{4}b = -\frac{1}{3}ax + \frac{1}{4}b$, $a \neq 0$.

11. $\frac{3y}{r_1} + \frac{r_1}{r_2{}^2} = 4$ $y - r_1$, $r_2 \neq 0$, $r_1 \neq \frac{3}{4}$, $r_1 \neq 0$.

12. $\frac{az}{b} + \frac{b}{a} = \frac{bz}{a} + \frac{a}{b}$, $b^2 - a^2 \neq 0$, $ab \neq 0$.

Read Examples 4 and 5 before working Exercises 13–16. In each of the attempted solutions over R, is the third equation redundant or defective relative to the given equation, and why?

13.
$$\frac{1}{x} + \frac{1}{x-3} = \frac{2x-3}{x(x-3)},$$

$$\left(\frac{1}{x} + \frac{1}{x-3}\right) x(x-3) = \frac{2x-3}{x(x-3)} x(x-3),$$

$$2x - 3 = 2x - 3.$$

14. $3x^2 = 6x$,

$$3x^2 \cdot \frac{1}{x} = 6x \cdot \frac{1}{x},$$

$$x = 2.$$

15. $ay^2 - by = 0$, $ab \neq 0$,

$$ay^2 \cdot \frac{1}{y} = by \cdot \frac{1}{y},$$

$$y = \frac{b}{a}.$$

16.
$$\frac{1}{u-a} + \frac{1}{u-b} = \frac{2u - (a+b)}{(u-a)(u-b)}, \quad a, b \in R,$$

$$\left(\frac{1}{u-a} + \frac{1}{u-b}\right)(u-a)(u-b)$$

$$= \frac{2u - (a+b)}{(u-a)(u-b)} (u-a)(u-b)$$

$$2u - (a+b) = 2u - (a+b).$$

Study Example 6 before working Exercises 17–20. In each of these, (a) find a subset K_1 of C such that $K_1 \Big\downarrow \dfrac{p(x)}{q(x)}$ holds,

(b) find a subset K_2 of C such that $K_2 \Big\uparrow \dfrac{p(x)}{q(x)}$ holds.

17. $p(x)$: $x^2 = -4$; $q(x)$: $x^2 + \dfrac{1}{x-1} = -4 + \dfrac{1}{x-1}$.

18. $p(x): x^2 = -1; q(x): x^2(x-2) = -(x-2)$.

19. $p(y): y^2 = 4; q(y): y^2(y+3) = 4(y+3)$.

20. $p(z): 7z^2 = 2; q(z): 7z^3 = 2z$.

7.3 EQUATIONS — FURTHER BASIC FACTS AND TECHNIQUES

Recall from Section 4.1 that if $a, b \in Q$, then $(x-a)(x-b) = 0$ if and only if $x = a \lor x = b$. This result is also valid if a and b are complex numbers. More generally, if $c_1, c_2, \cdots, c_k$ are complex numbers and x is a variable with replacement set C, then $(x-c_1) \cdot (x-c_2) \cdot \ldots \cdot (x-c_k) = 0$ if and only if $x = c_1 \lor x = c_2 \lor \cdots \lor x = c_k$.

Example 1. Solve the equation $x^2 = x$ over R.

Solution:

$$x^2 = x.$$
$$x^2 - x = 0.$$
$$x(x-1) = 0.$$
$$x = 0 \quad \lor \quad x - 1 = 0.$$
$$x = 0 \quad \lor \quad x = 1.$$

Hence the solution set is $\{0, 1\}$.

Example 2. Solve the equation $3x^3 + 10x^2 + 9x + 2 = 0$ over R.

Solution: We factor $3x^3 + 10x^2 + 9x + 2$ as follows:

$$3x^3 + 10x^2 + 9x + 2 = (3x^3 + 9x^2 + 6x) + (x^2 + 3x + 2)$$
$$= 3x(x^2 + 3x + 2) + (x^2 + 3x + 2)$$
$$= (3x + 1)(x^2 + 3x + 2)$$
$$= (3x + 1)(x + 1)(x + 2).$$

Now we have the following chain of equivalent equations:

$$3x^3 + 10x^2 + 9x + 2 = 0.$$
$$(3x + 1)(x + 1)(x + 2) = 0.$$
$$3x + 1 = 0 \quad \lor \quad x + 1 = 0 \quad \lor \quad x + 2 = 0.$$
$$x = -\tfrac{1}{3} \quad \lor \quad x = -1 \quad \lor \quad x = -2.$$

Therefore the solution set is $\{-2, -1, -\tfrac{1}{3}\}$.

The student should observe that in the last example the problem became very simple indeed, after we had factored the left-hand member of the given equation. Thus knowledge of factorization techniques appears to be an important prerequisite towards the mastery of solving equations.

Although factoring was covered in Chapter 5, we are mindful that at this time the student may need a brief review. The following short list of factorization formulas will provide a good starting point. If $a, b \in C$, then

(1) $a^2 - b^2 = (a - b)(a + b)$,
(2) $a^2 + 2ab + b^2 = (a + b)^2$,
(3) $a^2 - 2ab + b^2 = (a - b)^2$,
(4) $a^3 - b^3 = (a - b)(a^2 + ab + b^2)$,
(5) $a^3 + b^3 = (a + b)(a^2 - ab + b^2)$,
(6) $a^n - b^n = (a - b)(a^{n-1} + a^{n-2} b + \cdots + ab^{n-2} + b^{n-1})$,
  if $n \in N$ and $n \geq 2$.

Whenever a student tries to factor an expression, he should also keep the following techniques in mind:

factoring out a common factor,
factoring by grouping terms (associating terms),
factoring certain quadratic polynomials by inspection.

We illustrate these techniques with an example.

Example 3. Factor the following expressions:

(a) $c^2(c - 2) + (3c + 1)(c - 2)$,
(b) $a^2 + 2ab + b^2 + 2a + 2b + 1$,
(c) $3x^2 - 24x - 27$,
(d) $(a - b)^2 - 12(a - b) + 36$.

Solution: (a) Observe that the factor $c - 2$ is common to both terms. Thus,

$$c^2(c - 2) + (3c + 1)(c - 2) = (c^2 + 3c + 1)(c - 2).$$

(b) We first group the terms using the associative property of addition, and then factor.

$$a^2 + 2ab + b^2 + 2a + 2b + 1 = (a^2 + 2ab + b^2) + (2a + 2b) + 1$$
$$= (a + b)^2 + 2(a + b) + 1$$
$$= (a + b + 1)^2.$$

(c) $3x^2 - 24x - 27 \quad = 3(x^2 - 8x - 9)$
$$= 3(x + 1)(x - 9).$$

(d) Letting $c = a - b$, we get $c^2 - 12c + 36$, which is equal to $(c - 6)^2$. Thus

$$(a - b)^2 - 12(a - b) + 36 = (a - b - 6)^2.$$

An equation of the form

$$a_n x^n + a_{n-1} x^{n-1} + \cdots + a_1 x + a_0 = 0, \qquad \text{where } a_n \neq 0,$$

polynomial is called a *polynomial equation of degree n in one variable*. If
equation $n = 1$, we have a linear equation in one variable (this type of equation was completely solved in Example 2, Section 7.2). In the present section we proceed to solve polynomial equations of degree higher than one by factoring the left-hand member. Some other techniques, as well as some other types of equations, will be discussed in the subsequent sections of this chapter.

Example 4. Solve the equation $x^4 - 16 = 0$ (a) over R, (b) over C.

Solution: We write $x^4 - 16 = (x - 2)(x + 2)(x^2 + 4)$.

(a) Solving the given equation over R, we have

$$x^4 - 16 = 0.$$
$$R \qquad (x - 2)(x + 2)(x^2 + 4) = 0.$$
$$x = 2 \quad \vee \quad x = -2 \quad \vee \quad x^2 = -4.$$

Hence the solution set is $\{-2, 2\} \cup \{x \mid x^2 = -4\}$. Recalling from Section 6.1 that the square of any real number is a nonnegative real number, we conclude that $\{x \in R \mid x^2 = -4\} = \varnothing$. Thus the solution set of the equation $x^4 - 16 = 0$ over R is $\{-2, 2\}$.

(b) We note that $x^2 + 4 = (x + 2i)(x - 2i)$. Solving the given equation over C, we have

$$x^4 - 16 = 0.$$
$$C \qquad (x - 2)(x + 2)(x + 2i)(x - 2i) = 0.$$
$$x = 2 \quad \vee \quad x = -2 \quad \vee \quad x = -2i \quad \vee \quad x = 2i.$$

Hence the solution set over C is $\{-2, 2, -2i, 2i\}$.

If u and z are complex numbers and $u^2 = z$, then we say that
square root the number u is a *square root* of z.

Example 5. If $c, d \in C$ and $d^2 = c$ (that is, d is a square root of c), solve the equation $x^2 - c = 0$ over C.

Solution:

$$(x - d)(x + d) = x^2 - d^2$$
$$= x^2 - c.$$

Thus, $x^2 - c = (x - d)(x + d)$. Now we have

$$x^2 - c = 0.$$

$$C \quad (x - d)(x + d) = 0.$$

$$x = d \quad \lor \quad x = -d.$$

We conclude that the solution set is $\{d, -d\}$.

Exercise 7.3

Review Examples 1 and 2 before working Exercises 1–6. Solve each equation over R.

1. $3x^2 = 6x$.
2. $y^3 - y^2 - 4y + 4 = 0$.
3. $x^2 = -b^2$, $b \in R$.
4. $ax^2 - bx = 0$, $a, b \in R \land a \neq 0$.
*5. $2t^3 + 11t^2 + 17t + 6 = 0$.
*6. $6z^3 - 13z^2 + z + 2 = 0$.

Read Example 3 before doing Exercises 7–10. In each of these, factor the given expressions.

7. $ab + 2a + (3m + n)(b + 2)$.
8. $x^2 + 4xy + 4y^2 + 4x + 8y + 4$.
9. $21t - 24 + 3t^2$.
10. $(w - a)^2 - 14w + 14a + 49$.

Before doing Exercises 11–16, study Example 4. Solve each of the given equations (a) over R, (b) over C.

11. $V^2 = -k^2$, $k \in R$, $k \neq 0$.
12. $x^4 - 81 = 0$.
13. $b^4y^4 = c^4$, $b, c \in R \land bc \neq 0$.
14. $t^4 = 10^{-8}$.

15. $\dfrac{u^4}{a^4} - \dfrac{b^4}{c^4} = 0$, $a, b, c \in R \land abc \neq 0$.

16. $m^8 - n^8y^4 = 0$, $m, n \in R \land mn \neq 0$.

Study Example 5 before working Exercises 17–20.

17. If $a, \dfrac{b}{c} \in C$ and $\left(\dfrac{b}{c}\right)^2 = a$, solve $x^2 - a = 0$ over C.

18. If $a, b, c \in C$, and $(a + b)^2 = c$, solve $v^2 - c = 0$ over C.
19. If $m, -bn \in C$ and $(-bn)^2 = m$, solve $y^2 - m = 0$ over C.
20. Solve $x^2 + 4 = 0$ over C. [*Hint:* $x^2 + 4 = x^2 - (2i)^2$.]

7.4 QUADRATIC EQUATIONS, SOLVED BY COMPLETING THE SQUARE

In the present section we wish to consider equations of the form $ax^2 + bx + c = 0$, where a, b, and c are complex numbers and

quadratic $a \neq 0$. An equation of this form is called a *quadratic equation.*
equation Our purpose here is to develop a method for solving such equations. From Section 7.3 we know that if we can find complex numbers r_1 and r_2 such that $ax^2 + bx + c = a(x - r_1)(x - r_2)$, then the solution set of the quadratic equation $ax^2 + bx + c = 0$ over C is the set $\{r_1, r_2\}$. To appreciate the difficulty, the student should spend a few minutes trying to determine numbers r_1 and r_2 such that $2x^2 + x - 2 = 2(x - r_1)(x - r_2)$.

In Section 7.3, we introduced the notion of a square root of a complex number. Using Example 5 in Section 7.3, it is not hard to establish that (1) if d is a square root of a complex number c, so is $-d$, and (2) no complex number can have more than two distinct square roots.

One may inquire whether every complex number has a square root. It turns out that our ability to solve quadratic equations depends on the answer to this question. So let us continue with the discussion of square roots.

We quote now a theorem that allows us to gain new insights in this matter.

Theorem. For every non-negative real number a there is a unique non-negative real number b such that $b^2 = a$. This number b is

principal called *the principal square root of a* (or simply the square root
square root of a) and is denoted by $\sqrt{a}$ (also $a^{1/2}$).

The proof of this theorem depends very strongly on the completeness property of the real numbers and is too advanced to be presented here.

We wish to emphasize that if a is a non-negative real number, then $\sqrt{a} \geq 0$ and $(\sqrt{a})^2 = a$. To illustrate, $\sqrt{4} = 2$, since $2 \geq 0$ and $2^2 = 4$. Note that $\sqrt{4} \neq -2$, since $-2 < 0$. Thus while the number -2 is *a* square root of 4, it is not *the* (principal) square root of 4.

The following theorem is an easy consequence of the foregoing remarks.

Theorem. If a and b are non-negative real numbers, then

(1) $\sqrt{a} \cdot \sqrt{b} = \sqrt{a \cdot b}$,

(2) $\dfrac{\sqrt{a}}{\sqrt{b}} = \sqrt{\dfrac{a}{b}}$, provided $b \neq 0$.

Proof: (1) Since $\sqrt{a} \geq 0$ and $\sqrt{b} \geq 0$, $\sqrt{a} \cdot \sqrt{b} \geq 0$. Further,

$$(\sqrt{a} \cdot \sqrt{b})^2 = (\sqrt{a})^2 (\sqrt{b})^2$$

$$= a \cdot b.$$

Thus $\sqrt{a} \cdot \sqrt{b}$ is the unique non-negative number, whose square is the non-negative number ab. We conclude that $\sqrt{a} \cdot \sqrt{b} = \sqrt{a \cdot b}$.

The proof of part (2) is left to the student.

Example 1.

$$\sqrt{3} \cdot \sqrt{2} = \sqrt{3 \cdot 2}$$

$$= \sqrt{6}.$$

$$\frac{\sqrt{12}}{\sqrt{6}} = \sqrt{\frac{12}{6}}$$

$$= \sqrt{2}.$$

$$\sqrt{8} = \sqrt{4 \cdot 2}$$

$$= \sqrt{4} \cdot \sqrt{2}$$

$$= 2\sqrt{2}.$$

If $a \in R$, $\sqrt{a^2} = |a|$,   since $|a| \geq 0$ and $|a|^2 = a^2$.

Next we assume that a is a negative real number. Then $-a$ is a positive real number, hence $\sqrt{-a}$ is a unique positive real number. By direct computation we find that

$$(\sqrt{-a} \cdot i)^2 = (\sqrt{-a})^2 \cdot i^2$$

$$= (-a)(-1)$$

$$= a.$$

Thus the imaginary number $\sqrt{-a} \cdot i$ is a square root of a. *imaginary* (Recall that the members of $C - R$ are called *imaginary num-* *numbers* *bers*.) We extend the notion of the principal square root of a real number a by defining it to be $\sqrt{-a} \cdot i$ if a is negative. Thus we have:

If $a \geq 0$, $\sqrt{a}$ is the unique non-negative real number whose square is a.
If $a < 0$, $\sqrt{a} = \sqrt{-a} \cdot i$.

It is clear now that every real number has a square root among the complex numbers.

We say that we have completed the square of a quadratic polynomial $ax^2 + bx + c$ if this polynomial is written in the form

completing the
square

$a(x+h)^2 + k$. The next example illustrates the technique of completing the square.

Example 2. Complete the square of (a) $x^2 + 3x + 1$, (b) $2x^2 - x + 2$.

Solution: (a) We recall the formula $(x+h)^2 = x^2 + 2hx + h^2$. Note that $h^2 = \left(\dfrac{2h}{2}\right)^2$, that is, the third term is equal to one-half of the coefficient of x, squared. Motivated by this observation, we write

$$x^2 + 3x + 1 = x^2 + 3x + \left(\frac{3}{2}\right)^2 + 1 - \left(\frac{3}{2}\right)^2$$

$$= \left(x + \frac{3}{2}\right)^2 - \frac{5}{4}.$$

(b) $2x^2 - x + 2 = 2\left(x^2 - \dfrac{1}{2} \cdot x + 1\right)$

$$= 2\left(x^2 - \frac{1}{2}x + \left(\frac{-1}{4}\right)^2 + 1 - \left(\frac{-1}{4}\right)^2\right)$$

$$= 2\left(\left(x - \frac{1}{4}\right)^2 + \frac{15}{16}\right)$$

$$= 2\left(x - \frac{1}{4}\right)^2 + \frac{15}{8}.$$

Example 3. Solve the following equations over C:

(a) $x^2 + 3x + 1 = 0$,
(b) $2x^2 - x + 2 = 0$.

Solution: From Example 2,

$$x^2 + 3x + 1 = \left(x + \frac{3}{2}\right)^2 - \frac{5}{4}.$$

Further,

$$\left(x + \frac{3}{2}\right)^2 - \frac{5}{4} = \left(x + \frac{3}{2}\right)^2 - \left(\sqrt{\frac{5}{4}}\right)^2$$

$$= \left(x + \frac{3}{2} - \sqrt{\frac{5}{4}}\right)\left(x + \frac{3}{2} + \sqrt{\frac{5}{4}}\right)$$

$$= \left(x + \frac{3 - \sqrt{5}}{2}\right)\left(x + \frac{3 + \sqrt{5}}{2}\right).$$

Therefore,

$$x^2 + 3x + 1 = \left(x + \frac{3 - \sqrt{5}}{2}\right)\left(x + \frac{3 + \sqrt{5}}{2}\right).$$

Now we have

$$x^2 + 3x + 1 = 0.$$

$$\left(x + \frac{3 - \sqrt{5}}{2}\right)\left(x + \frac{3 + \sqrt{5}}{2}\right) = 0.$$

$$x + \frac{3 - \sqrt{5}}{2} = 0 \quad \text{or} \quad x + \frac{3 + \sqrt{5}}{2} = 0.$$

$$x = \frac{-3 + \sqrt{5}}{2} \quad \text{or} \quad x = \frac{-3 - \sqrt{5}}{2}.$$

Hence the solution set is $\left\{\dfrac{-3 + \sqrt{5}}{2}, \dfrac{-3 - \sqrt{5}}{2}\right\}$.

(b) From Example 2,

$$2x^2 - x + 2 = 2\left(x - \frac{1}{4}\right)^2 + \frac{15}{8}.$$

Now we have

$$2x^2 - x + 2 = 0.$$

$$2\left(x - \frac{1}{4}\right)^2 + \frac{15}{8} = 0.$$

$$\left(x - \frac{1}{4}\right)^2 + \frac{15}{16} = 0.$$

$$\left(x - \frac{1}{4}\right)^2 - \frac{-15}{16} = 0.$$

$$\left(x - \frac{1}{4}\right)^2 - \left(\sqrt{\frac{-15}{16}}\right)^2 = 0.$$

$$\left(x - \frac{1}{4}\right)^2 - \left(\frac{\sqrt{15}}{4}i\right)^2 = 0.$$

$$\left(x - \frac{1}{4} - \frac{\sqrt{15}}{4}i\right)\left(x - \frac{1}{4} + \frac{\sqrt{15}}{4}i\right) = 0.$$

$$x = \frac{1 + \sqrt{15}\,i}{4} \quad \vee \quad x = \frac{1 - \sqrt{15}\,i}{4}.$$

Hence the solution set is $\left\{\dfrac{1 + \sqrt{15}\,i}{4}, \dfrac{1 - \sqrt{15}\,i}{4}\right\}$.

The technique displayed in Example 3 can be shortened if we observe that if c is any real number, then the open sentences $x^2 = c$, $x = \sqrt{c} \lor x = -\sqrt{c}$, are equivalent relative to the replacement set C (the proof of this statement is left as an exercise). The next example illustrates a solution of a quadratic equation using this fact and the technique of *completing the square*.

Example 4. Solve $5x^2 + 2x + 1 = 0$ over C.

Solution:

$$5x^2 + 2x + 1 = 0.$$

$$x^2 + \frac{2}{5}x + \frac{1}{5} = 0.$$

$$x^2 + \frac{2}{5}x = -\frac{1}{5}.$$

$$x^2 + \frac{2}{5}x + \left(\frac{1}{5}\right)^2 = -\frac{1}{5} + \left(\frac{1}{5}\right)^2.$$

$$\left(x + \frac{1}{5}\right)^2 = -\frac{4}{25}.$$

$$x + \frac{1}{5} = \sqrt{-\frac{4}{25}} \lor x + \frac{1}{5} = -\sqrt{\frac{-4}{25}}.$$

$$x = \frac{-1 + 2i}{5} \lor x = \frac{-1 - 2i}{5}.$$

Thus the solution set is $\left\{\dfrac{-1 + 2i}{5}, \dfrac{-1 - 2i}{5}\right\}$.

Exercise 7.4

Study Example 1 before doing Exercises 1–6. In each of these, simplify the given expression.

1. $\sqrt{5} \cdot \sqrt{20}$.

2. $\dfrac{\sqrt{60}}{\sqrt{3}}$.

3. $\dfrac{\sqrt{3}(\sqrt{21} + \sqrt{63})}{\sqrt{189}}$.

4. $\sqrt{-4} \cdot \sqrt{-25}$.

5. $\sqrt{x^2 - 2x + 1}$, $x \in R$. 6. $\sqrt{2y^2 + 8y + 8}$, $y \in R$.

Study Example 2 before doing Exercises 7–12. In each of these, complete the square of the given expression.

7. $y^2 + 5y + 1$.

8. $x^2 - 5x - 3$.

9. $3x^2 - x + 3.$ **10.** $3z^2 + 4z + 2.$
11. $ax^2 - bx + 1, a \neq 0.$ *12.* $ax^2 + bx + c, a \neq 0.$

Study Example 3 before doing Exercises 13–18. In each of these, solve the given equation over C.

13. $x^2 + 5x + 1 = 0.$ **14.** $z^2 - 5z - 3 = 0.$
15. $3 + 3v^2 = v.$ **16.** $-2 = 4u + 3u^2.$
17. $ay^2 = by - 1, a \neq 0.$ *18.* $ax^2 + bx + c = 0, a \neq 0.$

Before doing Exercises 19–22, study Example 4. In each of these, solve the given equation as illustrated in Example 4.

19. $8t^2 + 4t + 5 = 0.$ **20.** $25z^2 = 30z - 11.$
21. $5 = 8y^2 + 4y.$ **22.** $4x^2 - 9 - 5x = 0.$

23. Prove that if $c \in R$, then relative to the replacement set C we have $x^2 = c \leftrightarrow x = \sqrt{c} \lor x = -\sqrt{c}.$
24. Solve the inequality $x^2 + 3x + 1 > 0.$ [*Hint:* Factor $x^2 + 3x + 1.$]
25. Solve by factoring:

(a) $(3x - 1)^2 - 4(3x - 1) - 12 = 0,$
(b) $x^3 + 3x^2 - x - 3 = 0.$

26. Write a quadratic equation with integral coefficients and solution set $\{-\frac{1}{3}, \frac{1}{2}\}.$
27. Find complex numbers r_1 and r_2 such that $2x^2 + x - 2 = 2(x - r_1)(x - r_2).$
28. Prove: If d is a square root of a complex number c, so is $-d.$
29. Prove that no complex number can have more than two distinct square roots. [*Hint:* Show that the equation $x^2 = c$ has at most two solutions.]
30. Give an example of a complex number that has only one square root.

In Exercises 31–36, simplify the given expression.

31. $\sqrt{-4} \cdot \sqrt{4}.$ **32.** $\sqrt{-9} \cdot \sqrt{-4}.$

33. $\dfrac{\sqrt{-5}}{\sqrt{5}}.$ **34.** $\dfrac{\sqrt{12}}{\sqrt{-3}}.$

35. $(\sqrt{7} - \sqrt{-3})(\sqrt{7} + \sqrt{-3}).$
36. $\sqrt{x^4 + 2\sqrt{x^2} + 1}.$

37. Prove: If $a \geqslant 0$ and $b > 0$, then $\dfrac{\sqrt{a}}{\sqrt{b}} = \sqrt{\dfrac{a}{b}}.$

*38. Let a and b be real numbers such that $\sqrt{a}\sqrt{b} = \sqrt{ab}$ and $a < 0$. What can you conclude about the number b? Justify your answer.

7.5 THE QUADRATIC FORMULA

Let us again consider the quadratic equation $ax^2 + bx + c = 0$, where a, b, c are complex numbers. We shall show in Chapter 11 that every complex number has a quare root. So, let d be a square root of $b^2 - 4ac$. We solve the quadratic equation $ax^2 + bx + c = 0$ over C as follows:

$$ax^2 + bx + c = 0.$$

$$x^2 + \frac{b}{a}x + \frac{c}{a} = 0.$$

$$x^2 + \frac{b}{a}x + \left(\frac{b}{2a}\right)^2 + \frac{c}{a} - \left(\frac{b}{2a}\right)^2 = 0.$$

$$\left(x + \frac{b}{2a}\right)^2 + \frac{4ac - b^2}{4a^2} = 0 \qquad \text{(why?)}.$$

$$\left(x + \frac{b}{2a}\right)^2 - \frac{b^2 - 4ac}{4a^2} = 0 \qquad \text{(why?)}.$$

$$\left(x + \frac{b}{2a}\right)^2 - \left(\frac{d}{2a}\right)^2 = 0 \qquad \begin{array}{l}\text{(since } d \text{ is a square} \\ \text{root of } b^2 - 4ac).\end{array}$$

$$\left(x + \frac{b}{2a} - \frac{d}{2a}\right)\left(x + \frac{b}{2a} + \frac{d}{2a}\right) = 0.$$

$$x + \frac{b}{2a} - \frac{d}{2a} = 0 \quad \vee \quad x + \frac{b}{2a} + \frac{d}{2a} = 0.$$

$$x = \frac{-b + d}{2a} \quad \vee \quad x = \frac{-b - d}{2a}.$$

Thus the solution set is $\left\{\dfrac{-b + d}{2a}, \dfrac{-b - d}{2a}\right\}$, where d is a square root of $b^2 - 4ac$.

In particular, if a, b, c are real numbers, then we may choose d to be the principal square root of $b^2 - 4ac$. In this instance the following equivalence holds:

$$ax^2 + bx + c = 0.$$

$$x = \frac{-b + \sqrt{b^2 - 4ac}}{2a} \quad \vee \quad x = \frac{-b - \sqrt{b^2 - 4ac}}{2a}.$$

The statement

$$x = \frac{-b + \sqrt{b^2 - 4ac}}{2a} \quad \lor \quad x = \frac{-b - \sqrt{b^2 - 4ac}}{2a}$$

is frequently abbreviated by writing

$$x = \frac{-b \pm \sqrt{b^2 - 4ac}}{2a}.$$

quadratic   The last expression is called the *quadratic formula*. Note
formula that the quadratic formula is an open sentence that is equivalent
to $ax^2 + bx + c = 0$ $(a \neq 0)$.

Example 1. Solve the equation $3x^2 - 2x + 1 = 0$ over C, using
the quadratic formula.

Solution: For the equation $ax^2 + bx + c = 0$ $(a \neq 0)$, the quadratic formula is

$$x = \frac{-b \pm \sqrt{b^2 - 4ac}}{2a}.$$

In our problem $a = 3$, $b = -2$, and $c = 1$. Thus $3x^2 - 2x + 1 = 0$
is equivalent to

$$x = \frac{-(-2) \pm \sqrt{(-2)^2 - 4 \cdot 3 \cdot 1}}{2 \cdot 3},$$

that is,

$$x = \frac{2 \pm \sqrt{-8}}{2 \cdot 3}.$$

Noting that $\sqrt{-8} = \sqrt{8}\,i = 2\sqrt{2}\,i$, we can write equivalently

$$x = \frac{1 \pm \sqrt{2}\,i}{3}.$$

Hence the solution set is $\left\{ \dfrac{1 + \sqrt{2}\,i}{3}, \dfrac{1 - \sqrt{2}\,i}{3} \right\}$.

Remark. Given a quadratic equation $ax^2 + bx + c = 0$, the quan-
discriminant tity $b^2 - 4ac$ is called the *discriminant of the equation*. If $a, b,$
$c \in R$, the discriminant determines the nature of the solutions
as follows:

(1) If $b^2 - 4ac \neq 0$, the equation has two distinct complex-
valued solutions. If $b^2 - 4ac > 0$, both solutions are real
numbers. If $b^2 - 4ac < 0$, both solutions are imaginary
numbers.
(2) If $b^2 - 4ac = 0$, then the equation has only one solution.
This solution is a real number.

Example 2. Determine the real values of k for which the solutions of $x^2 + 2x + k = 0$ are (a) real numbers, (b) imaginary numbers.

Solution: The discriminant is $2^2 - 4 \cdot 1 \cdot k$, that is, $4 - 4k$. By the foregoing remark, every solution is a real number if and only if $4 - 4k \geq 0$, that is, if and only if $k \leq 1$. Also, the solutions are imaginary numbers if and only if $4 - 4k < 0$, that is, if and only if $k > 1$. Thus, if $k < 1$ the equation has two distinct real-valued solutions; if $k = 1$, the equation has one real-valued solution; and if $k > 1$, there are two distinct imaginary solutions.

Example 3. Write the polynomial $3x^2 - 2x + 1$ as a product of two polynomials, each of degree one.

Solution: According to Example 1, the solutions of the equation $3x^2 - 2x + 1 = 0$ over C are

$$\frac{1 + \sqrt{2}\,i}{3}, \quad \frac{1 - \sqrt{2}\,i}{3}.$$

The factor theorem (see Section 5.4) allows us to conclude that

$$x - \frac{1 + \sqrt{2}\,i}{3} \quad \text{and} \quad x - \frac{1 - \sqrt{2}\,i}{3}$$

are factors of $3x^2 - 2x + 1$. It is easy to verify that

$$3x^2 - 2x + 1 = 3\left(x - \frac{1 + \sqrt{2}\,i}{3}\right)\left(x - \frac{1 - \sqrt{2}\,i}{3}\right),$$

which can be simplified to

$$3x^2 - 2x + 1 = (3x - 1 - \sqrt{2}\,i)\left(x - \frac{1 - \sqrt{2}\,i}{3}\right).$$

Exercise 7.5

Read Example 1 before doing Exercises 1–8. In each of these solve the given equation over C using the quadratic formula.

1. $2x^2 - 2x + 1 = 0$.   2. $5y^2 - 2 - 2y = 0$.
3. $3t - 7t^2 = 0$.   4. $\frac{1}{3}w^2 - \frac{1}{2}w + 1 = 0$.
5. $mz^2 + (m + n)z + n = 0$, $m \neq 0$, $n \neq 0$, $\{m, n\} \subset R$.
6. $ay^2 - (a - b)y - b = 0$, $a \neq 0$, $\{a, b\} \subset R$.
7. $x^2 + yx - 2y^2 = 0$ (solve for x), $y \in R$.
8. $12u^2 + V^3u - V^6 = 0$ (solve for u), $V \in R$.

Review Example 2 before working Exercises 9–16. For each of these, determine the real values of k for which the solutions of the given equation are (a) real numbers, (b) imaginary numbers.

9. $x^2 + 3x + k = 0$. 10. $y^2 - 2y - k = 0$.
11. $ky^2 - 3ky + 2 = 0$. 12. $3x^2 + 4kx + k = 0$.
13. $t^2 + 2kt - k = 0$. 14. $2w^2 - kw + 3k = 0$.
15. $3ky^2 - ky + 1 = 0$. 16. $5kx^2 + 2kx - 1 = 0$.

Study Example 3 before working Exercises 17–22. In each of these, write the given polynomial as a product of polynomials, each of degree one.

17. $3x^2 - 4x - 1$. 18. $4z^2 - 2z + \frac{7}{16}$.
19. $7u^2 - \sqrt{2}u - 1$. *20. $8x^3 - 27$.
*21. $y^3 - \frac{1}{2}y + \frac{1}{2}$. [*Hint:* $y + 1$ is a factor.]
*22. $x^6 - b^6$, $b \in R$, $b \neq 0$.

Write each of the following polynomials as a product of linear polynomials with integral coefficients whenever this is possible.

23. $2x^2 - x - 1$. 24. $3y^2 - 1$.
25. $\frac{1}{16}u^4 - 1$. 26. $12y^2 - 5y - 3$.
*27. If the solution set of $ax^2 + bx + c = 0$ $(a \neq 0)$ is $\{r_1, r_2\}$,
show that $r_1 + r_2 = \dfrac{b}{a}$ and $r_1 \cdot r_2 = \dfrac{c}{a}$.

**28. (a) Let $p(x)$ be the polynomial $ax^2 + bx + c$, where $\{a, b, c\} \subset R$ and $a < 0$. Show that $p\left(\dfrac{-b}{2a}\right) \geqslant p(x)$, whenever $x \in R$.

(b) A ball thrown vertically upward obeys the law $h(t) = 32t - 16t^2$, where t is the time in seconds after the throw and $h(t)$ is the height of the ball in feet. How high will the ball rise and how long will it take the ball to reach its maximum height?

*29. Prove: A quadratic equation has one and only one solution if and only if its discriminant is 0.

30. For what value or values of k does the quadratic equation $3x^2 + kx + 2x = 1 - k$ have one and only one solution?

*31. Solve: (a) $x^2 - (3 + 2i)x + (1 + 3i) = 0$, (b) $x^2 + (4 + 3i)x + (1 + 6i) = 0$.

*32. Factor the expression $a^2 + ab + b^2$, where $a, b \in R$.
[*Hint:* Complete the square.]

7.6 SOME OTHER TYPES OF EQUATIONS

In Sections 7.4 and 7.5 we showed how to solve quadratic equations. This knowledge sometimes enables us to solve other types of equations as well. We illustrate the ideas involved with the following examples.

Example 1. Solve the equation $(x-1)^4+2(x-1)^2-3=0$ over C.

Solution: The given equation is not quadratic in x; however, it is quadratic in the variable $(x-1)^2$. We let $y=(x-1)^2$ and consider the simpler equation $y^2+2y-3=0$. The solution set of this equation is found by methods of the preceding sections to be $\{-3, 1\}$. Now we have

$$\updownarrow \quad \begin{array}{c} (x-1)^4+2(x-1)^2-3=0. \\ (x-1)^2=-3 \quad \vee \quad (x-1)^2=1. \\ x-1=\pm\sqrt{-3} \quad \vee \quad x-1=\pm\sqrt{1}. \\ x=1\pm\sqrt{3}\,i \quad \vee \quad x=1\pm\sqrt{1}. \end{array}$$

Hence, the solution set is $\{0, 2, 1+\sqrt{3}\,i, 1-\sqrt{3}\,i\}$.

Example 2. Solve the equation $|x-3|=4$ over R.

Solution: We present two solutions: one has a geometric flavor, while the other is algebraic.

Solution 1: Consider the number line X as shown in Figure 7.1. We reason as follows. A number c is in the solution set of $|x-3|=4$ if and only if the distance between the points $P(c)$ and $P(3)$ is 4 units, that is, if and only if $c=-1$ or $c=7$ (see Figure 7.1). Thus the solution set is $\{-1, 7\}$.

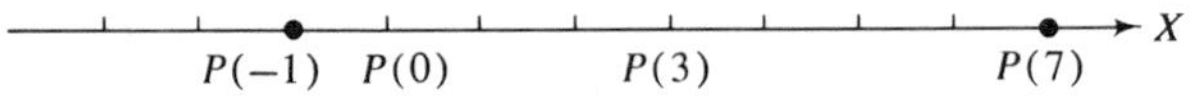

Figure 7.1

Solution 2:

$$\updownarrow \quad \begin{array}{c} |x-3|=4. \\ |x-3|^2=16. \\ (x-3)^2=16. \\ x-3=\pm\sqrt{16} \qquad \text{(why?)}. \\ x=3\pm 4. \\ x=7 \quad \vee \quad x=-1. \end{array}$$

Again, the solution set is $\{-1, 7\}$.

Example 3. Solve the equation $x-1=\sqrt{x-1}$ over R.

Solution:

$$x - 1 = \sqrt{x - 1},$$
$$(x - 1)^2 = (\sqrt{x - 1})^2,$$
$$(x - 1)^2 - (x - 1) = 0,$$
$$(x - 1)(x - 1 - 1) = 0,$$
$$x = 1 \quad \vee \quad x = 2.$$

Thus, the solution set of the given equation is a subset of $\{1, 2\}$. Next we check the numbers 1 and 2 to see whether they are solutions of the given equation.

CHECK. $1 - 1 = \sqrt{1-1}$ is a true proposition and so is $2 - 1 = \sqrt{2-1}$. We conclude that the solution set is $\{1, 2\}$.

Example 4. Solve the equation $x - 1 = \sqrt{x + 5}$ over R.

Solution:

$$x - 1 = \sqrt{x + 5},$$
$$(x - 1)^2 = (\sqrt{x + 5})^2,$$
$$x^2 - 2x + 1 = x + 5,$$
$$x^2 - 3x - 4 = 0,$$
$$(x - 4)(x + 1) = 0,$$
$$x = 4 \quad \vee \quad x = -1.$$

CHECK. $4 - 1 = \sqrt{4 + 5}$ is true while $-1 - 1 = \sqrt{-1 + 5}$ is false. Hence the solution set is $\{4\}$.

Example 4 should alert the student to the hazards of squaring each member of an equation: the new equation may be redundant relative to the given equation. We give another example to illustrate this fact.

Example 5. Solve the equation $|x - 1| = 2x + 1$ over R.

Solution:

$$|x - 1| = 2x + 1,$$
$$|x - 1|^2 = (2x + 1)^2,$$
$$x^2 - 2x + 1 = 4x^2 + 4x + 1,$$
$$3x^2 + 6x = 0,$$
$$3x(x + 2) = 0,$$
$$x = 0 \quad \vee \quad x = -2.$$

CHECK. $|0 - 1| = 2 \cdot 0 + 1$ is true while $|-2 - 1| = 2(-2) + 1$ is false. Hence the solution set is $\{0\}$.

Exercise 7.6

Study Example 1 before working Exercises 1–6. In each of these, solve the given equation over C.

1. $(x + 7)^4 - 5(x + 7)^2 + 6 = 0$.
2. $(y - 6)^4 - 9(y - 6)^2 + 20 = 0$.
3. $(w - 2)^4 + (w - 2)^2 - 6 = 0$.
4. $(r + 5)^4 - 2(r + 5)^2 - 3 = 0$.
5. $(y - a)^4 - b^2 = 0$, $a, b \in R$, $b > 0$.
6. $[(t + a)^2]^2 - (a^2 - 2ab + b^2)^2 = 0$, $a, b \in R$.

Before working Exercises 7–12, study Example 2. In each of these, solve the given equation over R.

7. $|x - 1| = 2$.   8. $|y + 3| = \sqrt{2}$.
9. $|x + 1| = -1$.   10. $|3u - a| = b$, $b > 0$.
11. $|ax + b| = c$, $a \neq 0$, $c > 0$.
12. $\left| \dfrac{y}{r_1} - \dfrac{1}{r_2^2} \right| = \dfrac{1}{r_1}$, $r_1 > 0$.

Study Examples 3, 4, and 5 before doing Exercises 13–25. In each of these, solve the given equation over R.

13. $t + 1 = \sqrt{1 + t}$.   14. $2r - \sqrt{4r - 3} = 1$.
15. $x - \sqrt{x + 1} = 1$.   16. $2 - z = \sqrt{8 - z}$.
17. $y + 1 = \sqrt{y - 1}$.   18. $x = 3 + \sqrt{1 - 2x}$.
19. $\sqrt{x - 1} + \sqrt{x + 1} = 2$.   20. $\sqrt{2y - 3} + \sqrt{y + 2} = 1$.
21. $\sqrt{x + 2} - \sqrt{x^2 - 3x + 5} = 0$.
22. $\sqrt{t^2 - a^2 - 2ab} = b$, $b > 0$, $a \in R$.
*23. $x + mn = \sqrt{(m + n)^2 x - 2n^2(x - m^2)}$,
    where $m > n > 0$.
24. $|x - 2| = 3x - 8$.   25. $|5 - 2x| = 3x - 5$.

7.7 SOME OTHER TYPES OF EQUATIONS (CONTINUED)

fractional equations We present here examples of the so called *fractional equations* (equations in which the variable appears in the denominator of a fraction).

Example 1. Solve the equation $\dfrac{x}{x + 2} - \dfrac{4}{x + 1} = \dfrac{-2}{x + 2}$ over C.

Solution: We note that the replacement set for the given equation is $C - \{-1, -2\}$. Since the solution set of $(x + 1)(x + 2) = 0$ over $C - \{-1, -2\}$ is $\varnothing$, from the theorem in Section 7.2, we conclude that relative to the replacement set $C - \{-1, -2\}$ the equations

$$\frac{x}{x + 2} - \frac{4}{x + 1} = \frac{-2}{x + 2}$$

and

$$(x + 1)(x + 2)\left(\frac{x}{x + 2} - \frac{4}{x + 1}\right) = (x + 1)(x + 2) \cdot \frac{-2}{x + 2}$$

are equivalent. Thus we have

$$C - \{-1, -2\}$$
$$\frac{x}{x + 2} - \frac{4}{x + 1} = \frac{-2}{x + 2}.$$

$$(x + 1)(x + 2)\left(\frac{x}{x + 2} - \frac{4}{x + 1}\right)$$

$$= (x + 1)(x + 2) \cdot \frac{-2}{x + 2}.$$

$$(x + 1)x - 4(x + 2) = -2(x + 1).$$

$$x^2 - x - 6 = 0.$$

$$(x - 3)(x + 2) = 0.$$

$$x = 3 \quad \vee \quad x = -2.$$

Hence the solution set is $(C - \{-1, -2\}) \cap \{-2, 3\} = \{3\}$.

Example 2. Solve $\dfrac{2x + 1}{x^2 + 5x + 6} + \dfrac{3}{x + 2} + \dfrac{1}{x + 3} = 0$ over R.

Solution: It is clear that the replacement set for the given equation is $R - \{-2, -3\}$.

$$R - \{-2, -3\}$$
$$\frac{2x + 1}{x^2 + 5x + 6} + \frac{3}{x + 2} + \frac{1}{x + 3} = 0.$$

$$(x + 2)(x + 3)\left(\frac{2x + 1}{x^2 + 5x + 6} + \frac{3}{x + 2} + \frac{1}{x + 3}\right)$$

$$= (x + 2)(x + 3) \cdot 0.$$

$$2x + 1 + 3(x + 3) + x + 2 = 0.$$

$$6x + 12 = 0.$$

$$x = -2.$$

Thus the solution set is $[R - \{-2, -3\}] \cap \{-2\} = \varnothing$.
We conclude this section with a stated problem.

Example 3. The numerator of a fraction is a positive number and exceeds the denominator by 1. If the numerator and the denominator are interchanged, the sum of this new fraction and the original one is $\frac{61}{30}$. Find the original fraction.

Solution: Let x denote the unknown numerator of the original fraction. Then the original fraction is $\dfrac{x}{x - 1}$. The problem states that

$$\frac{x}{x - 1} + \frac{x - 1}{x} = \frac{61}{30}.$$

Let R^+ denote the set of positive real numbers. From the context the replacement set for the foregoing equation is $R^+ - \{1\}$.

$$R^+ - \{1\} \begin{cases} \dfrac{x}{x - 1} + \dfrac{x - 1}{x} = \dfrac{61}{30}. \\[2mm] x^2 + (x - 1)^2 = \tfrac{61}{30}x(x - 1). \\[2mm] x^2 - x - 30 = 0. \\[2mm] (x + 5)(x - 6) = 0. \\[2mm] x = -5 \quad \vee \quad x = 6. \end{cases}$$

Thus the solution set of the foregoing equation is

$$(R^+ - \{1\}) \cap \{-5, 6\} = \{6\}.$$

Consequently, the original fraction is $\frac{6}{5}$. Indeed, $\frac{6}{5} + \frac{5}{6} = \frac{61}{30}$ and $6 = 5 + 1$.

Exercise 7.7

Study Examples 1 and 2 before working Exercises 1–6. In each of these, solve the given equation over C.

1. $\dfrac{x}{x + 3} + \dfrac{2x}{3x + 1} = 0.$

2. $\dfrac{2x}{x + 2} + \dfrac{2}{x - 1} = \dfrac{12}{x + 2}.$

3. $\dfrac{4x}{x^2 + x - 6} + \dfrac{3x - 3}{x + 3} + \dfrac{x - 3}{x - 2} = 3.$

4. $\dfrac{1}{x} + \dfrac{x}{x - 1} = \dfrac{x + 1}{x}.$

5. $\dfrac{x}{x+3} + \dfrac{2}{x+4} = \dfrac{-3}{x+3}.$

*6. Solve $\dfrac{x^2+3}{x+1} - 3\dfrac{x+1}{x^2+3} = 2$ over R.

7. A sports car that is 12 ft long overtakes a truck that is 32 ft long and that is traveling at the rate of 40 mph. How fast must the sports car be traveling in order to pass the truck in two seconds?

8. If each edge of a certain cube were increased by 2 in., the volume would be increased by $12\frac{2}{3}$ cu in. Find the length of the edge of this cube.

9. A baseball team won 24 and lost 5 of its first 29 games. If it won half and lost half of the remaining games and had an average of 60 percent wins at the end of the season, how many games were there?

10. The sum s of the first n consecutive positive integers $1, 2, 3, \cdots, n$ is given by the formula $s = \frac{1}{2}n(n+1)$. How many consecutive positive integers, starting with 1, must be added to obtain 2211?

In Exercises 11–20, solve the given equations over C.

11. $\dfrac{x}{x} = 2.$

16. $\dfrac{-2}{x} + \dfrac{x+6}{x^2-2x} = \dfrac{4}{x-2}.$

12. $\dfrac{a}{x-1} + \dfrac{a}{x+1} = 0,\ a \neq 0.$

17. $\dfrac{a^2}{x-a} = b,\ ab \neq 0.$

13. $\dfrac{10-3y^2}{y} = \dfrac{5-12y}{4}.$

18. $\dfrac{5}{3u} - \dfrac{10}{7u} = \dfrac{1}{21}.$

14. $\dfrac{x}{x} = 0.$

19. $\dfrac{2y}{y-3} + 5 = 0.$

15. $\dfrac{7}{u-1} = 2.$

20. $\dfrac{x(x-1)}{x^2-1} = 1.$

REVIEW EXERCISES

1. Which fundamental property of equality is used in each of the following statements?

    (a) If $a^2 + 2ab + b^2 = (a+b)^2$, then $(a+b)^2 = a^2 + 2ab + b^2$.

    (b) If $x^2 + 3x + 2 = (x+2)(x+1)$ and $(x+2)(x+1) = x(x+3) + 2$, then $x^2 + 3x + 2 = x(x+3) + 2$.

    (c) $4 = 4$.

In Exercises 2–6 a pair of equations is given along with a specified replacement set. In each of these, determine whether the equations are equivalent relative to the given replacement set.

2. $x^2 = 0$, $x = 0$. Replacement set is R.

3. $x^2 + 9x + 18 = 0$, $x + 6 = 0$. Replacement set is R.

4. Repeat Exercise 3 for the replacement set N (the set of natural numbers).

5. $(x^2 + 1)(x - 2) = 0$, $x - 2 = 0$. Replacement set is C.

6. Repeat Exercise 5 for the replacement set R.

7. (a) Verify that $1 + i$ is a square root of $2i$.

    (b) Solve the equation $x^2 - 2i = 0$ over C.

8. Let a and b be negative real numbers. Show that $\sqrt{ab} \neq \sqrt{a}\sqrt{b}$.

9. Complete the square of

    (a) $x^2 + 5x + 4$,

    (b) $3x^2 + 2x + 1$.

In each of Exercises 10–15 solve the given quadratic equation over C in two ways: (a) by completing the square of the left-hand member of the equation and then factoring, (b) by using the quadratic formula.

10. $6x^2 + x - 1 = 0$.   **11.** $4x^2 - 5x - 6 = 0$.

12. $x^2 - 4x + 2 = 0$.   **13.** $x^2 - \sqrt{5}x + 1 = 0$.

14. $2x^2 - x - 1 = 0$.   **15.** $2y^2 + y + 1 = 0$.

16. Factor completely $6x^2 - x - 1$.

17. Factor completely $4x^2 - 5x - 6$.

18. Factor completely $x^3 + x^2 + 4x + 4$. [*Hint:* $x + 1$ is a factor.]

19. Determine the real values of K for which the solution set of $x^2 + Kx + K = 0$ has (a) exactly one member, (b) two distinct members that are real numbers, (c) two distinct members that are imaginary numbers.

20. Solve the equation $x + 1 = \sqrt{(x + 1)^2}$ over R.

21. Solve the equation $x - 1 = |x + 1|$ over R.

22. Solve the equation $y^4 - 3y^2 - 4 = 0$ over C.

23. Solve the equation $\dfrac{x + 1}{x - 2} + \dfrac{x + 5}{x + 1} = 6$ over R.

24. Solve the equation $\dfrac{2x + 1}{2x} + \dfrac{1}{x} = \dfrac{5}{2}$ over R.

25. If $a, b \in R$, show that the open sentences $x = a \pm b$, $x = a \pm |b|$ are equivalent relative to the replacement set R.

8

STATED
PROBLEMS

8.1 INTRODUCTION

A written or verbal statement expressing some condition or conditions of equality that exist among some quantities, among which *stated* at least one is unknown, is called a *stated problem*.

problem   In the present section we outline the general procedure one should follow in solving such a problem. There are, however, certain types of problems that occur frequently and whose solution depends on the proper use of a particular formula or idea. We shall discuss some of these types in detail, pointing out, in each case, the formula or idea to be used.

The general procedure is as follows:

STEP 1. Read the problem carefully, and write a short summary of the problem on scratch paper if necessary.

STEP 2. Write down the particular idea or formula to be used in the problem. For example, in a motion problem, distance = velocity · time.

STEP 3. Ask yourself: "What are the quantities involved that are unknown?"

STEP 4. Represent one of these quantities by some symbol, say x, and be specific in writing down in what units the quantities are to be expressed.

We remark that, on occasion, denoting one of the unknown quantities x rather than some other unknown quantity may yield a simpler equation. The student will learn to make the "better" choice with practice, and in any event the solution of the problem does not depend on what he chooses to call x at this time. Therefore, it seems natural at the beginning to call x the unknown that is to be found in the particular problem (unless a better choice is obvious).

STEP 5. Write all unknown quantities in terms of x (or whatever symbol was used in step 4).

STEP 6. Form an equation that according to the statement of the problem, expresses the relation between the quantities introduced in step 5 and the known quantities of the problem.

STEP 7. Solve the equation.

STEP 8. Check your results.

We must stress that to check the correctness of the solution the student must verify that his results satisfy the conditions stated in the original problem. If he only checks that his results are solutions of the equation obtained in step 6, he has checked the correctness of his work in solving the equation but not the accuracy of his analysis of the problem.

8.2 MOTION PROBLEMS

We illustrate the basic idea of motion problems in two examples.

Example 1. A boy left town A to go to town B 21 miles away. He walked part of the way and then got a ride. If we know that the boy walked at a speed of 3 mph and that the car speed was 45 mph, how long did he walk if this trip took 2 hr and 20 min?

Solution: The fundamental formula is

$$\text{distance} = \text{speed} \cdot \text{time},$$

and the basic idea is illustrated in Figure 8.1.

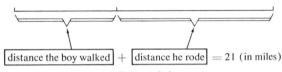

$$\boxed{\text{distance the boy walked}} + \boxed{\text{distance he rode}} = 21 \text{ (in miles)}$$

Figure 8.1

Let x hr be the time the boy walked. Note that

$$2 \text{ hr } 20 \text{ min} = (2 + \tfrac{1}{3}) \text{ hr} = \tfrac{7}{3} \text{ hr}.$$

Thus, the boy rode $(\tfrac{7}{3} - x)$ hr, since certainly the time he rode is equal to the total time minus the time he walked.

Using the fundamental formula, we conclude that the boy walked $3x$ mi and he rode $45 \cdot (\frac{7}{3} - x)$ mi. Thus, using the basic idea as stated above, we obtain

$$3x + 45 \cdot \left(\frac{7 - 3x}{3}\right) = 21.$$

We now proceed to solve this equation:

$$3x + 45 \cdot \left(\frac{7 - 3x}{3}\right) = 21.$$
$$3x + 15 \cdot (7 - 3x) = 21.$$
$$3x + 105 - 45x = 21.$$
$$-42x = -84.$$
$$x = 2.$$

Hence, the boy walked 2 hr and he rode 20 min.

To check, note that in 2 hr the boy would have walked 6 mi and in 20 min the car would have taken him 15 mi. Thus, he could reach town B in 2 hr and 20 min.

To illustrate the remark made in Section 8.1 concerning step 4, we give another solution of the same problem choosing x to represent some other unknown quantity.

Let x mi be the distance the boy walked. Then he must have ridden $(21 - x)$ mi, since the total distance is 21 mi.

Using the fundamental formula (distance $=$ speed $\cdot$ time or, equivalently, time $= \dfrac{\text{distance}}{\text{speed}}$), we get that the boy walked for $\dfrac{x}{3}$ hr and rode $\dfrac{21 - x}{45}$ hr. Since the total time was 2 hr 20 min ($\frac{7}{3}$ hr), we get

$$\frac{x}{3} + \frac{21 - x}{45} = \frac{7}{3}.$$

Proceeding to solve this equation, we obtain

$$15x + 21 - x = 105.$$
$$14x = 84.$$
$$x = 6.$$

Thus, the boy walked 6 mi and rode 15 mi. Since he walked at a speed of 3 mph, obviously he walked for 2 hr. The car speed was 45 mph, thus the boy rode $\frac{1}{3}$ hr or 20 min. Note that these results are the same as those obtained in the first solution.

Example 2. A car leaves Portland, Oregon, toward Vancouver, B.C., at 10:00 A.M. and travels at a speed of 70 mph. Another car leaves Seattle, Washington, for Vancouver, B.C., at noon of the same day and travels at 50 mph. The driver in the first car stops 30 min in Seattle and then continues his trip. At what time will he catch up with the second car? Seattle is 180 mi north of Portland.

Solution: The fundamental formula is

$$\text{distance} = \text{speed} \cdot \text{time},$$

and the basic idea is: when the first car passes the second car, then

$$\left(\begin{array}{l}\text{distance traveled}\\ \text{by first car}\end{array}\right) - \left(\begin{array}{l}\text{distance traveled}\\ \text{by second car}\end{array}\right) = 180 \text{ mi}$$

(see Figure 8.2).

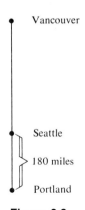

Vancouver

Seattle

180 miles

Portland

Figure 8.2

Let x hr be the time the first car traveled. It is clear that the second car left 2 hr after the first car started. However, the first driver stopped 30 min in Seattle while the second car was traveling. Hence, the first car traveled only $2 \text{ hr} - \frac{1}{2} \text{ hr} = \frac{3}{2} \text{ hr}$ more than the second car. Hence, the second car traveled $(x - \frac{3}{2})$ hr. Therefore, the distances traveled by the first and by the second car are, respectively, $70x$ mi and $50 \cdot (x - \frac{3}{2})$ mi. Note that $70x$ must be larger than $50 \cdot (x - \frac{3}{2})$, since the first car went a longer distance before they met. It follows that

$$70x - 50 \cdot (x - \tfrac{3}{2}) = 180.$$

Therefore,

$$70x - 50x + 75 = 180.$$

$$20x = 105.$$

$$x = \frac{105}{20}.$$

$$x = 5\tfrac{1}{4}.$$

Hence, the first car traveled 5 hr 15 min, and they met 5 hr 45 min after it started, since the driver stopped 30 min in Seattle. Therefore, the first car passed the second car at 3:45 P.M.

To check our result, note that the first car traveled $(70)\left(\tfrac{21}{4}\right)$ mi and the second car traveled $(50)\left(3 + \tfrac{3}{4}\right)$ mi. The difference is

$$\left[(70)\left(\tfrac{21}{4}\right) - (50)\left(\tfrac{15}{4}\right)\right] \text{ mi} = \tfrac{10}{4}(147 - 75) \text{ mi}$$

$$= \tfrac{5}{2}(72) \text{ mi}$$

$$= 180 \text{ mi}.$$

Since the distance from Portland to Seattle is 180 mi, we conclude that our result was correct.

Exercise 8.2

Read Examples 1 and 2 before working Exercises 1–10.

1. A boy left town A to go to town B, 15 mi away. He walked part of the way and then got a ride. If we know that the boy walked at a speed of 2.5 mph and that the car speed was 40 mph, how long did he walk if the trip took 2 hr and 15 min?

2. A man left town A in his car at 1:00 P.M. to go to town B, 350 mi away. The first part of the trip he traveled at 80 mph. A policeman stopped him for speeding and gave him a ticket. The man started again 15 min later and drove the remaining part of his trip at the more conservative speed of 50 mph. He arrived in town B at 6:45 P.M. How far from town A did the policeman stop him?

3. Two cars leave the same place at the same time and travel in opposite directions, one of them traveling 10 mph faster than the other. After 6 hr they are 300 mi apart. How fast is each car traveling?

4. A man walked into town at the rate of 4 mph. He rode back on a bus, over the same route, at a speed of 36 mph. If he spent 40 min in town and was back 4 hr later, how far did he walk?

5. Two jets left Chicago traveling in opposite directions. One traveled 50 mph slower than the other, and at the end of 2 hr they were 2100 mi apart. How fast was each jet flying?

6. A passenger train and a freight train leave towns 640 mi apart and travel toward each other. If the freight train travels 30 mph slower than the passenger train and they meet in 6 hr, what are their respective speeds?

7. At 7:00 A.M. a train traveling 70 mph leaves San Francisco for Seattle. The distance between the two cities is 900 mi. At 10:00 A.M., a train traveling 50 mph leaves Seattle for San Francisco. At what time will these two trains pass each other?

8. A cyclist started to ride to a town 20 mi away at the rate of 9 mph. After he had been riding for a time, his bicycle broke down, and he walked the remainder of the distance at a rate of 3 mph. If he reached town 2 hr and 40 min after he started, how far did he walk?

9. One car went 20 mi farther when traveling 50 mph than a second one, which traveled 4 hr longer at the speed of 40 mph. How far did each travel?

10. A motorist drove at an average speed of 45 mph outside the city limits and 20 mph inside the city limits. In 1 hr 40 min he traveled 50 mi. What was the distance he traveled outside the city limits?

8.3 MIXTURE AND INVESTMENT PROBLEMS

The nature and solution of this type of problem are best illustrated with examples.

Example 1. A grocer has two kinds of chocolate. The first kind sells at 60 cents a pound, the second kind at 90 cents a pound. How much should he take of each kind to get 100 lb of a mixture that he could sell at 80 cents a pound?

Solution: The fundamental formula is

weight in pounds · price per pound = total value.

The idea of this problem and of all mixture problems is to keep in mind that the total weight of the mixture is equal to the sum of the weights of the components and that the total value of the mixture is equal to the sum of the values of the components.

Let x pounds be the weight of the first kind of chocolate. Since the total weight of the mixture must be 100 lb, the grocer must take $(100 - x)$ lb of the second kind of chocolate. Therefore, the total value of the cheaper kind of chocolate used is $60x$ cents, and $90(100 - x)$ cents is the value of the other kind. But we know that the total value of the mixture is $(80) \cdot (100)$ cents $= 8000$ cents. Hence,

$$60x + 90(100 - x) = 8000.$$
$$60x + 9000 - 90x = 8000.$$
$$-30x = -1000.$$
$$x = \frac{100}{3}.$$
$$x = 33\tfrac{1}{3}.$$

We conclude that the grocer should use $33\tfrac{1}{3}$ lb of the first kind of chocolate and $66\tfrac{2}{3}$ lb of the second kind.

The student should check that these results are correct.

Remark. In analyzing a problem of this type, the student may find it helpful to tabulate the given information as follows:

	Weight in lb · price per pound = total value		
First kind		60 cents	
Second kind		90 cents	
Mixture	100 lb	80 cents	8000 cents

It is clear that the unknown quantities are those corresponding to the blank spaces in the foregoing table. Hence, we write x lb to represent the weight of the first kind of chocolate and complete the table as illustrated below.

	Weight in lb · price per pound = total value		
First kind	x lb	60 cents	$60x$ cents
Second kind	$(100 - x)$ lb	90 cents	$90(100 - x)$ cents
Mixture	100 lb	80 cents	8000 cents

Example 2. A man has $12,000 and wishes to earn exactly $530 interest in one year. He wants to invest all of his money in two banks. Bank A pays 5 percent interest and bank B pays 4 percent. How much should he invest in each bank?

Solution: Let x dollars be the amount the man will invest in bank A. Then he must invest $(12{,}000 - x)$ dollars in bank B. The interest he will earn from bank A is $\dfrac{5x}{100}$ dollars. On the other hand, he will earn $\dfrac{4 \cdot (12000 - x)}{100}$ dollars from bank B. Since the total interest for one year is 530 dollars, we obtain the following equation:

$$\frac{5x}{100} + \frac{4(12,000 - x)}{100} = 530,$$

$$5x + 48,000 - 4x = 53,000,$$

$$x = 5000.$$

Therefore, the man should invest $5000 in bank A and $7000 in bank B. Clearly his earned interest will be $250 in bank A and $280 in bank B, which total to $530 as required.

Exercise 8.3

Read Example 1 before working Exercises 1–7.

1. A grocer blended 70-cents-a-pound coffee with 90-cents-a-pound coffee so as to obtain 60 lb worth 85 cents a pound. How much of each did he take?

2. How much cream that is 25 percent butterfat should be mixed with milk that is 4 percent butterfat in order to obtain 20 gallons of light cream that is 21 percent butterfat?

3. A car radiator with 30-qt capacity has been prepared for fall driving with 3 qt of antifreeze, but winter driving requires a 20 percent solution of antifreeze. How many quarts of the fall mixture must be withdrawn and replaced with pure antifreeze?

4. A candy dealer has 30 lb of chocolates worth 90 cents a pound. He wants to mix with it bonbons that sell at 65 cents a pound to obtain a combination worth 80 cents a pound. How many pounds of bonbons should he use?

5. How much pure alcohol should be added to 52 oz. of a 55 percent solution to obtain a 75 percent solution?

6. A solution contains 3 parts alcohol to 4 parts water and another contains 5 parts alcohol to 3 parts water. How much of each solution should be taken in order to obtain 40 quarts of a solution that is 50 percent alcohol?

7. A grocer has a 100-lb mixture of nuts that sell at 70 cents a pound. He wants to reduce this price to 60 cents a pound by adding some nuts that sell at 45 cents a pound. How many pounds of this cheaper kind of nuts should he add?

Read Example 2 before working Exercises 8–12.

8. A sum of $7600 is invested part at 5 percent and the

remainder at 4 percent. A total interest of $338 is earned in one year. How much was invested at each rate?

9. Mr. Mason has three times as much money invested in 4 percent bonds as he has in stocks that pay 6 percent. His annual income from his stocks and bonds is $1080. How much has he invested in each?

10. A man has $3000 more invested in 4 percent bonds than he has in 6 percent stocks. If his total annual income from these two investments is $320, how much does he have invested in each?

11. A man invested part of $42,000 at 7 percent and the remainder at 4 percent. His annual income in one year from these two investments was the same as if he had invested the whole sum at 6 percent. How much did he invest at each rate?

12. A man has $10,000 invested in three bonds that pay different rates of interest. He has three times as much invested in a 4 percent bond as he has in a 5 percent bond, and the remainder is invested in a 6 percent bond. His annual interest from these three investments is $530. How much did he invest in each bond?

8.4 WORK PROBLEMS

Sometimes one is asked how long it would take several people to do a certain job given how long it takes each of these people to do the job. We illustrate this type of problem with an example.

Example 1. If John can do a job in 5 hr and Pete can do it in 7 hr, how long will it take John and Pete together to do that job?

Solution: The fundamental formula is

$$\text{(rate of work)} \cdot \text{(time)} = \text{fraction of work done.}$$

Since when the whole job is finished the fraction of work done is unity, one can easily find the rate of work using the following:

$$\text{rate of work} = \frac{1}{\text{time to do the whole job}}.$$

Therefore, John's rate of work is $\frac{1}{5}$ of the job per hour and Pete's is $\frac{1}{7}$ of the job per hour. Hence, together they can do

$$\tfrac{1}{5} + \tfrac{1}{7} \text{ of the job per hour.}$$

Let x hr be the time it will take both John and Pete to do the job. We obtain the equation

$$(\tfrac{1}{5} + \tfrac{1}{7})x = 1.$$

Therefore,

$$\frac{12x}{35} = 1,$$

$$12x = 35,$$

$$x = \tfrac{35}{12},$$

$$x = 2 + \tfrac{11}{12}.$$

Thus, John and Pete can do the job in 2 hr 55 min (since $\tfrac{11}{12}$ of an hour is 55 min). The student is asked to verify the correctness of our solution.

Example 2. A tank is equipped with a pipe that drains it continuously. Suppose that a faucet can fill the tank in 6 hr when it is turned on full force and in 24 hr when it is turned on only half speed. How long does it take the drain to empty the full tank?

Solution: The fundamental formula is

(rate of work) · (time) = fraction of work done.

Let x hr be the time it would take the drain to empty the tank.
    When the drain is open, and the faucet is turned on full force, $\frac{1}{6}$ of the tank is filled in 1 hr. Since $\frac{1}{x}$ of the tank was drained out, we must have had $\left(\frac{1}{6} + \frac{1}{x}\right)$ of the tank flowing from the faucet in one hour.
    Similarly, when the faucet is turned on half speed, we have $\left(\frac{1}{24} + \frac{1}{x}\right)$ of the tank flowing in from the faucet in 1 hr.
    Thus we have

$$\frac{1}{2}\left(\frac{1}{6} + \frac{1}{x}\right) = \frac{1}{24} + \frac{1}{x}.$$

$$\frac{1}{12} + \frac{1}{2x} = \frac{1}{24} + \frac{1}{x}.$$

$R - \{0\}$

$$\frac{1}{12} - \frac{1}{24} = \frac{1}{2x}.$$

$$24 = 2x.$$

$$x = 12.$$

Hence, it would take the drain 12 hr to empty the tank. The verification of the correctness of our result is left to the student.

Exercise 8.4

Read Example 1 before working Exercises 1–6.

1. If Betty can do a job in 4 hr and Shirley can do the same job in 5 hr, how long will it take Shirley and Betty together to do the job?

2. Frank and Roy can do a job in 3 hr. Roy and Ralph can do the same job in 4 hr, but it would take Frank and Ralph 5 hr to do that job. How long would it take Roy if he worked alone?

3. A boy can mow a lawn in 3 hr and another can mow it in 2 hr and 45 min. If both work together, how long will it take to do the job?

4. George can do a job in two-thirds the time required by Lloyd, and they can do it together in 45 min. How long would it take Lloyd to do the job by himself?

5. A man can do a piece of work in 1 hr and 45 min, his wife can do it in 2 hr, and their son can do it in 2 hr and 15 min. If all three work together for 20 min and then the man and his wife complete the work, how long will it take?

6. A man can do a job in 7 days and his older son can do it in 10 days. The man and his older son worked together for 2 days and then the man and his younger son finish the job in 3 days. How long would it take the younger son to do the job if he worked alone?

Read Example 2 before working Exercises 7–11.

7. One pipe can fill a tank in 3 hr, a second pipe takes 2 hr to fill it, and a third pipe takes 5 hr. If all three pipes are open, how long will it take to fill the tank?

8. A tank can be filled by one hose in 10 hr and by another hose in 15 hr. How long will it take to fill the tank using both hoses?

9. Two pipes are used for running water into a tank. When both pipes are used, the tank can be filled in 48 min. The larger pipe alone will fill the tank in 40 min less time than the smaller one. Find the number of minutes required for each pipe separately to fill the tank.

10. A pipe can fill a tank in 2 hr. After this pipe has been running for 30 min, it is shut off and a second pipe is opened. It takes 2 hr and 15 min for that pipe to finish filling the tank. How long would it have taken the second pipe alone to fill the tank?

11. A boy can watch a TV program in a half hour and his girl friend can also watch the same program in a half hour. How long will it take them together to watch this program?

12. It takes a pipe 20 min to fill a tank when the drain is closed, and 45 min when the drain is open. If the tank is three-quarters full, how long would it take the drain to empty it?

8.5 AGE PROBLEMS

The following example illustrates this type of problem.

Example 1. A man, being asked his age, replied: "If you take one year from my present age, the result will be three times my son's age, and three years ago my age was twice what his will be in four years." Find his age.

Solution: Let the father's age be x years. From the first statement we conclude that the son is $\dfrac{x-1}{3}$ years old. Three years ago the father was $(x-3)$ years old and in four years the son will be $\left(\dfrac{x-1}{3}+4\right)$ years old. From the second statement of the problem we can write the equation.

$$\underbrace{x-3}_{\substack{\text{Father's age} \\ \text{three years ago}}} = 2\underbrace{\left(\frac{x-1}{3}+4\right)}_{\substack{\text{the son's age} \\ \text{in four years.}}}$$

twice

We solve the foregoing equation as follows:

$$x - 3 = 2\left(\frac{x-1}{3} + 4\right).$$

$$x - 3 = \frac{2x - 2}{3} + 8.$$

$$3x - 9 = 2x - 2 + 24.$$

$$x = 31.$$

Hence, the father is 31 years of age.

To check, we see from the first statement that the son is 10 years old. Three years ago the father was 28 years of age and in four years the son will be 14 years old, which agrees with the father's second statement.

Exercise 8.5

Read Example 1 before working Exercise 1–8.

1. A man is six times as old as his son, and in 20 years he will be twice as old as his son. How old is each now?

2. Steven is one-third as old as his father and in 4 years he will be as old as his father was 20 years ago. How old is Steven?

3. A woman has two sons, one twice as old as the other. The woman's age now is four times that of her older son, and in 3 years she will be five times as old as her younger son. Find their present ages.

4. Carolyn is 1 year older than Jean, and Carolyn's age in 2 years will be twice what Jean's age was 2 years ago. Find their present ages.

5. A 36-year-old man has a son who is 12 years old. In how many years will the man be twice as old as his son?

6. A man 36 years old has three sons. The older son is 1 year older than the next son, who is himself 5 years older than the younger son. In two years the four ages will add up to 76 years. How old are the sons now?

7. A woman is 30 years older than her twin sons. In 6 years her age will be twice the combined ages of the twins. How old are they now?

8. John is twice as old as Louis, and in 4 years he will be three times as old as Pete, who is 6 years younger than Louis. How old are they now?

8.6 INTEGER PROBLEMS

Sometimes a statement is made concerning one or more integers, at least one of which is unknown, and one is asked to find the unknown integer.

The basic facts to remember for such problems are listed below.

(a) If x is an integer, $x + 1$ is the next larger integer.

(b) If x is an even integer, $x + 2$ is the next larger even integer.

(c) If x is an odd integer, $x + 2$ is the next larger odd integer.

(d) In general, any even integer can be represented by $2x$, where x is an integer. Similarly, any odd integer can be represented by $2x + 1$.

(e) If h is the hundreds digit, t the tens digit, and u the units digit of a three-digit number, the number itself is $100h + 10t + u$ and the sum of its digits is $h + t + u$.

The following examples will illustrate the basic ideas.

Example 1. The sum of three consecutive odd integers is 3 less than four times the first of these three numbers. What are the numbers?

Solution: If x is the first of the three consecutive odd integers, $x + 2$ and $(x + 2) + 2$ will be the next two odd integers.
    Hence,

$$\underbrace{x + (x + 2) + (x + 4)}_{\text{The sum of the three integers}} \overset{\uparrow}{=} \underbrace{4x - 3.}_{\substack{\text{3 less than four times} \\ \text{the first}}}$$

We solve the foregoing equation as follows:

$$x + (x + 2) + (x + 4) = 4x - 3.$$
$$3x + 6 = 4x - 3.$$
$$6 + 3 = 4x - 3x.$$
$$9 = x.$$

Hence, the three consecutive odd integers are 9, 11, and 13.
    To check our result, we note that the sum of these three integers is 33, which is 3 less than $4 \cdot 9$.

Example 2. In a two-digit number the units digit is 3 less than the tens digit. If three times the sum of the digits is 4 less than the number formed by reversing the order of the digits, find the number.

Solution: Let x be the tens digit. Then $x - 3$ is the units digit. Therefore, the number is $10x + (x - 3)$. The sum of the digits is $x + (x - 3)$ and, in the number formed by reversing the order of the digits, x is the units digit while $(x - 3)$ is the tens digit. Hence, the new number is $10(x - 3) + x$. From the statement given in the problem, we get the equation

$$3[x + (x - 3)] = 10(x - 3) + x - 4.$$

We solve the foregoing equation as follows:

$$3[x + (x - 3)] = 10(x - 3) + x - 4.$$
$$3(2x - 3) = 10x - 30 + x - 4.$$
$$6x - 9 = 11x - 34.$$
$$25 = 5x.$$
$$x = 5.$$

Therefore, the tens digit is 5 and the units digit is 2. The number is 52.

Note that the sum of the digits is 7 and that reversing the order of the digits we obtain 25. Further, $3 \cdot 7 = 21$, which is 4 less than 25.

Example 3. Establish the following rule. To square a natural number of two or more digits that ends in 5 proceed as follows: Delete the 5 and multiply the number so obtained by the sum of that number and 1. Multiply this product by 100 and add 25.

Solution: Before establishing the rule, we give an example. To square 85, we delete the 5 and obtain 8. Now $8 + 1 = 9$ and $8 \cdot 9 = 72$. Therefore, $85^2 = 7200 + 25 = 7225$.

To establish the rule, we suppose that m is a natural number that ends in 5 and suppose that k is the number obtained by deleting the 5. Then, $m = 10k + 5$ (for example, $85 = 10 \cdot 8 + 5$). Therefore,

$$
\begin{aligned}
m^2 &= (10k + 5)^2 \\
&= (10k)^2 + 2(10k) \cdot (5) + 5^2 \\
&= 100k^2 + 100k + 25 \\
&= 100k(k + 1) + 25 \\
&= k \cdot (k + 1) \cdot (100) + 25.
\end{aligned}
$$

Example 4. Square the number 95.

Solution: Here $k = 9$. Hence, $95^2 = 9 \cdot (9 + 1) \cdot (100) + 25 = 9025$.

Exercise 8.6

Read Example 1 before working Exercises 1–3.

1. The sum of two consecutive even integers is 46. Find the integers.

2. The sum of three consecutive odd integers is 45. Find the integers.

3. The sum of four consecutive even integers is 100. What are these integers?

Read Example 2 before working Exercises 4–9.

4. In a two-digit number the units digit is 2 less than the tens digit. Four times the sum of the digits is 9 less than the number obtained by reversing the order of the digits. What is the number?

5. The sum of the digits of a two-digit number is 5. If we interchange the digits and add 9 to the new number, we obtain the original number. What is the number?

6. The sum of the digits of a two-digit number is 13. If the digits are interchanged, the new number is 31 less than twice the original number. Find the number.

7. The tens digit of a two-digit number is 5 less than the units digit, and the number is three times the sum of the digits. What is the number?

8. In a three-digit number the tens digit is twice the hundreds digit and the units digit is twice the tens digit. If we reverse the order of the digits, we obtain a number that is 594 more than the original number. Find the number.

9. In a three-digit number the hundreds and units digits are equal and each is 3 more than the tens digit. The number itself is 46 times the sum of the digits. What is the number?

10. Prove that the square of any odd number is odd.

11. Prove that the square of any even number is divisible by 4.

Read Examples 3 and 4 before doing Exercises 12–14. In each of these, find the square of the given number using the technique illustrated in Example 4.

12. (a) 25, (b) 35, (c) 45.
13. (a) 55, (b) 65, (c) 75.
14. (a) 85, (b) 105, (c) 115.

8.7 VALUE PROBLEMS

The following examples illustrate this type of problem.

Example 1. A sum of money amounting to 20 dollars consists of nickels, dimes, and quarters. There are ten times as many dimes as quarters, and the number of nickels exceeds twice the number of dimes and quarters by 24. Find the number of coins of each denomination.

Solution: The fundamental formula is

(value of one coin) $\cdot$ (number of coins) = total value.

Let x be the number of quarters. Then, there are $10x$ dimes. Since the number of nickels exceeds twice the number of dimes and quarters by 24, the number of nickels is $2(10x + x) + 24$; that is, there are $(22x + 24)$ nickels. The value of x quarters is $25x$

cents. The value of $10x$ dimes is $100x$ cents and that of $(22x + 24)$ nickels is $5(22x + 24)$ cents. Since 20 dollars is 2000 cents, we can write

$$25x + 100x + 5(22x + 24) = 2000.$$

Solving the foregoing equation, we get

$$x = 8.$$

Hence, there are 8 quarters, 80 dimes, and 200 nickels. This clearly amounts to 20 dollars.

Example 2. A garage owner buys a certain number of spark plugs at the cost of 3 for a dollar. He sells half of them at 40 cents apiece, and to attract business he puts the other half on sale at 30 cents apiece. When they are all sold, he has realized a 5-dollar profit on the spark plugs. How many did he buy?

Solution: Let x be the number of spark plugs he bought. His cost was $\frac{x}{3}$ dollars. He sold the first half for $\frac{x}{2} \cdot \frac{40}{100}$ dollars and the second half for $\frac{x}{2} \cdot \frac{30}{100}$ dollars. Hence, he sold the spark plugs for

$$\left(\frac{x}{2} \cdot \frac{40}{100} + \frac{x}{2} \cdot \frac{30}{100} \right) \text{ dollars.}$$

Since his profit was 5 dollars, we obtain the equation

$$\left(\frac{x}{2} \cdot \frac{40}{100} + \frac{x}{2} \cdot \frac{30}{100} \right) - \frac{x}{3} = 5.$$

We solve the foregoing equation as follows:

$$\frac{40x}{200} + \frac{30x}{200} - \frac{x}{3} = 5.$$

$$\frac{x}{5} + \frac{3x}{20} - \frac{x}{3} = 5.$$

$$60 \left(\frac{x}{5} + \frac{3x}{20} - \frac{x}{3} \right) = 60 \cdot 5.$$

$$12x + 9x - 20x = 300.$$

$$x = 300.$$

We conclude that the garage owner bought 300 spark plugs.

CHECK. The owner's cost was 100 dollars. He sold 150 spark

plugs for 60 dollars and the last 150 for 45 dollars. Hence, he received 105 dollars, which gives him a 5-dollar profit.

Exercise 8.7

Read Examples 1 and 2 before doing Exercises 1–9.

1. A woman has three times as many dimes as quarters in her purse. If the dimes were quarters and the quarters were dimes, she would have 90 cents more. How many of each does she have?

2. A pile of 72 coins consisting of nickels and quarters is worth $6. Find how many coins of each denomination there are.

3. A sum of money amounting to $16.50 consists of nickels, dimes, and quarters. There are 10 more dimes than nickels, and the number of quarters is equal to the number of nickels and dimes put together. How many coins of each denomination are there?

4. A lady bought 100 Christmas cards for $18. For some she paid 15 cents each, and the others cost 25 cents each. How many of each kind did she get?

5. A farmer bought some cows at $200 each. Through disease he lost 5 percent of them, and he decided to sell the remainder for $250 each. He made a profit of $18,750 on the transaction. How many cows had he bought?

6. The admission charges for a basketball game are as follows: $1.50 for general admission, $2.50 for reserved seats, and $3.50 for box seats. The seating capacity of the auditorium is 16,000. There are five times as many general admission seats as there are box seats. If we know that the receipts for a capacity crowd are $36,000, find how many seats of each kind there are.

7. An ice cream vendor sells both 10-cent and 15-cent ice cream cones. On a certain day 650 cones were sold and he collected $85. How many of each kind did he sell?

8. A boy has 20 cents more than he needs to buy 12 ping pong balls. If each ball were to cost 5 cents more, he would be 10 cents short to buy 11 balls. How much money does he have?

9. The gate receipts at a baseball game were $53,030 from 21,510 paid admissions. If the bleacher tickets sold for $2 each and the grandstand tickets for $3 each, how many tickets of each kind were sold?

10. An auditorium has 720 seats. Each row contains the same number of seats. If 6 seats were removed from each row, 4 more rows would have to be added to maintain the same seating capacity. How many seats are presently in each row?

REVIEW EXERCISES

1. Two motorists start toward each other at 4:30 P.M. from towns 225 mi apart. If their respective average speeds are 30 mph and 45 mph, at what time will they meet?

2. A boy leaves town and starts walking toward home at the speed of 4 mph. When he is part of the way home, he begins to run at the speed of 8 mph. He reaches home 1 hr after he left town. How long did he walk and how long did he run, if his home is 5 mi from town?

3. Find two integers such that their sum is 215 and one of their quotients is $\frac{2}{3}$.

4. A can mow a lawn in 20 min and B can do it in 30 min. How long does it take the two to mow the lawn together, if they use two lawnmowers?

5. A man 5 ft 9 in. tall casts a shadow 6 ft 3 in. long when he stands 20 ft from a street light. How high is the light?

6. Find two consecutive positive integers whose product is 156.

7. The area of a rectangular field is 10,500 sq ft. If the length of the field exceeds the width by 5 ft, find the dimensions of the field.

8. A collection of 24 coins, worth $3.85, consists of nickels, dimes, and quarters. If the number of nickels exceeds the number of dimes by 2, how many nickels, quarters, and dimes are in the collection?

9. One faucet can fill a tank in 6 min less than it takes a second faucet to fill the same tank. Both faucets together can fill the tank in 4 min. How long does it take each faucet to fill the tank?

10. A car leaves a town traveling at a constant speed of 30 mph. Three hours later a second car leaves the same town and travels along the same route at the constant speed of 60 mph. How long will it take the second car to catch up with the first car?

11. How many gallons of water must be added to 5 gal of a solution that is 60 percent alcohol to obtain a solution that is 40 percent alcohol?

12. The distance from the ground of a body thrown vertically upward with the initial velocity of 128 ft/sec is given by the formula $h = 128t - 16t^2$, where t is the time in seconds and h is the distance in feet above the ground t sec after the throw. In how many seconds will the body return to the ground?

RADICALS AND FRACTIONAL EXPONENTS

9.1 INTRODUCTION

In Chapter 5 we introduced the following laws of integral exponents: If a, b are complex numbers, m, n are integers, and $ab \neq 0$, then

(1) $a^n a^m = a^{n+m}$,

(2) $(a^n)^m = a^{nm}$,

(3) $(ab)^n = a^n b^n$,

(4) $\dfrac{a^n}{a^m} = a^{n-m}$,

(5) $\dfrac{a^n}{b^n} = \left(\dfrac{a}{b}\right)^n$.

In the present chapter we discuss further the concept of exponent and we introduce the notion of radicals in detail.

It can be shown that if a is a nonzero complex number and n is a positive integer, the equation $x^n = a$ has exactly n distinct solutions in C.

Suppose that we wanted to denote any of these solutions by a^y, and further assume that a law such as law (2) above should hold. Then we would have

$$(a^y)^n = a.$$

Therefore,

$$a^{yn} = a^1.$$

We would be led to assume that $yn = 1$ — that is, $y = \dfrac{1}{n}$. Thus, it seems desirable to denote a solution of the equation $x^n = a$ by $a^{1/n}$. Unfortunately, this would lead to ambiguity, since the symbol $a^{1/n}$ would then represent n distinct numbers. To avoid this unpleasant state of affairs, we agree to reserve the symbol $a^{1/n}$ for a single one of the n solutions of the equation $x^n = a$.

principal
nth root

If a is a non-negative real number and n is an integer greater than 1, the foregoing equation has a unique non-negative real solution. It is called *the principal nth root of a* (or simply *the nth root of a*) and it is denoted by either of the two symbols $a^{1/n}$, $\sqrt[n]{a}$.

Example 1. $\sqrt[4]{16} = 2$. Note that $\{-2, 2, 2i, -2i\}$ is the solution set of the equation $x^4 = 16$. There is only one positive solution, namely, 2. Thus, $\sqrt[4]{16} = 2$. We could also have written $16^{1/4} = 2$.

radical
index
order
radicand
quadratic radical

The symbol $\sqrt[n]{a}$, denoting the principal nth root of the non-negative number a, is called a *radical*. The symbol $\sqrt{}$ is called a radical sign; the number n is called the *index* or *order* of the radical, and a is called the *radicand*.

If $n = 2$, the index is not usually written and the radical is said to be a *quadratic radical*.

Example 2. The solutions of the equation $x^2 = 9$ are -3 and 3. Since 3 is the positive solution, we write

$$9^{1/2} = \sqrt{9} = 3.$$

If we wanted to indicate that -3 is a solution also, we would say that $-\sqrt{9}$ is also a solution of the equation $x^2 = 9$.

principal
square root
principal
cube root

We remark that $\sqrt{a}$ is called the *principal square root of a*, and $\sqrt[3]{a}$ is called the *principal cube root of a* (rather than second and third roots, respectively).

It can also be shown that if n is an odd positive integer and a is a negative real number, then the equation $x^n = a$ has a unique real solution. This solution is negative, and we also call it the principal nth root of a and denote it by $\sqrt[n]{a}$ or by $a^{1/n}$.

Example 3. $\sqrt[3]{-8} = -2$ and $\sqrt[5]{-1} = -1$.

Exercise 9.1

Read Examples 1 and 3 before doing Exercises 1–18. In each of these, find the indicated principal nth root.

1. $\sqrt{4}$.

2. $\sqrt[3]{-27}$.

3. $64^{1/6}$.

4. $(-32)^{1/5}$.

5. $\sqrt{16}$.

6. $27^{1/3}$.

7. $121^{1/2}$.

8. $\sqrt[7]{-1}$.

9. $\left(\dfrac{81}{625}\right)^{1/4}$.

10. $\left(\dfrac{-32}{243}\right)^{1/5}$.

11. $(-0.001)^{1/3}$.

12. $\sqrt[4]{0.0001}$.

13. $\sqrt[4]{a^4}$ $(a \in R \wedge a > 0)$.

14. $\sqrt{a^4 + 2a^2b^2 + b^4}$ $(a, b \in R)$.

15. $\sqrt[3]{1/8}$.

16. $\sqrt[5]{-1/32}$.

17. $\sqrt{(x + 1)^2}$, $x \in R$.

18. $\sqrt[7]{-128}$.

Read Example 2 before working Exercises 19–26. In each of these, an equation is given. In each case find the solution that is the principal nth root of a number.

19. $x^2 = 4$.

20. $y^3 + 8 = 0$.

21. $z^4 - 256 = 0$.

22. $t^5 + 0.00001 = 0$.

23. $x^3 - \frac{216}{516} = 0$.

24. $r^4 = 0.0625$.

25. $v^3 - 27 = 0$.

26. $x^{17} + 1 = 0$.

9.2 FRACTIONAL EXPONENTS

In the preceding section we defined the principal nth root of a number a for the following two cases:

(1) if $a \geqslant 0$ and n is any integer greater than 1,
(2) if $a < 0$ and n is an odd integer greater than 1.

In both cases the (principal) nth root of a is a solution of the equation $x^n = a$. In the first case it is the unique non-negative solution; in the second case it is the unique negative solution. for example,

$$8^{1/3} = \sqrt[3]{8} = 2, \qquad (-32)^{1/5} = \sqrt[5]{-32} = -2, \qquad 81^{1/4} = \sqrt[4]{81} = 3.$$

Thus, the symbol $a^{1/n}$ has been defined at least for the two cases listed above. It can be defined for other cases as well, but in order to simplify the discussion, we shall limit ourselves to the above two cases.

If m and n are positive integers, we know that $\dfrac{m}{n} = \dfrac{1}{n}m$

$= m\dfrac{1}{n}.$

Thus, we would like the following three symbols $a^{m/n}$, $a^{\frac{1}{n} \cdot m}$, and $a^{m \cdot \frac{1}{n}}$ to represent the same number. If we wish the familiar laws of exponents to hold, the second symbol should represent $(a^{1/n})^m$ and the third $(a^m)^{1/n}$. That is, we should have

$$a^{\frac{m}{n}} = a^{\frac{1}{n} \cdot m} = (\sqrt[n]{a})^m \qquad \text{and} \qquad a^{\frac{m}{n}} = (a^m)^{\frac{1}{n}} = \sqrt[n]{a^m}.$$

Thus we should like to define $a^{\frac{m}{n}}$ to represent both the mth power of the principal nth root of a and the principal nth root of a^m.

If this is to make sense, we must have

$$(\sqrt[n]{a})^m = \sqrt[n]{a^m}.$$

If $a < 0$ and n is even, the foregoing equality may be false (see Example 3 below). Thus, we exclude that case and prove the following:

Theorem. If a is a real number, m, n are integers greater than 1, and the proposition "$a < 0 \wedge n$ is even" is false, then $(\sqrt[n]{a})^m = \sqrt[n]{a^m}$.

Proof: If $a \geqslant 0$, both sides of the foregoing equality are defined and are non-negative real numbers. It is easy to verify that both are solutions of the equation $x^n = a^m$. Since this equation has a unique non-negative solution, we conclude that the equality holds if $a \geqslant 0$.

If $a < 0$, n must be odd by hypothesis. Therefore, $\sqrt[n]{a}$ is negative. If m is even, both $(\sqrt[n]{a})^m$ and $\sqrt[n]{a^m}$ are positive, and if m is odd, they are both negative. Again, since both of the numbers $(\sqrt[n]{a})^m$ and $\sqrt[n]{a^m}$ are solutions of the equation $x^n = a^m$ and since this equation cannot have two distinct positive (or two distinct negative) solutions, we conclude that if $a < 0$ and n is odd, $(\sqrt[n]{a})^m = \sqrt[n]{a^m}$.

We now give the following definition formally.

Definition. If a is a real number and m, n are integers greater than 1, the symbol $a^{m/n}$ represents either of the two equal numbers $\sqrt[n]{a^m}$, $(\sqrt[n]{a})^m$ except in the case where $a < 0$ and n is even. In that case $a^{m/n}$ is not defined.

Question: Suppose m, n, p, q are positive integers and a is a real number. Further suppose that $\dfrac{m}{n} = \dfrac{p}{q}$ and that both $a^{m/n}$, $a^{p/q}$ are defined. Should we verify that $a^{m/n} = a^{p/q}$? To help the student see the significance of this question, we might ask: Is it obvious that $\sqrt[3]{4}$ and $\sqrt[6]{4^2}$ are equal? (See Exercise 19.)

Example 1. $8^{2/3} = \sqrt[3]{8^2} = \sqrt[3]{64} = 4$. Also, $8^{2/3} = (\sqrt[3]{8})^2 = 2^2 = 4$.

It can be proved that the laws of integral exponents can be generalized for rational exponents. That is, if a and b are real numbers and r, s are rational numbers, then

(1) $a^r a^s = a^{r+s}$,

(2) $(a^r)^s = a^{rs}$,

(3) $(ab)^r = a^r b^r$,

(4) $\dfrac{a^r}{a^s} = a^{r-s}$,

(5) $\dfrac{a^r}{b^r} = \left(\dfrac{a}{b}\right)^r$,

laws of rational exponents whenever each of the expressions above is defined. We shall not prove these *laws of rational exponents* here, but we give an example to illustrate their use.

Example 2. Evaluate each of the following:

(a) $\left(\dfrac{8}{27}\right)^{\frac{2}{3}}$,   (b) $\dfrac{2^{5/4}}{2^{1/4}}$,   (c) $(0.027)^{\frac{1}{3}}$,   (d) $\left(\dfrac{8}{81}\right)^{-\frac{2}{3}}$.

Solution:

(a) $\left(\dfrac{8}{27}\right)^{\frac{2}{3}} = \left(\dfrac{2^3}{3^3}\right)^{\frac{2}{3}} = \left[\left(\dfrac{2}{3}\right)^3\right]^{\frac{2}{3}} = \left(\dfrac{2}{3}\right)^2 = \dfrac{2^2}{3^2} = \dfrac{4}{9}$.

(b) $\dfrac{2^{5/4}}{2^{1/4}} = 2^{\frac{5}{4}-\frac{1}{4}} = 2^{4/4} = 2^1 = 2$.

(c) $(0.027)^{1/3} = \left(\dfrac{27}{1000}\right)^{1/3} = \left(\dfrac{3^3}{10^3}\right)^{1/3} = \left[\left(\dfrac{3}{10}\right)^3\right]^{1/3} = \dfrac{3}{10} = 0.3$.

(d) $\left(\dfrac{8}{81}\right)^{-\frac{2}{3}} = \left(\dfrac{8}{81}\right)^{0-\frac{2}{3}} = \dfrac{(8/81)^0}{(8/81)^{2/3}} = \dfrac{1}{\left(\dfrac{2^3}{3^4}\right)^{2/3}} = \dfrac{1}{\dfrac{(2^3)^{2/3}}{(3^4)^{2/3}}}$

$= \dfrac{(3^4)^{2/3}}{(2^3)^{2/3}} = \dfrac{3^{8/3}}{2^2} = \dfrac{3^{2+\frac{2}{3}}}{4} = \dfrac{3^2 3^{2/3}}{4} = \dfrac{9}{4}(3^2)^{1/3}$

$= \dfrac{9}{4}\sqrt[3]{9}$.

Example 3. Find both $\sqrt[n]{a^m}$ and $(\sqrt[n]{a})^m$ if $a = -4$, $m = 3$, and $n = 2$.

Solution: $\sqrt{(-4)^3} = \sqrt{-64} = \sqrt{(-1)64} = \sqrt{8^2}\,i = 8i$. Also,

$(\sqrt{-4})^3 = \sqrt{(-1)4})^3 = (\sqrt{2^2}\,i)^3 = (2i)^3 = 2^3 i^3$

$= 8i^2 i = 8(-1)i = -8i$.

Note that in this case $\sqrt[n]{a^m} \neq (\sqrt[n]{a})^m$. This example illustrates why the restriction "'$a < 0$ and n is even' is false" was necessary in the theorem of the present section.

Exercise 9.2

Read Examples 1 and 3 before doing Exercises 1–8. In each of these, *a*, *m*, and *n* are given. Find both $\sqrt[n]{a^m}$ and $(\sqrt[n]{a})^m$.

1. $a = 8$, $m = 4$, $n = 3$.
2. $a = 27$, $m = 2$, $n = 3$.
3. $a = -0.001$, $m = 5$, $n = 3$.
4. $a = 0.00032$, $m = 2$, $n = 5$.
5. $a = x^2 + 2xy + y^2$ (where $x + y \geq 0$), $m = 3$, $n = 2$.
6. $a = -9$, $m = 3$, $n = 2$.
7. $a = -16$, $m = 3$, $n = 2$.
8. $a = -16$, $m = 5$, $n = 2$.

Read Example 2 before doing Exercises 9–18. In each of these simplify the given expression.

9. $\left(\dfrac{125}{343}\right)^{2/3}$.

10. $\left(\dfrac{16}{81}\right)^{5/4}$.

11. $\dfrac{5^{53/50}}{25^{3/100}}$.

12. $\dfrac{3^{8/7}}{3^{1/7}}$.

13. $(0.0256)^{1/4}$.

14. $\dfrac{(4ab)^{1/2}}{(27a^{3/2})^{1/3}}$, $a > 0$, $b > 0$.

15. $\dfrac{(8x^{5/2})^{2/3}}{4x^2 y^{-1/3}}$, $x > 0$, $y \neq 0$.

16. $\dfrac{(x + y)^{5/2}}{x + y}$, $x > 0$, $y > 0$.

17. $\left(\dfrac{0.00243}{0.00032}\right)^{2/5}$.

18. $\dfrac{[(a + b)^3 (x^4 + 2x^2 y^2 + y^4)]^{5/6}}{[(a + b)^{1/4}(x^2 + y^2)^{1/6}]^{10}}$, where $a + b \neq 0$, $xy \neq 0$.

19. Show that if *m*, *n*, *p*, *q* are positive integers, $\dfrac{m}{n} = \dfrac{p}{q}$, and *a* is a positive real number, then $a^{m/n} = a^{p/q}$.

9.3 ARITHMETIC INVOLVING RADICALS

In some cases it is useful to be able to add, subtract, multiply, and divide numbers expressed in terms of radicals. In the present section we present the rules governing these operations.

The laws of radicals follow directly from the definitions and the laws of fractional exponents stated in the previous sections.

If a and b are real numbers and m, n are positive integers, then

(1) $(\sqrt[n]{a})^n = a$,

(2) $\sqrt[m]{\sqrt[n]{a}} = \sqrt[mn]{a}$,

(3) $\sqrt[n]{ab} = \sqrt[n]{a} \cdot \sqrt[n]{b}$,

(4) $\sqrt[n]{\dfrac{a}{b}} = \dfrac{\sqrt[n]{a}}{\sqrt[n]{b}}$ $(b \neq 0)$,

whenever each of the expressions above represents a real number. For example, we would not use the third equality above for $a = b = -1$ and $n = 2$, since $\sqrt{(-1)(-1)} = 1$ is real but $\sqrt{-1}$ does not represent a real number. Recalling that in Section 7.4 we defined $\sqrt{-1}$ to be the imaginary number i, we cannot use this here. Without this restriction we would have contradictions such as the following:

$$1 = \sqrt{1} = \sqrt{(-1)(-1)} = \sqrt{-1} \cdot \sqrt{-1} = i \cdot i = i^2 = -1.$$

Thus we must insist that a radical always represents a real number if we wish to use the four basic laws stated above.

If m and n are even natural numbers, and a is a real number, then the equalities $\sqrt[n]{a^m} = \sqrt[n]{|a|^m} = |a|^{m/n}$ follow from the laws of rational exponents. In particular, if m and n are even and equal natural numbers, then $\sqrt[n]{a^n} = |a|$ for any real number a.

For example, $\sqrt{(-3)^2} = \sqrt{9} = 3 = |-3|$. Thus, $\sqrt{(-3)^2} \neq -3$.

Example 1. A student was asked to simplify the expression

$$\sqrt{(x+1)^2} + \sqrt{(x-1)^2}.$$

He proceeded as follows:

$$\sqrt{(x+1)^2} + \sqrt{(x-1)^2} = (x+1) + (x-1) = 2x.$$

Show that his solution is incorrect.

Solution: Replacing x by 0 in the equation

$$\sqrt{(x+1)^2} + \sqrt{(x-1)^2} = 2x,$$

we obtain

$$\sqrt{1^2} + \sqrt{(-1)^2} = 2 \cdot 0.$$

That is,

$$1 + 1 = 0 \qquad \text{or} \qquad 2 = 0.$$

Since the last equality is obviously false, the student's simplification was in error.

The correct solution is

$$\sqrt{(x+1)^2} + \sqrt{(x-1)^2} = |x+1| + |x-1|.$$

Example 2. Simplify: (a) $\sqrt{18}$, (b) $\sqrt[4]{\frac{3}{16}}$, (c) $\sqrt[6]{16}$.

Solution:

(a) $\sqrt{18} = \sqrt{9 \cdot 2} = \sqrt{9} \cdot \sqrt{2} = 3\sqrt{2}.$

(b) $\sqrt[4]{\dfrac{3}{16}} = \dfrac{\sqrt[4]{3}}{\sqrt[4]{16}} = \dfrac{\sqrt[4]{3}}{2}.$

(c) $\sqrt[6]{16} = \sqrt[3]{\sqrt{16}} = \sqrt[3]{4}.$

simplest form   A radical $\sqrt[n]{a}$ is said to be in *simplest form* if

(a) no factor of a is a perfect nth power,
(b) $\sqrt[n]{a} = \sqrt[m]{b} \rightarrow n \leqslant m$,
(c) a does not involve fractions.

Example 3. Simplify the following expressions until all radicals are in simplest form: (a) $\sqrt[4]{32}$, (b) $\sqrt[6]{4}$, (c) $\sqrt[3]{\frac{5}{4}}$.

Solution:

(a) $\sqrt[4]{32} = \sqrt[4]{2^5} = \sqrt[4]{2^4 \cdot 2} = \sqrt[4]{2^4} \cdot \sqrt[4]{2} = 2\sqrt[4]{2}.$

(b) $\sqrt[6]{4} = \sqrt[6]{2^2} = 2^{2/6} = 2^{1/3} = \sqrt[3]{2}.$

(c) $\sqrt[3]{\dfrac{5}{4}} = \sqrt[3]{\dfrac{5}{2^2}} = \sqrt[3]{\dfrac{5 \cdot 2}{2^2 \cdot 2}} = \dfrac{\sqrt[3]{10}}{\sqrt[3]{2^3}} = \dfrac{1}{2}\sqrt[3]{10}.$

Example 4. Assuming a, b, c are positive real numbers, simplify $\sqrt[4]{64a^2 b^6 / c^2}$.

Solution: Exercises of this type are usually done with the help of fractional exponents. Thus,

$$\sqrt[4]{\frac{64a^2 b^6}{c^2}} = \left(\frac{2^6 a^2 b^6}{c^2}\right)^{1/4} = \frac{(2^6 a^2 b^6)^{1/4}}{(c^2)^{1/4}} = \frac{2^{6/4} a^{2/4} b^{6/4}}{c^{1/2}}$$

$$= \frac{2^{3/2} a^{1/2} b^{3/2}}{c^{1/2}} = \frac{2 \cdot 2^{1/2} \cdot a^{1/2} \cdot b \cdot b^{1/2} \cdot c^{1/2}}{c^{1/2} \cdot c^{1/2}}$$

$$= \frac{2b(2abc)^{1/2}}{c} = \frac{2b}{c}\sqrt{2abc}.$$

similar   The radicals $\sqrt[n]{a}$ and $\sqrt[m]{b}$ are said to be *similar* if, when each is reduced to simplest form, radicals with the same index and radicand result.

Example 5. $\sqrt{18}$ and $\sqrt{\frac{1}{2}}$ are similar since $\sqrt{18} = \sqrt{9 \cdot 2} = 3\sqrt{2}$ and

$$\sqrt{\frac{1}{2}} = \sqrt{\frac{2}{4}} = \frac{\sqrt{2}}{\sqrt{4}} = \frac{1}{2}\sqrt{2}.$$

In addition and subtraction involving radicals, we first reduce all radicals to simplest form and then combine those which are similar into single terms.

Example 6. Simplify, combining similar terms: (a) $6\sqrt{\frac{1}{3}} + \frac{5}{2}\sqrt{108} - 2\sqrt[4]{9}$, (b) $3\sqrt{12} + 7\sqrt{8} - 5\sqrt{48} - 2\sqrt{50}$.

Solution: (a)

$$\sqrt{\frac{1}{3}} = \sqrt{\frac{3}{9}} = \frac{\sqrt{3}}{\sqrt{9}} = \frac{\sqrt{3}}{3},$$

$$\sqrt{108} = \sqrt{3^3 2^2} = \sqrt{3 \cdot 3^2 \cdot 2^2} = \sqrt{3} \cdot \sqrt{3^2} \cdot \sqrt{2^2} = 6\sqrt{3},$$

$$\sqrt[4]{9} = \sqrt[4]{3^2} = 3^{2/4} = 3^{1/2} = \sqrt{3}.$$

Hence,

$$6\sqrt{\frac{1}{3}} + \frac{5}{2}\sqrt{108} - 2\sqrt[4]{9} = 6 \cdot \frac{\sqrt{3}}{3} + \frac{5}{2} \cdot 6\sqrt{3} - 2\sqrt{3}$$

$$= 2\sqrt{3} + 15\sqrt{3} - 2\sqrt{3}$$

$$= (2 + 15 - 2)\sqrt{3}$$

$$= 15\sqrt{3}.$$

(b)

$$3\sqrt{12} + 7\sqrt{8} - 5\sqrt{48} - 2\sqrt{50}$$

$$= 3\sqrt{4 \cdot 3} + 7\sqrt{4 \cdot 2} - 5\sqrt{16 \cdot 3} - 2\sqrt{25 \cdot 2}$$

$$= 3 \cdot 2\sqrt{3} + 7 \cdot 2\sqrt{2} - 5 \cdot 4\sqrt{3} - 2 \cdot 5\sqrt{2}$$

$$= 6\sqrt{3} + 14\sqrt{2} - 20\sqrt{3} - 10\sqrt{2}$$

$$= 14\sqrt{2} - 10\sqrt{2} + 6\sqrt{3} - 20\sqrt{3}$$

$$= (14 - 10)\sqrt{2} + (6 - 20)\sqrt{3}$$

$$= 4\sqrt{2} - 14\sqrt{3}.$$

Exercise 9.3

Read Example 1 before working Exercises 1–4. In each of these, a simplification is given: (a) show that this simplification is incorrect, (b) give a correct simplification.

1. $\sqrt{x^2 - 2x + 1} = \sqrt{(x-1)^2} = x - 1.$
2. $\sqrt{x^2} + \sqrt{(x-1)^2} = x + (x-1) = 2x - 1.$
3. $\sqrt{(t+1)^2} - \sqrt{4} = (t+1) - 2 = t - 1.$
4. $\sqrt{2(y+1)^2} - \sqrt{2(y-1)^2} = \sqrt{2}\,[(y+1) - (y-1)]$
$$= 2\sqrt{2}.$$

Study Example 2 before working Exercises 5–12. In each of these, simplify the given expression.

5. $\sqrt[3]{32}.$
6. $\sqrt{24}.$
7. $\sqrt[6]{\frac{64}{3}}.$
8. $\sqrt[5]{\frac{7}{32}}.$
9. $\sqrt[12]{8}.$
10. $\sqrt[5]{486x^{10}}.$

11. $\sqrt[4]{\dfrac{81x^8}{256y^{12}}},\ y > 0.$
12. $\sqrt[10]{1024\,(x^2 + 2xy + y^2)}.$

Study Example 3 before doing Exercises 13–24. In each of these, obtain radicals in simplest form.

13. $\sqrt[4]{243}.$
14. $\sqrt[6]{128}.$
15. $\sqrt[15]{125}.$
16. $\sqrt[3]{9}.$
17. $\sqrt[3]{625x^6}.$
18. $\sqrt[27]{343(x+y)^3}.$
19. $\sqrt[3]{14/3}.$
20. $\sqrt[3]{27/4}.$
21. $\sqrt[5]{64/27}.$
22. $\sqrt[3]{64/9}.$
23. $\sqrt{12}/\sqrt{3}.$
24. $\sqrt{2/3}.$

Study Example 4 before doing Exercises 25–30. In each of these, assume that the letters represent positive real numbers, and simplify the given expression.

25. $\sqrt[6]{\dfrac{27a^9c^3}{b^3}}.$
26. $\sqrt[4]{\dfrac{243a^8b^5}{c^2}}.$

27. $\sqrt[5]{\dfrac{25a^7x^{10}}{b^8c}}.$
28. $\sqrt[3]{\dfrac{32r^{12}s^6}{r^2t^{11}}}.$

29. $\sqrt[12]{\dfrac{5m^{17}n^{13}}{p^7}}.$
30. $\sqrt[5]{\dfrac{62a^{17}b^6}{c^{-8}d^4}}.$

Study Examples 5 and 6 before doing Exercises 31–37. In each of these, simplify the radicals and combine similar terms.

31. $\dfrac{6}{\sqrt{3}} + 5\sqrt[12]{729} - \sqrt{243}.$

32. $3\sqrt[4]{4} + 14\sqrt{\tfrac{1}{2}} - \sqrt{32}.$

33. $3\sqrt{54} - \dfrac{6}{\sqrt{6}} + \sqrt{8} - 7\sqrt{2} + \sqrt{45}.$

34. $5\sqrt{28} - 2\sqrt{45} - 3\sqrt{7} + 2\sqrt{80}.$

35. $2\sqrt[6]{a^2} - \dfrac{3a}{\sqrt[3]{a^2}} + \dfrac{7a}{\sqrt[9]{a^6}}$, where $a > 0$.

36. $6\sqrt[4]{a^2} - \dfrac{4a}{\sqrt{a}} + 7\sqrt[6]{b^3} - \dfrac{2b}{\sqrt[8]{b^4}}$, where $a > 0$, and $b > 0$.

37. $\dfrac{5x + 5y}{\sqrt{x + y}} - 3\sqrt[4]{x^2 + 2xy + y^2} + 2\sqrt[9]{(r + t)^3} - \dfrac{5(r + t)}{(\sqrt[3]{r + t})^2}$,

$$x + y > 0, \ r + t \neq 0.$$

9.4 ARITHMETIC INVOLVING RADICALS (CONTINUED)

We can multiply and divide radicals according to the laws stated in the preceding section. If the radicals have the same index, these operations can be performed readily. Otherwise, the radicals must first be converted to radicals with the same index. This is always possible, and we can do it best by converting the radicals to expressions having rational exponents.

We illustrate the techniques with several examples.

Example 1. Assuming that a and b are non-negative real numbers, find the product $\sqrt{30ab^3} \cdot \sqrt{15a^2b}$.

Solution:

$$\sqrt{30ab^3} \cdot \sqrt{15a^2b} = \sqrt{3 \cdot 5 \cdot 2ab^3 \cdot 3 \cdot 5 \cdot a^2b}$$
$$= \sqrt{3^2 \cdot 5^2 \cdot 2a \cdot a^2b^4}$$
$$= 3 \cdot 5 \cdot ab^2 \sqrt{2a}$$
$$= 15ab^2 \sqrt{2a}.$$

Example 2. Assuming that a is a non-zero real number, divide $\sqrt[4]{25}$ by $\sqrt{8a^2}$.

Solution:

$$\frac{\sqrt[4]{25}}{\sqrt{8a^2}} = \frac{\sqrt[4]{5^2}}{\sqrt{2 \cdot 2^2 a^2}} = \frac{\sqrt{5}}{2|a|\sqrt{2}} = \frac{1}{2|a|}\sqrt{\frac{5}{2}}.$$

If we wish to have the radical in simplest form, we proceed as follows:

$$\frac{1}{2|a|}\sqrt{\frac{5}{2}} = \frac{1}{2|a|}\sqrt{\frac{5}{2} \cdot \frac{2}{2}} = \frac{1}{2|a|}\frac{\sqrt{10}}{\sqrt{2^2}} = \frac{1}{4|a|}\sqrt{10}.$$

Example 3. Assuming that a and b are non-negative real numbers, multiply $\sqrt[3]{9a^4b^2}$ by $\sqrt{6a^3b}$.

Solution: By converting the radicals to fractional exponent forms, we obtain

$$(9a^4b^2)^{1/3}(6a^3b)^{1/2}.$$

The common denominator of the fractions $\frac{1}{3}, \frac{1}{2}$ is 6. Thus, we write the foregoing product as

$$(9a^4b^2)^{2/6}(6a^3b)^{3/6} = [(9a^4b^2)^2]^{1/6} \cdot [(6a^3b)^3]^{1/6}$$

$$= (3^4a^8b^4 \cdot 2^3 \cdot 3^3 \cdot a^9b^3)^{1/6}$$

$$= (2^33^7a^{17}b^7)^{1/6}$$

$$= (2^3 \cdot 3^6 \cdot 3a^{12}a^5b^6b)^{1/6}$$

$$= 3a^2b \sqrt[6]{8 \cdot 3 \cdot a^5b}$$

$$= 3a^2b \sqrt[6]{24a^5b}.$$

In many problems of arithmetic, an answer such as $1/\sqrt{2}$ is obtained. This is perfectly acceptable, unless it is desirable to approximate the result by a rational number. Knowing that $\sqrt{2}$ is approximately 1.414, we conclude that $\dfrac{1}{\sqrt{2}}$ is approximately $\dfrac{1}{1.414}$. However, if we had written $\dfrac{1}{\sqrt{2}} = \dfrac{\sqrt{2}}{2}$, it would be much easier to compute the approximation $\dfrac{1.414}{2}$. In fact, this can be done mentally, and we get 0.707.

The process of eliminating radicals from the denominator *rationalizing the* is called *rationalizing the denominator*. We illustrate this process *denominator* with an example.

Example 4. Rationalize the denominators of the following fractions:

(a) $\dfrac{\sqrt[3]{2}}{\sqrt{3}}$,

(b) $\dfrac{2 + \sqrt{3}}{\sqrt{7} - \sqrt{3}}$.

Solution: (a)

$$\frac{\sqrt[3]{2}}{\sqrt{3}} = \frac{\sqrt[3]{2}\sqrt{3}}{\sqrt{3}\cdot\sqrt{3}} = \frac{2^{1/3}3^{1/2}}{3}$$

$$= \frac{2^{2/6}3^{3/6}}{3} = \frac{4^{1/6}27^{1/6}}{3}$$

$$= \frac{(108)^{1/6}}{3} = \frac{\sqrt[6]{108}}{3}.$$

(b) Recalling the formula $(a - b)(a + b) = a^2 - b^2$, we multiply numerator and denominator of the given fraction by $(\sqrt{7} + \sqrt{3})$. We obtain

$$\frac{2 + \sqrt{3}}{\sqrt{7} - \sqrt{3}} = \frac{(2 + \sqrt{3})(\sqrt{7} + \sqrt{3})}{(\sqrt{7} - \sqrt{3})(\sqrt{7} + \sqrt{3})} = \frac{2\sqrt{7} + 2\sqrt{3} + \sqrt{21} + 3}{(\sqrt{7})^2 - (\sqrt{3})^2}$$

$$= \frac{3 + 2\sqrt{3} + 2\sqrt{7} + \sqrt{21}}{4}.$$

Exercise 9.4

Study Example 1 before doing Exercises 1–4. In each of these, assume that the given letters represent positive real numbers and find the product.

1. $\sqrt{3a^2b^3} \cdot \sqrt{6ab^4}$.
2. $\sqrt[3]{16ab^5} \cdot \sqrt[6]{4a^8b^2}$.
3. $\sqrt[3]{12(a + b)^2} \cdot \sqrt[15]{32(a + b)^{10}}$.
4. $\sqrt[4]{27a^{11}(a^2 + b^2)} \cdot \sqrt[12]{27(a^2 + b^2)^9}$.

Review Example 2 before doing Exercises 5–9. In each of these, perform the indicated division.

5. Divide $\sqrt{8(a - b)^2}$ by $\sqrt[4]{25b^{16}}$.
6. Divide $\sqrt[4]{162a^4}$ by $\sqrt[8]{48a^2b^8}$, $ab \neq 0$.
7. Divide $\sqrt[6]{125}$ by $\sqrt{8b^2}$, $b \neq 0$.
8. Divide $\sqrt[3]{16a^6}$ by $\sqrt[6]{64}$.
9. Divide $6y\sqrt{2x^3y}$ by $5x\sqrt{8xy^3}$, $xy > 0$.

Review Example 3 before working Exercises 10–14. In each of these, assume that the letters represent non-negative real numbers and perform the indicated multiplication.

10. Multiply $\sqrt[4]{8a^5b^7}$ by $\sqrt{6a^4b}$.
11. Multiply $\sqrt[3]{8a^2b^2}$ by $\sqrt[4]{2a^3b}$.
12. Multiply $\sqrt[5]{16a^7(x + y)^4}$ by $\sqrt[4]{32b^3(x + y)^5}$.
13. Multiply $\sqrt[m]{n}$ by $\sqrt[n]{m}$, where m and n are integers greater than 2.
14. Multiply $\sqrt[4]{2m^3n}$ by $\sqrt[12]{5m^5n^9}$.

Study Example 4 before doing Exercises 15–22. In each of these, rationalize the denominator of the given expression.

15. $\dfrac{\sqrt[3]{3}}{\sqrt{2}}$.

16. $\dfrac{\sqrt[4]{2}}{\sqrt{5}}$.

17. $\dfrac{\sqrt{2} + \sqrt{3}}{\sqrt{2} - \sqrt{3}}$.

18. $\dfrac{2 - \sqrt{2}}{4 + \sqrt{3}}$.

19. $\dfrac{\sqrt{4a}}{\sqrt[3]{6a}}$, $a > 0$.

20. $\dfrac{\sqrt{a + b} - \sqrt{a - b}}{\sqrt{a + b} + \sqrt{a - b}}$, $a + b > 0$, $a - b > 0$, $b \neq 0$.

21. $\dfrac{3\sqrt{2} - \sqrt{3}}{\sqrt{3} + \sqrt{2}}.$ **22.** $\dfrac{5\sqrt[3]{5}}{\sqrt[3]{5} - 1}.$

23. Simplify $\sqrt{7 + \sqrt{48}} + \sqrt{7 - \sqrt{48}}.$ [*Hint:* Set the expression equal to x and square.]

24. Simplify $\sqrt{12 + 2\sqrt{27}} + \sqrt{12 - 2\sqrt{27}}.$

9.5 EQUATIONS INVOLVING RADICALS

In Section 7.6, Example 4, we illustrated a technique for solving equations involving a quadratic radical. Since one of the steps was to square both sides in order to eliminate the radical, it was pointed out that the equation obtained may be redundant relative to the given one. Hence, in such cases, it is always necessary to check which of the solutions of the new equation are also solutions of the original one. We now give further examples of this technique.

Example 1. Solve the equation $\sqrt{x} + \sqrt{x + 5} = 5$ over R.

Solution:

$$\sqrt{x} + \sqrt{x + 5} = 5,$$
$$\sqrt{x + 5} = 5 - \sqrt{x},$$
$$x + 5 = 25 - 10\sqrt{x} + x,$$
$$\sqrt{x} = 2,$$
$$x = 4.$$

CHECK. $\sqrt{4} + \sqrt{4 + 5} = \sqrt{4} + \sqrt{9} = 2 + 3 = 5.$

Thus the solution set of the given equation is $\{4\}$.

Example 2. Solve the equation $\sqrt{2x + 9} = \sqrt{x + 4} + \sqrt{x + 1}$ over R.

Solution:

$$\sqrt{2x + 9} = \sqrt{x + 4} + \sqrt{x + 1},$$
$$2x + 9 = x + 4 + 2\sqrt{x + 4} \cdot \sqrt{x + 1} + x + 1,$$
$$2 = \sqrt{x + 4}\, \sqrt{x + 1},$$
$$4 = (x + 4)(x + 1),$$
$$x^2 + 5x = 0,$$
$$x(x + 5) = 0,$$
$$x = 0 \quad \vee \quad x = -5.$$

CHECK. $\sqrt{2 \cdot 0 + 9} = \sqrt{0 + 4} + \sqrt{0 + 1}$ is a true statement, since

$$\sqrt{9} = \sqrt{4} + \sqrt{1}$$

is true.

$$\sqrt{(2)(-5) + 9} = \sqrt{-5 + 4} + \sqrt{-5 + 1}$$

is not true.

Thus, the solution set of the given equation is $\{0\}$.

Example 3. Solve the equation $\sqrt{x + 5} - 1 = \dfrac{x}{\sqrt{x + 5}}$ over R.

Solution: Note that the replacement set for the given equation is $R - \{-5\}$. Thus, we have

$$R - \{-5\} \quad \left| \begin{array}{l} \sqrt{x + 5} - 1 = \dfrac{x}{\sqrt{x + 5}}, \\[2mm] x + 5 - \sqrt{x + 5} = x, \\[2mm] 5 = \sqrt{x + 5}, \\[2mm] 25 = x + 5, \\[2mm] x = 20. \end{array} \right.$$

CHECK. $\sqrt{20 + 5} - 1 = \dfrac{20}{\sqrt{20 + 5}}$ is true, since $5 - 1 = \frac{20}{5}$ is true. Thus the solution set of the given equation is $\{20\}$.

Example 4. Solve the equation $\sqrt[3]{x + 1} = x - 5$ over R.

Solution:

$$\left| \begin{array}{l} \sqrt[3]{x + 1} = x - 5, \\[2mm] x + 1 = (x - 5)^3, \\[2mm] x + 1 = x^3 - 15x^2 + 75x - 125, \\[2mm] 0 = x^3 - 15x^2 + 74x - 126, \\[2mm] 0 = (x - 7)(x^2 - 8x + 18), \\[2mm] x = 7 \ \lor \ x = 4 + \sqrt{2}\,i \ \lor \ x = 4 - \sqrt{2}\,i. \end{array} \right.$$

Thus, the solution set of the given equation (which must be a subset of R) is $\{7\}$, as we can see, since $\sqrt[3]{7 + 1} = 7 - 5$ is true.

Exercise 9.5

Study Examples 1 and 2 before doing Exercises 1–6. In each of these, solve the given equation over R.

1. $\sqrt{x+2} + \sqrt{x} = 2$.
2. $\sqrt{y} - \sqrt{y+2} = 1$.
3. $\sqrt{t+1} + \sqrt{t-1} = 1$.
4. $\sqrt{x+8} - \sqrt{x-2} = 2$.
5. $\sqrt{7x+27} = \sqrt{6x+12} + \sqrt{x+3}$.
6. $\sqrt{5y-10} + \sqrt{y+2} = \sqrt{6y+2}$.

Study Example 3 before doing Exercises 7–11. In each of these, solve the given equation over R.

7. $\sqrt{x-6} + 1 = \dfrac{x-3}{\sqrt{x-6}}$.

8. $\sqrt{2t-1} = \dfrac{2t+2}{\sqrt{2t-1}} - 1$.

9. $\dfrac{y-1}{\sqrt{y-3}} - 1 - \sqrt{y-3} = 0$.

10. $\sqrt{ax - a^2b^2} + c = \dfrac{ax + abc - a^2b^2}{\sqrt{ax - a^2b^2}}$, where $a > 0$,

$\qquad\qquad\qquad\qquad\qquad\qquad\qquad\qquad\qquad b > 0$, and $c > 0$.

11. Solve the equation $\sqrt{2x+1} - \sqrt{4x+6} = 2$ over R.

Study Example 4 before working Exercises 12–16. In each of these, solve the given equation over R.

12. $\sqrt[3]{2x+2} = x - 1$.
13. $\sqrt[3]{4t-40} - t - 2 = 0$.
14. $t + 1 - \sqrt[3]{2t-2} = 0$.
15. $\sqrt[3]{4x-2} = 2x + 1$.
16. $\sqrt[3]{9x-4} = 2\sqrt[3]{x-1}$.
∗17. $x^{2/3} - 2x^{1/3} - 8 = 0$.
∗18. Explain why we obtained a redundant equation in Example 2 of Section 9.5.

REVIEW EXERCISES

In Exercises 1–4 find the principal nth root.

1. $\sqrt{(-2)^2}$.   2. $\sqrt[3]{-64}$.
3. $\sqrt[4]{0.0001}$.   4. $\sqrt{(x-2)^2}$, $x \in R$.

In Exercises 5–20 simplify the given expressions.

5. $a^{2/3} \cdot a^{1/2} \cdot a^{-(1/3)}$, $a > 0$.

6. $\dfrac{h^3 r^2}{h^{5/2} r^{7/2}}$, $h > 0$, $r > 0$.

7. $\sqrt[4]{\dfrac{a^4b^{-4}}{16}}$, $a > 0$, $b > 0$.

8. $\sqrt[3]{\sqrt{64}}$.

9. $5\sqrt{2} + \sqrt{8} - 3\sqrt{32}$.

10. $\sqrt[3]{a + b} \cdot \sqrt[3]{(a + b)^2}$.

11. $\sqrt{4 + \sqrt{3}} + \sqrt{4 - \sqrt{3}}$.

12. $\dfrac{\sqrt[4]{a^8b^4}}{\sqrt{b^8}}$, $b \neq 0$.

13. $\dfrac{10}{\sqrt{13} - \sqrt{3}}$.

14. $\dfrac{\sqrt[3]{27a^6}}{\sqrt[6]{729}}$.

15. $\sqrt[3]{(a - b)} \ \sqrt[3]{a^2 + ab + b^2}$.

16. $3\sqrt[3]{81} - 2\sqrt[3]{648}$.

17. $\sqrt{3^3} - (\sqrt{3})^3$.

18. $\sqrt{x^3} - (\sqrt{x})^3$, $x \geqslant 0$.

19. $2\sqrt[4]{4} + 7\sqrt{\tfrac{1}{2}} - 3\sqrt{32}$.

20. $\dfrac{1}{\sqrt{3}} + 4\sqrt[12]{729} - 2\sqrt{243}$.

In Exercises 21–26 solve the equations over R.

21. $\sqrt{x - 3} = 1$.

22. $\sqrt{x - 3} = \sqrt{x + 5} - 2$.

23. $\sqrt{(x + 1)^2} = 2$.

24. $\sqrt{y + 6} + \sqrt{y + 1} = 5$.

25. $\sqrt{y - 6} = y - 12$.

26. $\sqrt{x^2 + a^2} = \sqrt{x^2} + \sqrt{a^2}$, $a \neq 0$.

COORDINATE GEOMETRY

O

10.1 INTRODUCTION

In the present chapter we introduce a technique that will enable us to solve certain geometric problems algebraically and, conversely, to give geometric solutions to certain algebraic problems.

The student is undoubtedly familiar with the fact that ordered pairs of real numbers can be represented by points in a plane. For completeness we describe the method very briefly. We draw a horizontal straight line, which we call the *X-axis*, and a vertical straight line, the *Y-axis*. Their point of intersection is called the *origin*.

X-axis
Y-axis
origin

Recall from Section 4.5 that there is a one-to-one correspondence between the set R of real numbers and a straight line. Let P_x and P_y be these correspondences between the real numbers and the X and Y axes, respectively—that is, if $c \in R$, then $P_x(c)$ is the image of c on the horizontal axis and $P_y(c)$ is the image of c on the vertical axis.

Further, assume the following:

(1) $P_x(0) = P_y(0) = O$ (the origin).
(2) If $c > 0$, $P_x(c)$ is to the right of the origin and $P_y(c)$ is above the origin.

(3) If $c < 0$, $P_x(c)$ is to the left of the origin and $P_y(c)$ is below the origin.

(4) If c_1, c_2, c_3, c_4 are real numbers, $|c_1 - c_2| = |c_3 - c_4|$ if and only if the segment $P_x(c_1) P_x(c_2)$ is congruent to the segment $P_y(c_3) P_y(c_4)$.

(See Figure 10.1.)

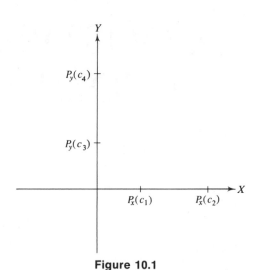

Figure 10.1

Now we establish a one-to-one correspondence between the set of all ordered pairs of real numbers and the plane in which the two axes have been drawn, as follows. The point Q corresponds to the ordered pair of real numbers (a, b) if and only if the projection of Q on the horizontal axis is $P_x(a)$ and the projection of Q on the vertical axis is $P_y(b)$. In this case, we say that a is the *abscissa* (X-coordinate, first coordinate) of Q and b is the *ordinate* (Y-coordinate, second coordinate) of Q. The ordered pair (a, b) is called *the ordered pair of coordinates of Q*, or simply the *coordinates of Q*.

Clearly, if the coordinates of a point are known, so is its position. Conversely, if a point is known, its coordinates can easily be found. The process of locating a point whose coordinates are given is called *plotting the point*. The plane together with the two axes is called the *coordinate plane*. (It is often called the *Cartesian coordinate plane*.)

The point with coordinates (a, b) is denoted $P(a, b)$ (see Figure 10.2) and is called the *image of (a, b)*.

abscissa

ordinate

ordered pair of coordinates

coordinates

plotting the point

coordinate plane

Cartesian coordinate plane

image

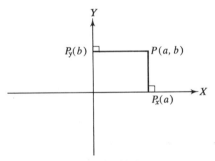

Figure 10.2

Example 1. Plot the points $P(2, 3)$, $P(-3, 2)$, $P(-2, -3)$, and $P(3, -2)$.

Solution: The four points are plotted in Figure 10.3.

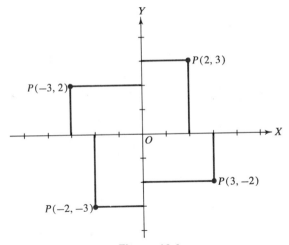

Figure 10.3

Note that the set of points of the plane that do not belong to any axis is divided into four subsets. Each of these subsets is quadrant called a *quadrant*. The first, second, third, and fourth quadrants are denoted Q_1, Q_2, Q_3, and Q_4, respectively, and are defined as follows:

$$Q_1 = \{P(a, b) \,|\, a > 0 \wedge b > 0\},$$
$$Q_2 = \{P(a, b) \,|\, a < 0 \wedge b > 0\},$$
$$Q_3 = \{P(a, b) \,|\, a < 0 \wedge b < 0\},$$
$$Q_4 = \{P(a, b) \,|\, a > 0 \wedge b < 0\}.$$

Note that $P(a, b)$ is on the X-axis if and only if $b = 0$. Similarly it is on the Y-axis if and only if $a = 0$. (See Figure 10.4.)

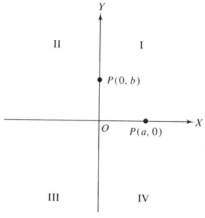

Figure 10.4

Exercise 10.1

Study Example 1 before doing Exercises 1–10. In Exercises 1–6 plot the given set of points.

1. $\{P(1, 3), P(-1, 3), P(1, -3), P(-1, -3)\}$.
2. $\{P(0, 2), P(2, 0), P(0, 0), P(0, -2), P(-2, 0)\}$.
3. $\{P(-2, 1), P(-1, 0), P(0, 1), P(1, 2), P(2, 3)\}$.
4. $\{P(3, 2), P(2, -1), P(1, 0), P(0, 1), P(2, -2)\}$.
5. $\{P(-2, 4), P(-\frac{3}{2}, \frac{9}{4}), P(-1, 1), P(0, 0), P(1, 1),$
   $P(\frac{3}{2}, \frac{9}{4}), P(2, 4)\}$.
6. $\{P(-2, 0), P(2, 0), P(0, -2), P(0, 2), P(1, -\sqrt{3}),$
   $P(1, \sqrt{3}), P(-1, -\sqrt{3}), P(-1, \sqrt{3})\}$.

In Exercises 7–15 determine in which quadrant each of the given points lies.

7. (a) $P(-1, 2)$,
   (b) $P(-3, -5)$,
   (c) $P(1, -4)$.
8. (a) $P(3, b), b < 0$,
   (b) $P(a, 6), a < 0$,
   (c) $P(-2, b), b > 0$.
9. (a) $P(-5, -3)$,
   (b) $P(-3, 2)$,
   (c) $P(6, 1)$.
10. (a) $P(a, b)$, where $ab > 0$,
    (b) $P(c, d)$, where $cd < 0$,
    (c) $P(a, b)$, where $ab = 0$.
11. $P(a, b)$, where $a \neq 0$ and $\dfrac{b}{a} < 0, b > 0$.
12. $P(a, b)$, where $b > 0, a < 0$.
13. $P(c, d)$, where $\dfrac{d}{c} > 0, c < 0$.

14. $P(x, y)$, where $xy \neq 0$ and $y = |y|$, $x = -|x|$.

15. $P(a, b)$, where $ab \neq 0$ and $a = |b|$, $b = -|a|$.

10.2 THE DISTANCE FORMULA

If A and B are two points in a coordinate plane, we are interested in finding the distance between them. If the coordinates of A are (x_1, y_1) and those of B are (x_2, y_2), then the distance d between A and B is given by the formula

$$d = \sqrt{(x_2 - x_1)^2 + (y_2 - y_1)^2}.$$

Since $(x_2 - x_1)^2 = (x_1 - x_2)^2$ and $(y_2 - y_1)^2 = (y_1 - y_2)^2$, it makes no difference which point is labeled (x_1, y_1); the distance between A and B is the same as the distance between B and A.

This formula can be proved in the general form just given. We do not wish to give the proof here, but we illustrate the idea with the example that follows. We first remind the student of the following well-known theorem from basic geometry.

Pythagorean Theorem. The square of the length of the hypotenuse of a right triangle is equal to the sum of the squares of the lengths of the other two sides.

Example 1. Find the distance between the points whose coordinates are $(2, 3)$ and $(-1, -1)$.

Solution: The two points are represented in Figure 10.5 and labeled A and B. Through A we draw a parallel to the Y-axis and through B a parallel to the X-axis. These two lines intersect at C to form the right triangle ACB. The length of the side BC is 3 units (note that $3 = 2 - (-1)$). Similarly the length of the side AC is 4 units ($4 = 3 - (-1)$). If we let d units be the length of the hypotenuse, the Pythagorean theorem leads to the following:

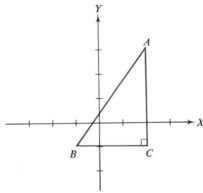

Figure 10.5

$$d^2 = 3^2 + 4^2.$$

Thus,

$$d = \sqrt{3^2 + 4^2}$$
$$= \sqrt{9 + 16}$$
$$= \sqrt{25}$$
$$= 5.$$

Note that $d = \sqrt{3^2 + 4^2}$ could have been written as

$$d = \sqrt{(2 - (-1))^2 + (3 - (-1))^2},$$

which agrees with the general formula given at the beginning of the present section.

Example 2. Find the distance between the points whose coordinates are $(1, 2)$ and $(-4, 5)$.

Solution: Using the distance formula and replacing x_1, y_1, x_2, y_2 by 1, 2, -4, and 5, respectively, we obtain

$$d = \sqrt{(-4 - 1)^2 + (5 - 2)^2}$$
$$= \sqrt{(-5)^2 + 3^2}$$
$$= \sqrt{25 + 9}$$
$$= \sqrt{34}.$$

Example 3. The points A, B, and C have coordinates $(-1, 7)$, $(1, 1)$, and $(5, 5)$ respectively. Show that the triangle ABC is isosceles.

Solution: The length of the side AB is

$$\sqrt{(1 - (-1))^2 + (1 - 7)^2} = \sqrt{2^2 + (-6)^2}$$
$$= \sqrt{4 + 36} = \sqrt{40} = 2\sqrt{10}.$$

The length of the side AC is

$$\sqrt{(5 - (-1))^2 + (5 - 7)^2} = \sqrt{6^2 + (-2)^2}$$
$$= \sqrt{36 + 4} = \sqrt{40} = 2\sqrt{10}.$$

Since the two sides AB and AC have equal lengths, the triangle is isosceles.

Example 4. Using the distance formula, prove the converse of the Pythagorean theorem.

Solution: We must prove that if the square of the length of one side of a triangle is equal to the sum of the squares of the lengths of the other two sides, then the triangle is a right triangle. Suppose such a triangle is given, and set it on a coordinate plane, as illustrated in Figure 10.6.

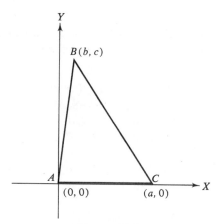

Figure 10.6

The triangle is such that the square of the distance from B to C is equal to the sum of the squares of the distances from A to B and to C. We shall show that $b = 0$ and therefore that the point B must be on the Y-axis. From this we shall conclude that the angle BAC is a right angle.

We have

$$\sqrt{(b-a)^2 + (c-0)^2}^2 = \sqrt{(b-0)^2 + (c-0)^2}^2 + \sqrt{(a-0)^2 + (0-0)^2}^2$$

—that is,

$$(b-a)^2 + c^2 = b^2 + c^2 + a^2.$$

Hence,

$$b^2 - 2ab + a^2 + c^2 = b^2 + c^2 + a^2.$$

Therefore,

$$2ab = 0.$$

Since both 2 and a are not zero, it follows that $b = 0$.

Exercise 10.2

Review Example 1 before working Exercises 1–7. In each of these, find the distance between the given pair of points using the Pythagorean theorem.

1. $P(6, 5)$, $P(2, 2)$.   **2.** $P(0, 3)$, $P(4, 0)$.
3. $P(-2, -2)$, $P(4, 3)$.   **4.** $P(-2, 3)$, $P(4, -5)$.
5. $P(-1, 3)$, $P(-5, 6)$.   **6.** $P(2, 4)$, $P(5, 9)$.
7. $P(-4, 0)$, $P(2, 3)$.

Review Example 2 before working Exercises 9–16. In each of these, use the distance formula to find the distance between the given pairs of points.

8. $P(-1, -3)$, $P(-4, -7)$.
9. $P(0, 5)$, $P(-8, 5)$.
10. $P(6, 7)$, $P(6, 0)$.
11. $P(3, 10)$, $P(-7, 2)$.
12. $P(a^2, a^2 + b^2)$, $P(-a^2, b^2)$.
13. $P(a, b)$, $P(b, a)$.
14. $P(x, k)$, $P(a, k)$.
15. $P(x + a, y)$, $P(x + a, b)$.
16. $P(a, a^2)$, $P(a + 1, (a + 1)^2)$.

Read Example 3 before doing Exercises 17–20.

17. The points A, B, C have coordinates $(-2, 0)$, $(0, 5)$, $(2, 0)$, respectively. Show that the triangle ABC is isosceles.

18. The points A, B, C have coordinates $(2, 3)$, $(10, 4)$, $(6, 10)$, respectively. Show that triangle ABC is isosceles.

19. The points A, B, C have coordinates $(1, 1)$, $(5, -3)$, $(4, 4)$, respectively. Determine whether triangle ABC is isosceles.

20. The points A, B, C have coordinates $(-4, -5)$, $(3, -3)$, $(-3, \frac{19}{4})$, respectively. Determine whether triangle ABC is isosceles.

21. Show that the points A, B, C, whose coordinates are $(-1, -1)$, $(2, 5)$, $(3, 7)$, respectively, are collinear. [*Hint:* Show that the distance from A to C is equal to the sum of the distances from B to C and to A.]

22. Repeat Exercise 21 with A, B, C having coordinates $(-2, 8)$, $(0, 2)$, $(4, -10)$, respectively.

23. Repeat Exercise 21 with A, B, C having coordinates $(-2, 3)$, $(4, 0)$, $(6, -1)$, respectively.

10.3 EQUATION OF A CIRCLE

Suppose a curve in a coordinate plane is given. This curve is a set of points in the plane, and to each of these points corresponds an ordered pair (x, y) of real numbers. Suppose that an equation can be found, relating x and y, such that a point with coordinates (a, b) belongs to the given curve if and only if the

equation of
the curve
graph of the
equation

equation becomes a true statement when x and y are replaced respectively by a and b. Then the equation is called an *equation of the curve*, and the curve is called the *graph of the equation*. In the present and the next two sections we shall be concerned with the problem of finding the equations of certain curves. The general technique to be used is the following.

Given the curve C (described geometrically), choose a point of C arbitrarily and call its coordinates (x, y). Then translate the geometric description of C into an algebraic statement involving x and y, thus getting an equation. Since the point of C was chosen arbitrarily, the same equation would have been obtained if we had chosen any other point. Thus, if the coordinates of any point of C are (a, b), then the equation becomes a true statement if x is replaced by a and y by b. However, to be able to conclude that the equation is the equation of the curve, we must also show that if (a, b) "satisfies" the equation (in the sense just described), then the point with coordinates (a, b) is on C. We illustrate this technique with an example.

Example 1. Find the equation of the circle whose center is the point $P(4, 2)$ and whose radius is 3.

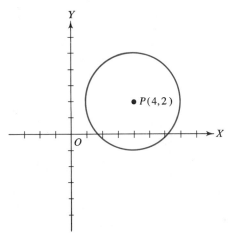

Figure 10.7

Solution: The circle is represented in Figure 10.7. Choose a point arbitrarily on the circle and call its coordinates (x, y). Since the radius of the circle is 3, we know that the distance from the center to that point is 3. By the distance formula we also know that the distance from the center to the chosen point is

$$\sqrt{(x - 4)^2 + (y - 2)^2}.$$

Thus, we have

$$\sqrt{(x-4)^2 + (y-2)^2} = 3.$$

Squaring both sides, we get

$$(x-4)^2 + (y-2)^2 = 9.$$

Conversely, if a point has coordinates (a, b) and the statement

$$(a-4)^2 + (b-2)^2 = 9$$

is true, then the distance from the center to the point $P(a, b)$ is 3, and the point is on the circle. Therefore, the equation of the given circle is

$$(x-4)^2 + (y-2)^2 = 9.$$

Using the procedure of the foregoing example, we obtain the following:

Theorem. If a point with coordinates (h, k) is the center of a circle with radius r, then the equation of the circle is

$$(x-h)^2 + (y-k)^2 = r^2.$$

Sometimes the equation of a circle is known and we need to find the center and radius of that circle. The process of completing squares is useful for that purpose. (See Example 2, Section 7.4.)

Example 2. An equation of a circle is $x^2 - 6x + y^2 + 4y - 3 = 0$. Find the center and radius of that circle.

Solution: We must add $\left(\dfrac{-6}{2}\right)^2$ to the expression $x^2 - 6x$ and $\left(\dfrac{4}{2}\right)^2$ to the expression $y^2 + 4y$ in order to get perfect squares. Hence, from the given equation we obtain

$$x^2 - 6x + \left(\frac{-6}{2}\right)^2 + y^2 + 4y + \left(\frac{4}{2}\right)^2 = 3 + \left(\frac{-6}{2}\right)^2 + \left(\frac{4}{2}\right)^2$$

—that is,

$$x^2 - 6x + 9 + y^2 + 4y + 4 = 3 + 9 + 4.$$

Hence,

$$(x-3)^2 + (y+2)^2 = 16.$$

The foregoing equation can be written

$$(x-3)^2 + (y-(-2))^2 = 4^2.$$

Therefore, the center of the circle has coordinates $(3, -2)$ and the circle has radius 4.

Exercise 10.3

Study Example 1 before working Exercises 1–9. In each of these, the coordinates of the center of a circle and the length of its radius are given; find the equation of the circle.

1. $(3, 3)$; 3.
2. $(-1, 2)$; 2.
3. $(-5, -3)$; 1.
4. $(7, -3)$; k $(k > 0)$.
5. $(0, 0)$; 1.
6. $(a + b, a - b)$; k $(k > 0)$.

7. Derive an equation for the circle with center at $(2, 1)$ and tangent to the X-axis. Sketch.

8. Write an equation of the circle with center at $(-1, 1)$ and passing through the point $(-1, 5)$.

9. How do the graphs of the equations $x^2 + y^2 - 25 = 0$ and $(x^2 + y^2) \cdot (x^2 + y^2 - 25) = 0$ differ?

Study Example 2 before doing Exercises 10–17. In each of these, an equation of a circle is given; find the center and radius of that circle.

10. $x^2 - 2x + y^2 + 6y + 6 = 0$.
11. $y^2 - 4y + x^2 + 4x - 8 = 0$.
12. $4x^2 + 4x + 4y^2 + 8y - 31 = 0$.
13. $x^2 - 4hx + y^2 - 4hy + 4h^2 = 0$ $(h > 0)$.
14. $x^2 + 2ax + y^2 + 2by = 0$.
15. $y^2 + 2aby + x^2 - 2abx = -(ab)^2$ $(ab > 0)$.
16. $6x^2 + 12x + 6y^2 - 24y - 12 = 0$.
17. $\frac{1}{2}x^2 + 6x + \frac{1}{2}y^2 - 4y - 1 = 0$.

10.4 EQUATION OF A STRAIGHT LINE

Suppose L is a straight line in a Cartesian coordinate plane. We should like to find its equation, using the technique outlined in the previous section. Thus we must find a way of describing the line geometrically. It is clear that the line is completely determined if we know that it passes through the point $P(a, b)$ and is parallel to the line that passes through the origin and the point $P(1, m)$ (see Figure 10.8). In this case we say that the number

slope m is the *slope* of the line L.

It can be shown that if the line L is not parallel to the Y-axis and $P(x_1, y_1)$, $P(x_2, y_2)$ are two distinct points of L, then the real number m is the slope of L if and only if

$$m = \frac{y_2 - y_1}{x_2 - x_1}.$$

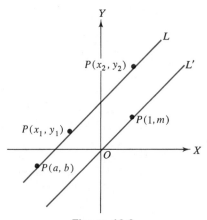

Figure 10.8

Note that the foregoing equation simply states that the slope of a straight line is the ratio between the change in ordinates and change in abscissas of any two distinct points on that line.

Example 1. The slope of the line that passes through the points $P(-2, 3)$ and $P(3, -4)$ is obtained as follows:

$$m = \frac{-4 - 3}{3 - (-2)} = \frac{-7}{5} = -1.4.$$

We remark that if m is a real number, then the line through the origin and the point $P(1, m)$ is not parallel to the Y-axis. On the other hand, if L is any line not parallel to the Y-axis, there is a unique line L' through the origin that is parallel to L. Further, there is a unique point on L' with abscissa 1. The ordinate m of that point is the slope of L. (See Figure 10.9.)

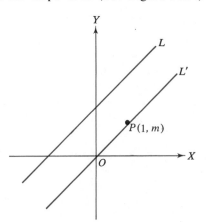

Figure 10.9

Hence, we conclude that *the slope of a line L is defined if and only if L is not parallel to the Y-axis.*

We are now ready to prove the following:

Theorem. Suppose that L_1 and L_2 are lines with slopes m_1 and m_2, respectively. Then L_1 is parallel to L_2 if and only if $m_1 = m_2$.

Proof: By definition of the slope of a line we know that L_1 is parallel to the line L_3 that passes through the origin and the point $P(1, m_1)$. Similarly, L_2 is parallel to the line L_4 that passes through the origin and the point $P(1, m_2)$. Thus, L_1 is parallel to L_2 if and only if L_3 is parallel to L_4. Since both L_3 and L_4 pass through the origin, L_1 is parallel to L_2 if and only if L_3 and L_4 coincide. But this will happen if and only if $(1, m_1) = (1, m_2)$ — that is, if and only if $m_1 = m_2$.

We also obtain the following:

Theorem. Suppose that L_1 and L_2 are lines with slopes m_1 and m_2, respectively. Then L_1 is perpendicular to L_2 if and only if $m_1m_2 = -1$.

Proof: Let L_3 and L_4 be described as in the proof of the preceding theorem. Then L_1 is perpendicular to L_2 if and only if L_3 is perpendicular to L_4 — that is, if and only if the origin and the points $P(1, m_1)$ and $P(1, m_2)$ are the vertices of a right triangle. (See Figure 10.10.) But, by the Pythagorean theorem and its converse, this will be true if and only if the following is true:

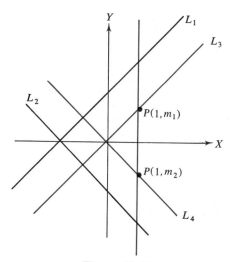

Figure 10.10

$$\sqrt{(1-0)^2 + (m_1-0)^2}^2 + \sqrt{(1-0)^2 + (m_2-0)^2}^2$$
$$= \sqrt{(1-1)^2 + (m_1-m_2)^2}^2.$$

The foregoing equation is equivalent to

$$1 + m_1^2 + 1 + m_2^2 = m_1^2 - 2m_1m_2 + m_2^2.$$

But the foregoing equations holds if and only if $m_1m_2 = -1$.
   We also obtain the following useful theorem.

Theorem. If a line L passes through the point $P(a, b)$ and its slope is m, then its equation is

$$y - b = m(x - a).$$

Proof: Let (x, y) be the coordinates of an arbitrary point of L. If $(x, y) = (a, b)$, then $x = a$ and $y = b$. But

$$b - b = m(a - a)$$

is a true statement.
   On the other hand, if $(x, y) \neq (a, b)$, we have

$$m = \frac{y - b}{x - a}. \qquad \text{(Why?)}$$

Therefore,

$$y - b = m(x - a).$$

We have shown that if (x, y) are the coordinates of any point of L, then

$$y - b = m(x - a).$$

Conversely, suppose that (c, d) is an ordered pair of real numbers and suppose that

$$d - b = m(c - a)$$

is true. We wish to show that the point $P(c, d)$ is on L.
   We consider two cases.

CASE 1. $c = a$. Then $d = b$, and therefore $(c, d) = (a, b)$. Since $P(a, b)$ is on L, so is $P(c, d)$.

CASE 2. $c \neq a$. Then we know that the line L' through the points $P(a, b)$ and $P(c, d)$ is not parallel to the Y-axis. The slope of L' is $(d - b)/(c - a)$. But we know that $d - b = m(c - a)$ and $c - a \neq 0$. Hence,

$$\frac{d - b}{c - a} = m.$$

We conclude that the slope of L' is m. Hence, L and L' have the same slope and therefore are parallel. Since they both pass through the point $P(a, b)$, they must coincide. Hence, the point $P(c, d)$, which is on L', must also be on L. (See Figure 10.11.)

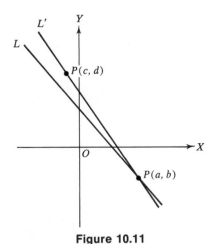

Figure 10.11

We conclude the present section with two more examples.

Example 2. Find the equation of the line passing through the points $P(1, 3)$ and $P(-2, -3)$.

Solution: The slope of the line is obtained as follows:

$$m = \frac{-3 - 3}{-2 - 1} = \frac{-6}{-3} = 2.$$

Since the line passes through the point $P(1, 3)$, we can use the preceding theorem and obtain the equation

$$y - 3 = 2(x - 1).$$

Note that if we had used the point $P(-2, -3)$ instead, we would have obtained the equation

$$y - (-3) = 2(x - (-2))$$

— that is,

$$y + 3 = 2(x + 2).$$

Are these two answers consistent? The following shows that they are.

$$\uparrow \begin{cases} y - 3 = 2(x - 1). \\ y - 3 = 2x - 2. \\ y = 2x + 1. \\ y + 3 = 2x + 4. \\ y + 3 = 2(x + 2). \end{cases}$$

Often, of the five foregoing equivalent equations, one would choose to use $y = 2x + 1$ as the simplest answer.

Example 3. Let L_1 be a line that contains the points $P(1, 3)$ and $P(-2, 9)$. Let L_2 and L_3 be lines that pass through the point $P(2, 5)$ and that are, respectively, parallel and perpendicular to L_1. Find their equations.

Solution: The slope of L_1 is obtained as follows:

$$m_1 = \frac{9 - 3}{-2 - 1} = \frac{6}{-3} = -2.$$

Since L_2 is parallel to L_1, the slope of L_2 is also -2. We also know that L_2 passes through the point $P(2, 5)$. Therefore, the equation of that line is

$$y - 5 = -2(x - 2).$$

Equivalently,

$$y = -2x + 9.$$

Let m be the slope of L_3. Since L_3 and L_1 are perpendicular, we have

$$m(-2) = -1.$$

Therefore,

$$m = \tfrac{1}{2}.$$

But L_3 passes through the point $P(2, 5)$ also. Hence, its equation is

$$y - 5 = \tfrac{1}{2}(x - 2)$$

—that is,

$$y = \tfrac{1}{2}x + 4.$$

Remark. If a line L is not parallel to the Y-axis, the following two propositions are equivalent:

(1) L has slope m and intersects the Y-axis at $P(0, b)$.
(2) An equation of the line L is $y = mx + b$.

Exercise 10.4

Read Example 1 before doing Exercises 1–9. In each of these find the slope of the line that passes through the given pairs of points.

1. $P(-1, 1)$, $P(4, 2)$.   **2.** $P(3, 4)$, $P(7, 11)$.
3. $P(4, 1)$, $P(-2, 3)$.   **4.** $P(\frac{7}{2}, \frac{1}{4})$, $P(\frac{1}{3}, \frac{3}{2})$.
5. $P(\frac{2}{3}, 4)$, $P(\frac{2}{3}, 5)$.   **6.** $P(k, k^2)$, $P(h, h^2)$, $k \neq h$.
7. $P(a, c)$, $P(b, c)$, $a \neq b$.
8. $P(b, b + 1)$, $P(c, c + 1)$, $b \neq c$.
9. $P(a, -a + k)$, $P(b, -b + k)$, $a \neq b$.

Read Example 2 before working Exercises 10–19. In each of these, find an equation of the line passing through the given two points.

10. $P(3, -5)$, $P(7, -13)$.   **11.** $P(-1, -4)$, $P(4, 6)$.
12. $P(1, -5)$, $P(2, -2)$.   **13.** $P(0, 2)$, $P(3, -2)$.
14. $P(3, 1)$, $P(-1, -1)$.   **15.** $P(1, 6)$, $P(2, 8)$.
16. $P(0, -3)$, $P(-4, 1)$.   **17.** $P(-1, -6)$, $P(-3, -7)$.
18. $P(0, 5)$, $P(5, 0)$.   **19.** $P(-3, -3)$, $P(-2, -2)$.

20. Obtain an equation of the line containing the centers of the circles whose equations are $(x - 4)^2 + (y - 1)^2 = 4$ and $(x + 1)^2 + (y - 5)^2 = 6$, respectively.

Read Example 3 before working Exercises 21–24.

21. Let L_1 be a line that contains the points $P(2, -1)$ and $P(-3, 2)$. Let L_2 and L_3 be lines that pass through $P(1, 1)$ and that are, respectively, parallel and perpendicular to L_1. Find their equations.

22. Let L_1 be a line that contains the points $P(3, -2)$ and $P(2, 3)$. Let L_2 and L_3 be lines that pass through $P(-5, -6)$, and that are, respectively, parallel and perpendicular to L_1. Find their equations.

23. Let L_1 be a line that contains the points $P(3, 4)$ and $P(4, 3)$. Let L_2 and L_3 be lines that pass through $P(-2, -2)$ and that are, respectively, parallel and perpendicular to L_1. Find their equations.

24. Let L_1 be a line that contains the points $P(a, b)$ and $P(b, a)$, where $a \neq b$. Let L_2 and L_3 be lines that pass through $P(a + b, a + b)$ and that are, respectively, parallel and perpendicular to L_1. Find their equations.

25. If a line through the points $P(1, 2)$ and $P(-1, -1)$ is perpendicular to a line through $P(-4, 1)$ and $P(c, -3)$, determine c.

26. If a line through the points $P(1, 2)$ and $P(-1, -1)$ is parallel to a line through $P(-4, 1)$ and $P(c, -3)$, determine c.

27. Graph the equation $x^2 = y^2$. [*Hint:* $x^2 = y^2$ is equivalent to $(y - x)(y + x) = 0$.]

28. (a) Write an equation of the line parallel to the Y-axis and passing through the point $P(-1, 0)$.

(b) Write an equation of the line parallel to the Y-axis and passing through the point $P(a, 0)$.

29. Show that $y = b$ is an equation of the line parallel to the X-axis and passing through the point $P(0, b)$.

10.5 THE PARABOLA

Suppose that D and F denote, respectively, a straight line and a point both contained in the same plane P and such that $F \notin D$. *parabola* Then the locus of all points of P that are equidistant from D *directrix* and F is called a *parabola* with *directrix* D and *focus* F. In the *focus* present section we illustrate how to find the equation of a parabola whose directrix is parallel to one of the axes of a Cartesian coordinate system.

Example 1. Find an equation of the parabola with focus the point $P(1, 3)$ and directrix the graph of the equation $y = 1$.

Solution: A point with coordinates (x, y) is on the parabola if and only if the distance between $P(x, y)$ and $P(1, 3)$ is equal to the distance between $P(x, y)$ and the graph of $y = 1$. The distance between $P(x, y)$ and $P(1, 3)$ is

$$\sqrt{(x - 1)^2 + (y - 3)^2}.$$

To find the distance between $P(x, y)$ and the graph of $y = 1$, we draw the perpendicular from that point to the straight line. The foot of the perpendicular is clearly the point $P(x, 1)$. (See Figure 10.12.) Hence, the distance from $P(x, y)$ to the directrix of the parabola is

$$\sqrt{(x - x)^2 + (y - 1)^2}.$$

Therefore, a point with coordinates (x, y) is on the parabola if and only if

$$\sqrt{(x - 1)^2 + (y - 3)^2} = \sqrt{(y - 1)^2}.$$

The foregoing equation is equivalent to

$$(x - 1)^2 + (y - 3)^2 = (y - 1)^2,$$

which in turn is equivalent to

$$x^2 - 2x + 1 + y^2 - 6y + 9 = y^2 - 2y + 1$$

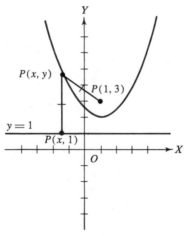

Figure 10.12

and to

$$y = \frac{x^2}{4} - \frac{x}{2} + \frac{9}{4}.$$

We expect the point $P(1, 2)$ to be on the parabola, since it is equidistant from $P(1, 3)$ and the graph of $y = 1$. Indeed we can verify that

$$2 = \frac{1^2}{4} - \frac{1}{2} + \frac{9}{4}$$

is true. Similarly, if

$$b = \frac{a^2}{4} - \frac{a}{2} + \frac{9}{4}$$

is true, then we can show (by reversing the steps above) that the distance from $P(a, b)$ to $P(1, 3)$ is equal to the distance from $P(a, b)$ to the graph of $y = 1$. Hence, $P(a, b)$ is on the parabola.

Exercise 10.5

Study Example 1 before working Exercises 1–10. In each of these, a point and the equation of a line are given. Find an equation of the parabola whose focus and directrix are, respectively, the given point and line.

1. $P(0, -4)$; $y = 4$.   2. $P(0, 2)$; $y = -2$.
3. $P(2, 5)$; $y = 2$.   4. $P(6, 2)$; $y = -3$.
5. $P(1, 1)$; $x = -1$.   6. $P(5, 2)$; $x = 3$.

7. $P(1, 3)$; $x = 5$.   **8.** $P(-3, 2)$; $x = 1$.

9. $P(-5, 6)$; $x = 3$.   **10.** $P(4, 1)$; $y = x$.

[*Hint — Exercise 10:* From a point $P(x, y)$ of the parabola draw a perpendicular to the graph of $y = x$. Find the coordinates of the intersection of these two lines and use this result to find the distance from $P(x, y)$ to the graph of $y = x$, and so on.]

10.6 THE ELLIPSE (OPTIONAL)

On occasion the student may have been asked to sketch a circle but did not have a compass at hand. At such times he could have used the following device. On a piece of paper lying on a flat (but old) table, choose two points. Insert a pin in the first point and hold the pencil tip at the other. Then tie a loop of string snugly around the pin and pencil tip and swing the pencil around, keeping the string taut until the circle is completed. (See Figure 10.13.)

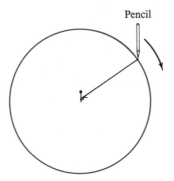

Pencil

Figure 10.13

A similar device may be used to sketch another closed curve, which may be thought of as a generalized circle. Choose three noncollinear points on the piece of paper; insert pins in the first two points and hold the pencil tip at the third. Now tie a loop of string snugly around the two pins and pencil tip. Finally, swing the pencil around, keeping the string taut until the closed curve is completed, as illustrated in Figure 10.14. The significant property of the curve so obtained is that the sum of the distances from any of its points to the two pins is constant. Such a curve is called an ellipse.

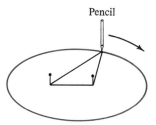

Figure 10.14

ellipse   An *ellipse* is a curve that is the locus of all points in a plane the sum of whose distances from two fixed points in the same *focus* plane is constant. Each of the two fixed points is called a *focus* of the ellipse.

Example 1. Let b and c be positive numbers. Find an equation of the ellipse that has as foci the points $P(-c, 0)$ and $P(c, 0)$ and that passes through the point $P(0, b)$.

Solution: Clearly the distance between $P(0, b)$ and $P(-c, 0)$ is equal to the distance between $P(0, b)$ and $P(c, 0)$. If we let a be this common distance, we see that the sum of the distances from $P(0, b)$ to $P(-c, 0)$ and to $P(c, 0)$ is $2a$. Thus, a point $P(x, y)$ is on the ellipse if and only if the sum of the distances from $P(x, y)$ to $P(-c, 0)$ and to $P(c, 0)$ is $2a$. (See Figure 10.15.)

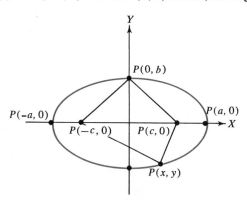

Figure 10.15

Thus, if $P(x, y)$ is on the ellipse, then

$$\sqrt{(x - (-c))^2 + (y - 0)^2} + \sqrt{(x - c)^2 + (y - 0)^2} = 2a.$$

Therefore,

$$\sqrt{(x + c)^2 + y^2} = 2a - \sqrt{(x - c)^2 + y^2}.$$

Squaring both sides of the foregoing equation, we get

$$(x + c)^2 + y^2 = 4a^2 - 4a \sqrt{(x - c)^2 + y^2} + (x - c)^2 + y^2.$$

Hence,

$$a\sqrt{(x - c)^2 + y^2} = a^2 - cx.$$

Again squaring both sides of the equation, we obtain

$$a^2(x^2 - 2cx + c^2 + y^2) = a^4 - 2a^2cx + c^2x^2$$

— that is,

$$a^2x^2 - c^2x^2 + a^2y^2 = a^4 - a^2c^2.$$

Therefore,

$$(a^2 - c^2)x^2 + a^2y^2 = a^2(a^2 - c^2).$$

Note that the points $P(0, b)$, $P(c, 0)$, and $P(0, 0)$ are the vertices of a right triangle. Hence,

$$a^2 = b^2 + c^2.$$

Therefore,

$$a^2 - c^2 = b^2.$$

We conclude that if the point $P(x, y)$ is on the ellipse, then

$$b^2x^2 + a^2y^2 = a^2b^2.$$

Dividing both sides of this equation by a^2b^2, we obtain

$$\frac{x^2}{a^2} + \frac{y^2}{b^2} = 1.$$

Conversely, we must show that if x_1 and y_1 are real numbers and

$$\frac{x_1^2}{a^2} + \frac{y_1^2}{b^2} = 1$$

is true, then the point $P(x_1, y_1)$ is on the ellipse.

Note that $-a \leq x_1 \leq a$, because otherwise $\frac{x_1^2}{a^2}$ would be larger than one, and since $\frac{y_1^2}{b^2} \geq 0$, the statement $\frac{x_1^2}{a^2} + \frac{y_1^2}{b^2} = 1$ could not be true.

If $x_1^2 = a^2$, then $y_1 = 0$, and it can be shown that the points $P(a, 0)$, $P(-a, 0)$ are both on the ellipse. (See Exercise 7 below.) In the case $-a < x_1 < a$, draw a perpendicular to the X-axis through the point $P(x_1, 0)$ as in Figure 10.16. Since the ellipse is a curve that can be drawn continuously without lifting the pencil, it is clear that this perpendicular intersects the ellipse at two points, $P(x_1, y_2)$ and $P(x_1, y_3)$. Since these two points are on the ellipse, we know that both statements

$$\frac{x_1^2}{a^2} + \frac{y_2^2}{b^2} = 1 \quad \text{and} \quad \frac{x_1^2}{a^2} + \frac{y_3^2}{b^2} = 1$$

are true. It was given that $\frac{x_1^2}{a^2} + \frac{y_1^2}{b^2} = 1$ is also true. Therefore, y_1 must be equal to y_2 or to y_3, for otherwise the quadratic equation (in y)

$$\frac{x_1^2}{a^2} + \frac{y^2}{b^2} = 1$$

would have the three distinct solutions $y_1, y_2,$ and y_3. But this is impossible. Thus the point $P(x_1, y_1)$ is on the ellipse.

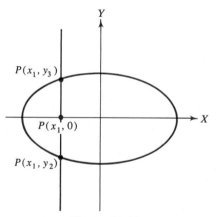

Figure 10.16

Example 2. Find an equation of the ellipse whose foci are the points $P(-2, 0)$ and $P(2, 0)$ and that passes through the point $P(2, 3)$.

Solution: The distance between $P(-2, 0)$ and $P(2, 3)$ is

$$\sqrt{(2 - (-2))^2 + (3 - 0)^2} = \sqrt{4^2 + 3^2} = \sqrt{16 + 9} = \sqrt{25} = 5.$$

The distance between $P(2, 0)$ and $P(2, 3)$ is

$$\sqrt{(2 - 2)^2 + (3 - 0)^2} = 3.$$

Thus, the sum of the distances from any point of the ellipse to $P(-2, 0)$ and $P(2, 0)$ is $5 + 3 = 8$. Comparing this example with Example 1, we see that $c = 2$, $2a = 8$, and therefore $a = 4$.

Further, $b^2 = a^2 - c^2 = 4^2 - 2^2 = 16 - 4 = 12$. Thus the equation of the ellipse is

$$\frac{x^2}{16} + \frac{y^2}{12} = 1.$$

(See Figure 10.17.)

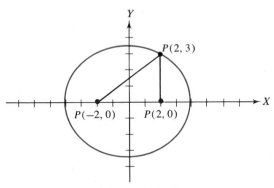

Figure 10.17

Exercise 10.6

Study Examples 1 and 2 before doing Exercises 1–6. In each of these, three points are given. Find an equation of the ellipse whose foci are the first two points and that passes through the third point.

1. $P(-\sqrt{5}, 0)$, $P(\sqrt{5}, 0)$, $P(0, 2)$.
2. $P(-3, 0)$, $P(3, 0)$, $P(-5, 0)$.
3. $P(-8, 0)$, $P(8, 0)$, $P(5, -3\sqrt{3})$.
4. $P(-2, 1)$, $P(2, 1)$, $P(0, 3)$.
5. $P(0, -2)$, $P(0, 2)$, $P(3, 2)$.
6. $P(0, -6)$, $P(0, 6)$, $P(-8, 0)$.

7. Refer to Figure 10.15 and show that the sum of the distances from $P(a, 0)$ to $P(c, 0)$ and $P(-c, 0)$ is $2a$. Hence $P(a, 0)$ is on the ellipse. Similarly, show that $P(-a, 0)$ is on the ellipse.

10.7 SYMMETRY

In plotting the graph of an equation, it is often time-saving to know certain geometric properties of the graph. For example, in plotting the graph of the equation $y = x^2$, it is useful to note that if $b = a^2$ is a true statement, then $b = (-a)^2$ is also true. It follows that if $P(a, b)$ is a point of the graph of the equation $y = x^2$, then $P(-a, b)$ is also a point of that graph. Noting that $P(-a, b)$ is the reflection of $P(a, b)$ through the Y-axis (see Figure 10.18), we could plot the part of the graph of the equation $y = x^2$ for $x \geq 0$, and obtain the other part by reflection through the Y-axis (see Figure 10.19).

symmetric   In general, we say that the points *P and Q are symmetric with respect to a line L* if that line is a perpendicular bisector of the line segment PQ. Similarly, we say that *P and Q are symmetric with respect to a point C* if C is the midpoint of the line segment PQ.

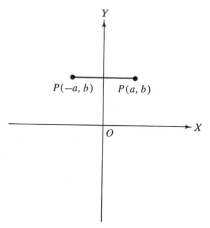

Figure 10.18

A set S is *symmetric with respect to a line L* if for each $P \in S - L$ there is a $Q \in S - L$ such that P and Q are symmetric with respect to L. Also, we say that S is *symmetric with respect to a point C* if for each $P \in S - \{C\}$ there is a $Q \in S - \{C\}$ such that P and Q are symmetric with respect to C. (See Figure 10.20.)

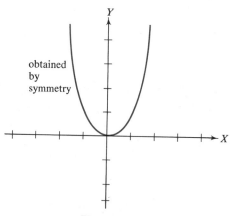

Figure 10.19

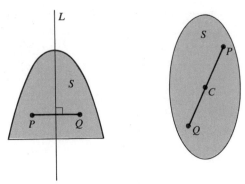

Figure 10.20

In the remaining portion of the present section and in the next section we shall consider only points in a Cartesian coordinate plane.

The following theorem can easily be proved.

Theorem. The points $P(a, b)$ and $P(c, d)$ are

(a) symmetric with respect to the X-axis if and only if $a = c$ and $b = -d$,

(b) symmetric with respect to the Y-axis if and only if $a = -c$ and $b = d$,

(c) symmetric with respect to the origin if and only if $a = -c$ and $b = -d$,

(d) symmetric with respect to the graph of $y = x$ if and only if $a = d$, $b = c$.

The proof is left as an exercise.

Example 1. Plot the following pairs of points: (a) $P(2, 3)$ and $P(2, -3)$, (b) $P(-3, 1)$ and $P(3, 1)$, (c) $P(-2, 4)$ and $P(2, -4)$, (d) $P(3, 2)$ and $P(2, 3)$.

Solution: The points are plotted in Figure 10.21. Note the symmetry.

The foregoing theorem may be used to test whether the graph of an equation is symmetric with respect to one of the axes, the origin, or the graph of the equation $y = x$. We illustrate the technique with two examples.

Example 2. Show that the graph of $x - y^2 = 4$ is symmetric with respect to the X-axis.

Solution: Let G be the graph of the equation $x - y^2 = 4$. Then we obtain the following equivalent propositions:

$$P(a, b) \in G.$$
$$a - b^2 = 4.$$
$$a - (-b)^2 = 4. \quad \text{(Why?)}$$
$$P(a, -b) \in G.$$

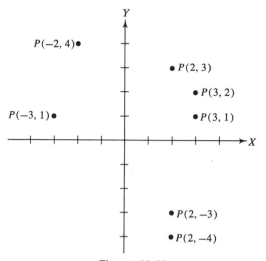

Figure 10.21

Since $P(a, b) \in G \leftrightarrow P(a, -b) \in G$ and the points $P(a, b)$ and $P(a, -b)$ are symmetric with respect to the X-axis, we conclude that G is symmetric with respect to the X-axis. (See Figure 10.22.)

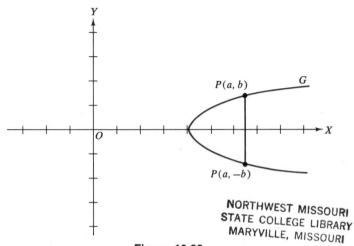

Figure 10.22

Example 3. Show that the graph of the equation $xy = 1$ is symmetric with respect to the graph of $y = x$.

Solution: Let G be the graph of the equation $xy = 1$. Then we have

$$P(a, b) \in G.$$
$$a \cdot b = 1.$$
$$b \cdot a = 1.$$
$$P(b, a) \in G.$$

Since $P(a, b) \in G \leftrightarrow P(b, a) \in G$ and the points $P(a, b)$ and $P(b, a)$ are symmetric with respect to the graph of $y = x$, then G is also symmetric with respect to that graph.

Note also that

$$P(a, b) \in G.$$
$$a \cdot b = 1.$$
$$(-1)(-1)ab = 1.$$
$$(-a)(-b) = 1.$$
$$P(-a, -b) \in G.$$

Thus, $P(a, b) \in G \leftrightarrow P(-a, -b) \in G$, from which we conclude that G is symmetric with respect to the origin. (See Figure 10.23.)

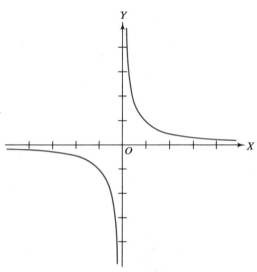

Figure 10.23

Exercise 10.7

Read Example 1 before doing Exercises 1–6. In each of these, plot the given pairs of points and note the symmetry.

1. (a) $P(-1, 2)$ and $P(-1, -2)$, (b) $P(-2, -2)$ and $P(2, -2)$, (c) $P(2, 3)$ and $P(-2, -3)$, (d) $P(-4, -2)$ and $P(-2, -4)$.

2. (a) $P(-2, 1)$ and $P(1, -2)$, (b) $P(-2, 1)$ and $P(2, 1)$, (c) $P(-3, 1)$ and $P(1, -3)$, (d) $P(-2, 7)$ and $P(7, -2)$.

3. (a) $P(5, 3)$ and $P(3, 5)$, (b) $P(-4, -1)$ and $P(4, 1)$, (c) $P(0, 2)$ and $P(0, -2)$.

4. (a) $P(-1, 0)$ and $P(1, 0)$, (b) $P(0, 4)$ and $P(4, 0)$, (c) $P(6, 5)$ and $P(5, 6)$.

5. (a) $P(a, b)$ and $P(a, -b)$, (b) $P(a, b)$ and $P(-a, b)$, (c) $P(a, b)$ and $P(-a, -b)$, (d) $P(a, b)$ and $P(b, a)$.

6. (a) $P(m, m - n)$ and $P(m, n - m)$, (b) $P(x, x^2)$ and $P(-x, x^2)$, (c) $P(x, \sqrt{r^2 - x^2}$ and $P(-x, -\sqrt{r^2 - x^2}), 0 < x < r$.

Study Examples 2 and 3 before working Exercises 7–16.

7. Show that the graph of $x^2 - y = 3$ is symmetric with respect to the Y-axis.

8. Show that the graph of $x - \sqrt{r^2 - y^2} = 0$ is symmetric with respect to the X-axis.

9. Show that the graph of $x - \dfrac{a}{b} = b^2 - y^2$ is symmetric with respect to the X-axis.

10. Show that the graph of $x^2 y = 3$ is symmetric with respect to the Y-axis.

11. Show that the graph of $|x| + |y| = 1$ is symmetric with respect to (a) the origin, (b) the graph of $y = x$, (c) the X-axis.

12. Determine the symmetry of the graph of $x - y = 0$.

13. Determine the symmetry of the graph of $y(x^2 + 1) = 1$.

14. Determine the symmetry of the graph of $x^2 + x - y^2 = k$, $k > 0$.

15. Determine the symmetry of the graph of $\dfrac{x^2}{4} + \dfrac{y^2}{2} = 1$.

16. Determine the symmetry of the graph of $x^2 + y^2 = r^2$, $r > 0$.

  ∗**17.** Graph the equation $x^2 + y^2 - xy - 6 = 0$.

18. Graph the equations (a) $y = |x|$, (b) $y = |x| - x$.

19. Prove part (a) of the theorem of this section.

20. Prove part (b) of the theorem of this section.

21. Prove part (c) of the theorem of this section.

22. Prove part (d) of the theorem of this section.

23. Find an equation of the line L, if L is symmetric with respect to (a) the X-axis and passes through $(1, 0)$, (b) the Y-axis and passes through $(0, 1)$, (c) the line $y = x$ and passes through $(-1, 1)$.

10.8 APPLICATIONS

Some geometric problems can be solved algebraically; conversely, some algebraic results can be obtained through geometric means. In the present section we give some examples to illustrate these techniques.

Example 1. Prove that if an angle is inscribed in a semicircle, it is a right angle.

Solution: The equation of a circle with center the origin and radius a is $x^2 + y^2 = a^2$. Thus, the equation of the upper semicircle shown in Figure 10.24 is $y = \sqrt{a^2 - x^2}$ (since $y \geqslant 0$).

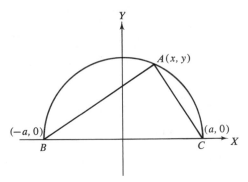

Figure 10.24

Consider the angle BAC and let the coordinates of the points B, A, C be $(-a, 0)$, (x, y), and $(a, 0)$, respectively. Note that $-a < x < a$. The slope of the line through the points B and A is

$$\frac{y - 0}{x - (-a)} = \frac{y}{x + a}.$$

The slope of the line that passes through A and C is

$$\frac{y - 0}{x - a} = \frac{y}{x - a}.$$

The product of these two slopes is

$$\frac{y}{x + a} \cdot \frac{y}{x - a} = \frac{y^2}{x^2 - a^2}.$$

Since the point A is on the circle, $y^2 = a^2 - x^2$. Therefore, the foregoing product is equal to

$$\frac{a^2 - x^2}{x^2 - a^2} = \frac{-(x^2 - a^2)}{x^2 - a^2} = -1.$$

Recalling that two lines are perpendicular if the product of their slopes is -1, we conclude that angle BAC is a right angle.

Example 2. Show that a parallelogram is a rectangle if and only if its two diagonals have equal length.

Solution: Draw the parallelogram as in Figure 10.25 and let d_1 be the distance from $P(0,0)$ to $P(a+b,c)$ and d_2 be the distance from $P(a,0)$ to $P(b,c)$.

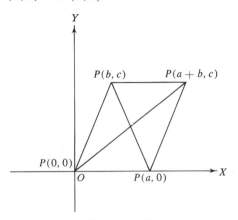

Figure 10.25

Now it is clear that the parallelogram is a rectangle if and only if $b = 0$. But we have the following equivalent propositions:

$$d_1 = d_2.$$
$$\sqrt{(a+b)^2 + c^2} = \sqrt{(a-b)^2 + c^2}.$$
$$(a+b)^2 + c^2 = (a-b)^2 + c^2.$$
$$2ab = -2ab.$$
$$4ab = 0.$$
$$b = 0.$$

Since $d_1 = d_2 \leftrightarrow b = 0$, we conclude that the parallelogram is a rectangle if and only if its diagonals have the same length.

Example 3. It can be shown that if the coordinates of the points A and B are, respectively, (a, b) and (c, d), then the coordinates of the midpoint of the line segment AB are $\left(\dfrac{a+c}{2}, \dfrac{b+d}{2}\right)$. Use this result to show that the length of the segment joining the midpoints of two sides of a triangle is half the length of the third side.

Solution: Set the triangle as in Figure 10.26. Then segment M_1M_2 has length

$$\sqrt{\left(\frac{a+b}{2} - \frac{b}{2}\right)^2 + \left(\frac{c}{2} - \frac{c}{2}\right)^2} = \sqrt{\left(\frac{a}{2}\right)^2} = \frac{a}{2} \qquad \text{(since } a > 0\text{)}.$$

Thus, the length of the segment M_1M_2 is half the length of the side AB of the triangle.

Also note that the line through M_1 and M_2 has slope

$$\frac{\dfrac{c}{2} - \dfrac{c}{2}}{\dfrac{a+b}{2} - \dfrac{b}{2}} = 0.$$

Thus, that line is parallel to the X-axis, and we have shown that M_1M_2 is parallel to AB.

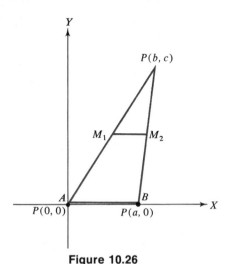

Figure 10.26

Example 4. If a and b are positive numbers, construct a line segment whose length is $\sqrt{ab}$.

Solution: Consider the circle with center the point $P\left(\dfrac{a+b}{2}, 0\right)$ and radius $\dfrac{a+b}{2}$. The equation of that circle is

$$\left(x - \frac{a+b}{2}\right)^2 + y^2 = \left(\frac{a+b}{2}\right)^2.$$

Through the point $P(a, 0)$ draw a perpendicular to the X-axis. This perpendicular intersects the upper half of the circle at the point $P(a, y)$. (See Figure 10.27.) We claim that $y = \sqrt{ab}$. To see this, note that $P(a, y)$ is on the circle. Thus, we must have

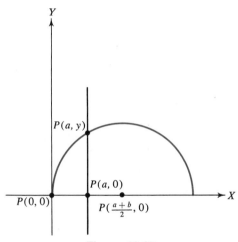

Figure 10.27

Hence,

$$\left(\frac{a-b}{2}\right)^2 + y^2 = \left(\frac{a+b}{2}\right)^2,$$

$$y^2 = \left(\frac{a+b}{2}\right)^2 - \left(\frac{a-b}{2}\right)^2,$$

$$y^2 = ab.$$

Since $P(a, y)$ is on the upper half circle, $y > 0$ and $y = \sqrt{ab}$.

Example 5. If a and b are positive numbers, $\dfrac{a+b}{2}$ and $\sqrt{ab}$ are, respectively, the arithmetic average and the geometric average of a and b. Give a geometric argument to show that $\sqrt{ab} \leqslant \dfrac{a+b}{2}$.

Solution: From Figure 10.27 we see that $\frac{a+b}{2}$ is the radius of the circle and hence the length of the hypotenuse of the right triangle whose vertices are $P(a, y)$, $P(a, 0)$, and $P\left(\frac{a+b}{2}, 0\right)$.

On the other hand, $\sqrt{ab}$ is the length of a side of that right triangle. Since the hypotenuse is the largest side of a right triangle, we always have $\sqrt{ab} \leqslant \frac{a+b}{2}$. Note also from the geometry that $\sqrt{ab} = \frac{a+b}{2}$ if and only if $a = b$.

Exercise 10.8

Read Example 4 before working Exercises 1 and 2.

1. Let a and b be positive numbers. Find the equation of the circle that has the segment from $P(-a, 0)$ to $P(b, 0)$ as one of its diameters. This circle intersects the positive Y-axis at the point $P(0, y)$. Prove that $y = \sqrt{ab}$.

2. Using the idea described in Exercise 1, give a geometric construction to find $\sqrt{5}$. [*Hint:* $5 = 1 \cdot 5$.]

Read Examples 2 and 3 before doing Exercises 3–7.

3. Prove that the sum of the squares of the sides of a parallelogram equals the sum of the squares of the diagonals.

4. Prove that two of the medians of an isosceles triangle have equal length. [*Hint:* First show that the triangle can be set so that its vertices are $P(0, a)$, $P(-b, 0)$, $P(b, 0)$.]

5. Prove that if two medians of a triangle have equal length, then the triangle is isosceles.

6. Prove that if a triangle is isosceles, one of its medians is perpendicular to one of its sides.

7. State and prove the converse of the theorem proved in Exercise 6.

8. The diagonals of a square lie along the coordinate axes. If the length of the side of the square is 4 units, find the coordinates of the vertices.

9. Prove: If a radius of a circle bisects a chord, then it is perpendicular to the chord.

10. Prove that the center of a circle, circumscribed about a right triangle, is the midpoint of the hypotenuse.

11. (a) Prove that every line segment that lies on the X-axis has a unique midpoint.

(b) Prove that the midpoint of the line segment determined by the points (a, b) and (c, d) has the coordinates $\left(\dfrac{a + c}{2}, \dfrac{b + d}{2}\right)$.

12. Justify why it was correct to set the parallelogram as we did in Figure 10.25. That is, show that if $P(0, 0)$, $P(a, 0)$, $P(b, c)$ are three vertices of a parallelogram with a, b, c positive numbers, then the fourth vertex is $P(a + b, c)$, or $P(b - a, c)$.

REVIEW EXERCISES

1. Plot the given points and name the quadrants in which they lie:
   (a) $P(1, 3)$,
   (b) $P(-1, 1)$,
   (c) $P(2, -1)$,
   (d) $P(-1, -2)$.

2. Using the distance formula, verify that $P\left(\dfrac{x_1 + x_2}{2}, \dfrac{y_1 + y_2}{2}\right)$ is the midpoint of the line segment joining the points $P(x_1, y_1)$ and $P(x_2, y_2)$.

3. Find the perimeter of the triangle whose vertices are $P(2, 3)$, $P(-2, 0)$, and $P(1, -4)$.

4. Show that the point $P(-3, 11)$ lies on the perpendicular bisector of the line segment joining the points $P(-2, -1)$ and $P(6, 3)$.

5. Find the coordinates of the midpoint of the line segment joining the points $P(2, 1)$ and $P(4, 3)$. [*Hint:* See Exercise 2.]

6. Find the center and the radius of the circle whose equation is

$$x^2 - 4x + y^2 + 6y + 9 = 0.$$

7. Obtain an equation of the circle with center at $P(1, -1)$ and radius 3.

8. Find the slope of the line whose equation is $2y + 3x + 1 = 0$.

9. Obtain an equation of the line that passes through the points $P(1, 3)$ and $P(-1, 2)$.

10. Let L_1 be the line that contains the points $P(-2, 2)$ and $P(1, 1)$. Let L_2 and L_3 be lines that pass through $P(1, 2)$ and that are, respectively, parallel and perpendicular to L_1. Find their equations.

11. Find the equation of the parabola with focus the point $P(-1,0)$ and directrix the graph of the equation $x = 1$.

12. Find the equation of the ellipse whose foci are the points $P(-3,0)$, $P(3,0)$ and that passes through the point $P(0,1)$.

In Exercises 13–18 sketch the graph of each equation and test whether it is symmetric with respect to the origin, the X-axis, the Y-axis, and the line $y = x$.

13. $y = x$.   **14.** $x^2 + y^2 = 1$.

15. $y = x + 1$.   **16.** $y = 4x^2$.

17. $4y^2 + x^2 = 1$.   **18.** $y^2 = x^2$.

19. Construct a line segment whose length is $\sqrt{6}$.

20. Prove that the diagonals of a square are perpendicular.

21. Prove that the diagonals of a rhombus are perpendicular.

22. Show that the points $P(-11,-7)$, $P(13,1)$, $P(15,2)$ do not lie on a straight line.

23. L_1 is a line with equation $ay + bx + c = 0$ and L_2 is a line with equation $by - ax + c = 0$, where $ab \neq 0$. Show that the two lines are perpendicular.

24. Find an equation of the circle that passes through the origin and has its center at $P(3,4)$.

25. Find an equation of the circle with radius 5 and center coinciding with the center of the circle whose equation is

$$x^2 + y^2 - 4x + 2y - 4 = 0.$$

SYSTEMS

OF

EQUATIONS

AND

INEQUALITIES

○

11.1 SYSTEMS OF TWO LINEAR EQUATIONS IN TWO VARIABLES

linear equation An equation of the form $ax + by = c$, where a, b, and c are real
in two numbers, is called a *linear equation in two variables*. We agree
variables here that x and y are variables with replacement set R. We say
solution that an ordered pair of real numbers (x_0, y_0) is a *solution* of
$ax + by = c$ if and only if $ax_0 + by_0 = c$ is a true proposition.
solution set The set of all solutions is called the *solution set*.

If $a_1x + b_1y = c_1$ and $a_2x + b_2y = c_2$ are linear equations in
two variables, then the open sentence $a_1x + b_1y = c_1 \wedge a_2x
+ b_2y = c_2$ is *called a system of two linear equations in two*
linear system *variables*, or simply *a linear system in two variables*. It is customary to write

$$\begin{cases} a_1x + b_1y = c_1, \\ a_2x + b_2y = c_2, \end{cases}$$

instead of $a_1x + b_1y = c_1 \wedge a_2x + b_2y = c_2$.

Example 1.

$$\begin{cases} x + y = 1, \\ x - y = 3, \end{cases}$$

is a linear system in two variables. Writing $x + y = 1$ in the equivalent form $y - 1 = (-1)(x - 0)$, we recognize that the graph of this equation is the straight line L_1 passing through the point $(0, 1)$ and with slope -1. Similarly, the graph of the equation $x - y = 3$ is the straight line L_2 passing through the point $(0, -3)$ and with slope 1. (See Figure 11.1.)

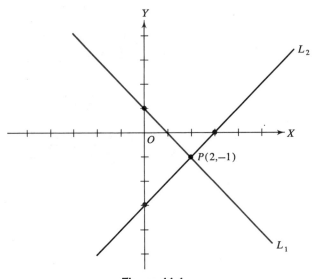

Figure 11.1

Since

$$\{(x, y) \,|\, x + y = 1 \wedge x - y = 3\} = \{(x, y) \,|\, x + y = 1\}$$
$$\cap \, \{(x, y) \,|\, x - y = 3\},$$

we know that the graph of the solution set of the given linear system consists of all those points which are common to both lines L_1 and L_2. Because the slopes of the lines L_1 and L_2 are -1 and 1, respectively, L_1 and L_2 must intersect at one and only one point. From Figure 11.1 we estimate that the point in $L_1 \cap L_2$ may have the coordinates $(2, -1)$. Since $2 + (-1) = 1 \wedge 2 - (-1) = 3$ is a true proposition, the ordered pair $(2, -1)$

is indeed a solution. In fact, it is the only solution, and the solution set is $\{(2, -1)\}$.

In the present section we wish to provide an algebraic method to solve systems of two linear equations in two variables (that is, to determine the solution sets of such systems). In Example 1 we were able to solve the given system by a graphical method. Since a graphical solution necessitates estimation of the coordinates of certain points, it is clear that a more accurate method is highly desirable.

Let us consider a general linear system of the form

$$\begin{cases} a_1x + b_1y = c_1, & a_1{}^2 + b_1{}^2 \neq 0, \\ a_2x + b_2y = c_2, & a_2{}^2 + b_2{}^2 \neq 0. \end{cases}$$

It can be shown that *the graph of each of the foregoing two equations in the Cartesian plane is a straight line.* Consequently, if L_k is the graph of $a_kx + b_ky = c_k$, for $k \in \{1, 2\}$, then we know that each L_k is a straight line. Hence, we have the following possibilities.

consistent and independent

    (a) L_1 and L_2 intersect at a unique point (see Figure 11.2(a)). In this case we say that the system is *consistent and independent*.

consistent and dependent

    (b) The lines L_1 and L_2 coincide. Then the coordinates of every point on L_1 are solutions of the given system, and consequently the solution set has infinitely many members (see Figure 11.2(b)). In this case we say that the system is *consistent and dependent*.

inconsistent

    (c) L_1 and L_2 are two distinct, parallel lines (see Figure 11.2(c)). It is clear that in this case the solution set is $\varnothing$. We say that the system is *inconsistent*.

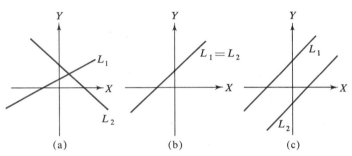

Figure 11.2

We solve a linear system algebraically by generating a chain of equivalent systems until a linear system is obtained whose solution set is obvious. The following theorem provides a criterion for equivalence of two linear systems.

Theorem. If K_1 and K_2 are real numbers and $K_2 \neq 0$, then the linear systems

$$\begin{cases} a_1x + b_1y = c_1, \\ a_2x + b_2y = c_2, \end{cases} \qquad \begin{cases} a_1x + b_1y = c_1, \\ K_1(a_1x + b_1y) + K_2(a_2x + b_2y) = K_1c_1 \\ \qquad\qquad\qquad\qquad\qquad\qquad + K_2c_2, \end{cases}$$

are equivalent.

The proof of this theorem is left as an exercise.

Example 2. Solve the following linear system algebraically:

$$\begin{cases} x + y = 0, \\ 3x + y = 1. \end{cases}$$

Solution:

$$\begin{cases} x + y = 0, \\ 3x + y = 1; \end{cases}$$

$$\begin{cases} x + y = 0, \\ -(x + y) + 3x + y = -0 + 1; \end{cases}$$

(Note that we are trying to obtain $0 \cdot y$ in the second equation.)

$$\begin{cases} x + y = 0, \\ 2x + 0 \cdot y = 1; \end{cases}$$

$$\begin{cases} x + y = 0, \\ x + 0 \cdot y = \tfrac{1}{2}; \end{cases}$$

$$\begin{cases} -(x + 0 \cdot y) + x + y = -\tfrac{1}{2} + 0, \\ x + 0 \cdot y = \tfrac{1}{2}; \end{cases}$$

(Note that we are trying to get $0 \cdot x$ in the first equation.)

$$\begin{cases} 0 \cdot x + y = -\tfrac{1}{2}; \\ x + 0 \cdot y = \tfrac{1}{2}. \end{cases}$$

Hence the solution set is

$$\{(x, y) \mid 0 \cdot x + y = -\tfrac{1}{2}\} \cap \{(x, y) \mid x + 0 \cdot y = \tfrac{1}{2}\}$$

$$= \{(x, -\tfrac{1}{2}) \mid x \in R\} \cap \{(\tfrac{1}{2}, y) \mid y \in R\},$$

or simply $\{(\frac{1}{2}, -\frac{1}{2})\}$. Note that the given system is consistent and independent.

Example 3. Solve the following linear system algebraically:

$$\begin{cases} -3x + y = 2, \\ x - \frac{1}{3}y = -\frac{2}{3}. \end{cases}$$

Solution:

$$\begin{cases} -3x + y = 2, \\ x - \frac{1}{3}y = -\frac{2}{3}; \end{cases}$$

$$\begin{cases} 3(x - \frac{1}{3}y) + (-3x) + y = 3(-\frac{2}{3}) + 2, \\ x - \frac{1}{3}y = -\frac{2}{3}; \end{cases} \quad \text{(Note that we are trying to obtain } 0 \cdot x \text{ in the first equation.)}$$

$$\begin{cases} 0 \cdot x + 0 \cdot y = 0, \\ x - \frac{1}{3}y = -\frac{2}{3}. \end{cases}$$

Hence, the solution set is

$$\{(x, y) \mid 0 \cdot x + 0 \cdot y = 0\} \cap \{(x, y) \mid x - \frac{1}{3}y = -\frac{2}{3}\},$$

which is the same as $\{(x, y) \mid x - \frac{1}{3}y = -\frac{2}{3}\}$. (Why?) Consequently, the given linear system is consistent and dependent.

Example 4. Solve algebraically

$$\begin{cases} x + y = 2, \\ -2x - 2y = -3. \end{cases}$$

Solution:

$$\begin{cases} x + y = 2, \\ -2x - 2y = -3. \end{cases}$$

$$\begin{cases} x + y = 2, \\ 2(x + y) + (-2x - 2y) = 2 \cdot 2 + (-3) \end{cases} \quad \text{(Note that we are trying to obtain } 0 \cdot x \text{ in the second equation.)}$$

$$\begin{cases} x + y = 2, \\ 0 \cdot x + 0 \cdot y = 1. \end{cases}$$

Therefore, the solution set is

$$\{(x, y) \mid x + y = 2\} \cap \{(x, y) \mid 0 \cdot x + 0 \cdot y = 1\}.$$

Noting that $\{(x, y) \mid 0 \cdot x + 0 \cdot y = 1\} = \varnothing$, we have now shown that the solution set of the given linear system is $\varnothing$. Thus this linear system is inconsistent.

In summary we remark that the basic procedure we have followed in solving a linear system of the form

$$\begin{cases} a_1x + b_1y = c_1, \\ a_2x + b_2y = c_2, \end{cases}$$

is to obtain an equivalent system of the form

$$\begin{cases} 0 \cdot x + d_1y = e_1, \\ d_2x + 0 \cdot y = e_2, \end{cases}$$

with $d_1d_2 \neq 0$. Then we can conclude that

$$x = \frac{e_2}{d_2} \quad \text{and} \quad y = \frac{e_1}{d_1}.$$

Thus, the solution set can readily be seen to be

$$\left\{ \left(\frac{e_2}{d_2}, \frac{e_1}{d_1} \right) \right\}.$$

elimination of variables by addition This procedure is often called *elimination of variables by addition*, since the foregoing system can be written

$$\begin{cases} d_1y = e_1, \\ d_2x = e_2, \end{cases}$$

where the variable x has been "eliminated" from the first equation and y from the second.

Exercise 11.1

Read Example 1 before doing Exercises 1–6. In each of these, a linear system in two variables is given. Let L_1 and L_2 denote the graph of the first and second equations, respectively. Then (a) write each equation in an equivalent form $y - y_1 = m(x - x_1)$, (b) from part (a) determine the slopes of L_1 and L_2, (c) sketch the graphs of L_1 and L_2 on the same coordinate system, (d) obtain the solution set of the system by considering the intersection of L_1 and L_2.

1. $\begin{cases} x - y = 1, \\ x + y = 2. \end{cases}$

2. $\begin{cases} x - y = 2, \\ x + y = \frac{1}{2}. \end{cases}$

3. $\begin{cases} 3x - y = 1, \\ x + 2y = 2. \end{cases}$ **4.** $\begin{cases} 3x + 2y = 3, \\ 2x + 5y = 1. \end{cases}$

5. $\begin{cases} 4x + 3y = -2, \\ 5x - 2y = 3. \end{cases}$ **6.** $\begin{cases} 5x - 7y = -1, \\ 3x + 4y = -3. \end{cases}$

Study Examples 2, 3, and 4 before working Exercises 7–22. In each of these, solve the given system algebraically.

7. $\begin{cases} x - y = 1, \\ 2x + y = 0. \end{cases}$ **8.** $\begin{cases} 3x + 4y = -1, \\ 2x + 3y = -1. \end{cases}$

9. $\begin{cases} 3u + 2v = 2, \\ 4u - v = 1. \end{cases}$ **10.** $\begin{cases} 2x - y = 5, \\ -x + \frac{1}{2}y = -\frac{5}{2}. \end{cases}$

11. $\begin{cases} x - \frac{1}{3}y = -1, \\ -3x + y = 3. \end{cases}$ **12.** $\begin{cases} 7x - 2y = -2, \\ -21x + 6y = 6. \end{cases}$

13. $\begin{cases} 6x - 5y = \frac{1}{2}, \\ 2x - \frac{5}{3}y = \frac{1}{6}. \end{cases}$ **14.** $\begin{cases} s + t = 1, \\ 2s + 2t = -2. \end{cases}$

15. $\begin{cases} 21x - 3y = 7, \\ 14x - 2y = 5. \end{cases}$ **16.** $\begin{cases} -4x + 2y = 8, \\ x - \frac{1}{2}y = 1. \end{cases}$

17. $\begin{cases} 3ax + 2y = 1, \\ -ax + y = 2, \ a \neq 0. \end{cases}$

18. $\begin{cases} au + bv = 2, \\ -au + 2bv = 1, \ \text{where both } a \text{ and } b \text{ are nonzero} \end{cases}$
                 constants.

19. $\begin{cases} au + bv = c, \\ bu + av = c, \ \text{where } a, b \text{ are constants with} \end{cases}$
                $(a^2 - b^2)(a^2 + b^2) \neq 0.$

20. $\begin{cases} \dfrac{x}{a} - \dfrac{y}{b} = 1, \\ -bx + ay = -ab, \ \text{where } a \text{ and } b \text{ are constants with} \end{cases}$
                $ab \neq 0.$

21. $\begin{cases} (a - b)y + (a + b)z = 1, \\ (a^2 - b^2)y + (a^2 + b^2)z = a + b, \ \text{where } a > b > 0. \end{cases}$

22.
$$\begin{cases} m^2s + n^2r = m^2n^2, \\ \dfrac{s}{n^2} + \dfrac{1}{m^2}r = -1, \text{ where both } m \text{ and } n \text{ are nonzero} \\ \hspace{4.5cm} \text{constants.} \end{cases}$$

23. Consider the linear system

$$\begin{cases} ax + by = 1, \\ bx + ay = 1, \quad \text{where } a \neq 0 \wedge b \neq 0. \end{cases}$$

Place conditions on a and b such that the linear system is (a) consistent and independent, (b) consistent and dependent, (c) inconsistent.

***24.** Consider the linear system

$$\begin{cases} a_1x + b_1y = c_1, \\ a_2x + b_2y = c_2, \quad \text{where } b_1b_2 \neq 0. \end{cases}$$

Place conditions on a_i, b_i, c_i, $i \in \{1, 2\}$, such that the linear system is (a) consistent and independent, (b) consistent and dependent, (c) inconsistent.

25. Solve the system

$$\begin{cases} \dfrac{2}{x} + \dfrac{3}{y} = 2, \\ \dfrac{1}{x} - \dfrac{1}{y} = \dfrac{1}{6}. \end{cases}$$

[*Hint:* Let $u = \dfrac{1}{x}$, $v = \dfrac{1}{y}$.]

***26.** Find the point equidistant from the points $P(-3, 1)$, $P(2, -4)$, and $P(1, 1)$. [*Hint:* This point is on each of the perpendicular bisectors of the segments whose endpoints are the given points taken two at a time.]

***27.** Let $(0, 0)$, (a, b), $(c, 0)$ be the coordinates of the vertices of a triangle. Show that the medians of the triangle intersect at a unique point and find the coordinates of this point.

***28.** Prove the theorem stated in Section 11.1. [*Hint:* Show that (a, b) is a solution of the first system if and only if it is a solution of the second system.]

***29.** Show that the distance d between $P(c_1, c_2)$ and the line L with equation $y = mx + b$ is given by the formula

$$d = \frac{|c_2 - mc_1 - b|}{\sqrt{m^2 + 1}}.$$

30. Use Exercise 29 to find the distance between the given points and lines: (a) $P(1, 2)$, L: $y = 2x + 1$; (b) $P(0, 0)$, L: $y = x + 3$; (c) $P(1, 1)$, L: $y = \frac{1}{2}x + \frac{1}{2}$; (d) $P(0, 2)$, L: $y = x - 1$.

11.2 CRAMER'S RULE (OPTIONAL)

In Chapter 7 we saw that the solution of any quadratic equation in one variable could be accomplished through the use of the quadratic formula. Let us now inquire whether it is possible to have a formula for the solution of certain linear systems.

Consider the linear system

$$\begin{cases} a_1x + b_1y = c_1, \\ a_2x + b_2y = c_2. \end{cases}$$

We make the additional assumptions that (1) $a_1b_2 - a_2b_1 \neq 0$ and (2) $a_2 \neq 0$. Now we have the following chain of equivalent systems:

$$\begin{cases} a_1x + b_1y = c_1, \\ a_2x + b_2y = c_2; \end{cases}$$

$$\begin{cases} a_1(a_2x + b_2y) - a_2(a_1x + b_1y) = -a_2c_1 + a_1c_2, \\ a_2x + b_2y = c_2; \end{cases}$$

$$\begin{cases} y = \dfrac{a_1c_2 - a_2c_1}{a_1b_2 - a_2b_1}, \\ a_2x + b_2y = c_2; \end{cases}$$ (We have left out the term $0 \cdot x$ on the left side of the first equation, since $0 \cdot x = 0$.)

$$\begin{cases} y = \dfrac{a_1c_2 - a_2c_1}{a_1b_2 - a_2b_1}, \\ -b_2y + a_2x + b_2y = (-b_2)\dfrac{a_1c_2 - a_2c_1}{a_1b_2 - a_2b_1} + c_2; \end{cases}$$

(A) $$\begin{cases} y = \dfrac{a_1c_2 - a_2c_1}{a_1b_2 - a_2b_1}, \\ x = \dfrac{c_1b_2 - c_2b_1}{a_1b_2 - a_2b_1}. \end{cases}$$

Next let us assume that $a_1b_2 - a_2b_1 \neq 0$ and $a_2 = 0$. Then, clearly, $a_1 \cdot b_2 \neq 0$.

We now have

$$\begin{cases} a_1x + b_1y = c_1, \\ a_2x + b_2y = c_2; \end{cases}$$

$$\begin{cases} a_1x + b_1y = c_1, \\ b_2y = c_2 \end{cases} \quad \text{(since } a_2 = 0\text{);}$$

$$\begin{cases} a_1x + b_1y = c_1, \\ y = \dfrac{c_2}{b_2}; \end{cases}$$

$$\begin{cases} -b_1y + a_1x + b_1y = c_1 - b_1 \cdot \dfrac{c_2}{b_2}, \\ y = \dfrac{c_2}{b_2}; \end{cases}$$

$$\begin{cases} x = \dfrac{c_1b_2 - c_2b_1}{a_1b_2}, \\ y = \dfrac{c_2}{b_2}; \end{cases}$$

$$\text{(B)} \quad \begin{cases} y = \dfrac{c_2}{b_2}, \\ x = \dfrac{c_1b_2 - c_2b_1}{a_1b_2}. \end{cases}$$

Comparing systems (A) and (B), we see that, provided $a_1b_2 - a_2b_1 \neq 0$, the following equivalence holds:

$$\begin{cases} a_1x + b_1y = c_1, \\ a_2x + b_2y = c_2. \end{cases} \longleftrightarrow \begin{cases} x = \dfrac{c_1b_2 - c_2b_1}{a_1b_2 - a_2b_1}, \\ y = \dfrac{a_1c_2 - a_2c_1}{a_1b_2 - a_2b_1}. \end{cases}$$

The last statement is readily remembered with the aid of square arrays of numbers of the form

$$\begin{pmatrix} a & b \\ c & d \end{pmatrix},$$

matrices called 2×2 (read "two-by-two") *matrices*. With each such *determinant* matrix we associate the number $ad - bc$, which we call the *determinant* of the matrix. We use the symbol

$$\begin{vmatrix} a & b \\ c & d \end{vmatrix}$$

to denote the determinant of the matrix

$$\begin{pmatrix} a & b \\ c & d \end{pmatrix}.$$

Thus,

$$\begin{vmatrix} a & b \\ c & d \end{vmatrix} = ad - cb.$$

Example 1.
$$\begin{vmatrix} 3 & -7 \\ 2 & 5 \end{vmatrix} = 3 \cdot 5 - 2 \cdot (-7) = 29.$$

$$a_1 b_2 - a_2 b_1 = \begin{vmatrix} a_1 & b_1 \\ a_2 & b_2 \end{vmatrix},$$

$$c_1 b_2 - c_2 b_1 = \begin{vmatrix} c_1 & b_1 \\ c_2 & b_2 \end{vmatrix},$$

$$a_1 c_2 - a_2 c_1 = \begin{vmatrix} a_1 & c_1 \\ a_2 & c_2 \end{vmatrix}.$$

We summarize the preceding discussion with the aid of determinants in the following theorem.

Theorem. If

$$\begin{vmatrix} a_1 & b_1 \\ a_2 & b_2 \end{vmatrix} \neq 0,$$

then the following linear systems are equivalent:

$$\begin{cases} a_1 x + b_1 y = c_1, \\ a_2 x + b_2 y = c_2, \end{cases} \qquad \begin{cases} x = \dfrac{Dx}{D}, \\ y = \dfrac{Dy}{D}, \end{cases}$$

where

$$D = \begin{vmatrix} a_1 & b_1 \\ a_2 & b_2 \end{vmatrix}, \qquad Dx = \begin{vmatrix} c_1 & b_1 \\ c_2 & b_2 \end{vmatrix}, \qquad Dy = \begin{vmatrix} a_1 & c_1 \\ a_2 & c_2 \end{vmatrix}.$$

Cramer's rule   We remark that this result is a particular case of an important theorem, called *Cramer's rule*. To keep our discussion simple, we do not intend to offer a more detailed treatment of either
Cramer's rule or determinants. The student who plans to continue his study of mathematics beyond this course is bound to
encounter these topics again. It will suffice to remark that the
significance of determinants does not lie in a notational convenience alone, as our usage here may suggest.

We conclude the present section with one more example.

Example 2. Solve the following system, using the theorem of
this section:

$$\begin{cases} 5x + 7y = -1, \\ 3x + 4y = \tfrac{1}{2}. \end{cases}$$

Solution:

$$D = \begin{vmatrix} 5 & 7 \\ 3 & 4 \end{vmatrix} = -1, \qquad Dx = \begin{vmatrix} -1 & 7 \\ \tfrac{1}{2} & 4 \end{vmatrix} = -\tfrac{15}{2} \qquad \text{and}$$

$$Dy = \begin{vmatrix} 5 & -1 \\ 3 & \tfrac{1}{2} \end{vmatrix} = \tfrac{11}{2}.$$

Hence,

$$\begin{cases} 5x + 7y = -1, \\ 3x + 4y = \tfrac{1}{2}; \end{cases}$$

$$\begin{cases} x = \dfrac{-\tfrac{15}{2}}{-1} = \tfrac{15}{2}, \\[2ex] y = \dfrac{\tfrac{11}{2}}{-1} = -\tfrac{11}{2}. \end{cases}$$

Thus the solution set is $\{(\tfrac{15}{2}, -\tfrac{11}{2})\}$.

Exercise 11.2

Study Example 1 before doing Exercises 1–14. In each of these,
evaluate the given determinant.

1. $\begin{vmatrix} 3 & 1 \\ -1 & 2 \end{vmatrix}$ 2. $\begin{vmatrix} 1 & 7 \\ 4 & 3 \end{vmatrix}$

3. $\begin{vmatrix} 0 & -3 \\ 2 & 0 \end{vmatrix}$ 4. $\begin{vmatrix} 0 & 0 \\ 1 & 4 \end{vmatrix}$

5. $\begin{vmatrix} 2 & 4 \\ 4 & 8 \end{vmatrix}$ 6. $\begin{vmatrix} a & b \\ b & a \end{vmatrix}$

7. $\begin{vmatrix} a & a \\ b & b \end{vmatrix}$ 8. $\begin{vmatrix} a_1 & b_1 \\ ka_2 & kb_2 \end{vmatrix}$

9. $\begin{vmatrix} a_1 & b_1 \\ a_2 + ka_1 & b_2 + kb_1 \end{vmatrix}$ 10. $\begin{vmatrix} a_1 & b_1 + ka_1 \\ a_2 & b_2 + ka_2 \end{vmatrix}$

11. $\begin{vmatrix} ax & ay \\ bx & by \end{vmatrix}$ 12. $\begin{vmatrix} 2a & 2 \\ -1 & 3 \end{vmatrix}$

13. $\begin{vmatrix} ax & d \\ c & b \end{vmatrix}$ 14. $\begin{vmatrix} x & 1 \\ 1 & x \end{vmatrix}$

Study Example 2 before doing Exercises 15–23. In each of these, solve the given linear system using determinants.

15. $\begin{cases} x - y = 2, \\ x + y = 1. \end{cases}$ 16. $\begin{cases} 2x + y = -1, \\ x - 2y = 1. \end{cases}$

17. $\begin{cases} 3x + 5y = 3, \\ -4x + 9y = 0. \end{cases}$ 18. $\begin{cases} 6x - 5y = 7, \\ 3x + 4y = \frac{11}{5}. \end{cases}$

19. $\begin{cases} \frac{1}{3}u + \frac{1}{2}v = \frac{1}{5}, \\ \frac{1}{2}u + \frac{1}{3}v = \frac{1}{5}. \end{cases}$ 20. $\begin{cases} 2x - \frac{3}{7}y = \frac{1}{2}, \\ 2x - \frac{3}{8}y = \frac{1}{2}. \end{cases}$

21. $\begin{cases} ar + bt = c, \\ br + at = c, \text{ where } a^2 - b^2 \neq 0. \end{cases}$

22. $\begin{cases} a_1x + a_2y = c_1, \\ -a_2x + a_1y = c_1, \text{ where } a_1^2 + a_2^2 \neq 0. \end{cases}$

23. $\begin{cases} mu + nv = m, \\ amu + bnv = cm, \text{ where } m, n, a, \text{ and } b \text{ are nonzero} \\ \qquad\qquad\qquad\quad \text{constants and } a \neq b. \end{cases}$

11.3 LINEAR SYSTEMS OF THREE EQUATIONS IN THREE VARIABLES

In the present section we wish to solve linear systems of the form

$$(1) \quad \begin{cases} a_1x + b_1y + c_1z = d_1, \\ a_2x + b_2y + c_2z = d_2, \\ a_3x + b_3y + c_3z = d_3, \end{cases}$$

where x, y, and z are variables with replacement set R, and for *ordered triple* $k \in \{1, 2, 3\}$ a_k, b_k, c_k, d_k are real numbers. An *ordered triple* (x_0, y_0, z_0) of real numbers is a solution if and only if

$$a_1x_0 + b_1y_0 + c_1z_0 = d_1 \wedge a_2x_0 + b_2y_0 + c_2z_0$$
$$= d_2 \wedge a_3x_0 + b_3y_0 + c_3z_0 = d_3$$

is a true proposition. To solve a system of this form means to determine the set of all solutions, called the solution set.

Although it is possible to interpret the solution set geometrically as the collection of all points common to three planes in a three-dimensional space, for simplicity we shall not pursue this idea. Instead, our attention will be focused on algebraic techniques.

The proof of the following theorem is sufficiently straightforward to be left as an exercise.

Theorem. If in the system (1) we multiply the members of any equation by some constant and then add them to the corresponding members of some other equation, the new system with one equation changed in this manner is equivalent to system (1). Further, if any equation in the system (1) is multiplied by a nonzero real number, the resultant system is equivalent to system (1).

We illustrate the application of this theorem with an example.

Example 1. Solve the system

$$\begin{cases} x + 2y - z = -3, \\ 2x - y + z = 5, \\ 3x + 2y - 2z = -3. \end{cases}$$

Solution: Repeated applications of the foregoing theorem yield the following chain of equivalent systems. To shorten the explanation, we introduce the notation $cE_j + E_k$ to represent the longer statement: the members of the jth equation have been multiplied by the number c and then added to the corresponding members of the kth equation. The notation sE_j means, of course, that the members of the jth equation have been multiplied by the constant s.

$$\begin{cases} x + 2y - z = -3, \\ 2x - y + z = 5, \\ 3x + 2y - 2z = -3. \end{cases} \qquad \textit{Explanation}$$

$$\begin{cases} x + 2y - z = -3, \\ 0 \cdot x - 5y + 3z = 11, \\ 0 \cdot x - 4y + z = 6. \end{cases} \qquad \begin{aligned} & \\ & -2E_1 + E_2 \\ & -3E_1 + E_3 \end{aligned}$$

$$\begin{cases} x - 2y + 0 \cdot z = 3 \\ 0 \cdot x + 7y + 0 \cdot z = -7, \\ 0 \cdot x - 4y + z = 6. \end{cases} \qquad \begin{aligned} & 1 \cdot E_3 + E_1 \\ & -3E_3 + E_2 \\ & \end{aligned}$$

$$\begin{cases} x - 2y + 0 \cdot z = 3, \\ 0 \cdot x + y + 0 \cdot z = -1, \\ 0 \cdot x - 4y + z = 6. \end{cases} \qquad \begin{aligned} & \\ & \tfrac{1}{7}E_2 \\ & \end{aligned}$$

$$\begin{cases} x + 0 \cdot y + 0 \cdot z = 1, \\ 0 \cdot x + y + 0 \cdot z = -1, \\ 0 \cdot x + 0 \cdot y + z = 2. \end{cases} \qquad \begin{aligned} & 2E_2 + E_1 \\ & \\ & 4E_2 + E_3 \end{aligned}$$

Hence, the solution set is

$$\{(x, y, z) \mid x + 0 \cdot y + 0 \cdot z = 1\} \cap \{(x, y, z) \mid 0 \cdot x + y$$
$$+ 0 \cdot z = -1\} \cap \{(x, y, z) \mid 0 \cdot x + 0 \cdot y + z = 2\},$$

which is equal to the set $\{(1, -1, 2)\}$.

The amount of writing can be reduced if we use the symbolic form

$$\begin{pmatrix} a_1 & b_1 & c_1 & d_1 \\ a_2 & b_2 & c_2 & d_2 \\ a_3 & b_3 & c_3 & d_3 \end{pmatrix}$$

augmented (called the *augmented matrix* of the linear system) to denote
matrix the system (1) described at the beginning of this section.

Using this notation, we can show the work in the foregoing example as follows:

Explanation

$$\begin{pmatrix} 1 & 2 & -1 & -3 \\ 2 & -1 & 1 & 5 \\ 3 & 2 & -2 & -3 \end{pmatrix}$$

$$\begin{pmatrix} 1 & 2 & -1 & -3 \\ 0 & -5 & 3 & 11 \\ 0 & -4 & 1 & 6 \end{pmatrix} \qquad \begin{array}{l} -2E_1 + E_2 \\ -3E_1 + E_3 \end{array}$$

$$\begin{pmatrix} 1 & -2 & 0 & 3 \\ 0 & 7 & 0 & -7 \\ 0 & -4 & 1 & 6 \end{pmatrix} \qquad \begin{array}{l} 1 \cdot E_3 + E_1 \\ -3E_3 + E_2 \end{array}$$

$$\begin{pmatrix} 1 & -2 & 0 & 3 \\ 0 & 1 & 0 & -1 \\ 0 & -4 & 1 & 6 \end{pmatrix} \qquad \tfrac{1}{7}E_2$$

$$\begin{pmatrix} 1 & 0 & 0 & 1 \\ 0 & 1 & 0 & -1 \\ 0 & 0 & 1 & 2 \end{pmatrix} \qquad \begin{array}{l} 2E_2 + E_1 \\ 4E_2 + E_3 \end{array}$$

Note that the entries in the fourth column of the last augmented matrix reveal the solution set to be $\{(1, -1, 2)\}$.

Example 2. Solve the linear system

$$\begin{cases} 3x + y + z = 0, \\ x + y + z = 1, \\ 7x + 3y + 3z = 2. \end{cases}$$

Solution: Using augmented matrices, we have

Explanation

$$\begin{pmatrix} 3 & 1 & 1 & 0 \\ 1 & 1 & 1 & 1 \\ 7 & 3 & 3 & 2 \end{pmatrix}$$

$$\begin{pmatrix} 3 & 1 & 1 & 0 \\ 1 & 1 & 1 & 1 \\ 1 & 1 & 1 & 2 \end{pmatrix} \qquad -2E_1 + E_3$$

$$\begin{pmatrix} 3 & 1 & 1 & 0 \\ 1 & 1 & 1 & 1 \\ 0 & 0 & 0 & 1 \end{pmatrix} \qquad (-1)E_2 + E_3$$

Thus the given system is equivalent to

$$\begin{cases} 3x + y + z = 0, \\ x + y + z = 1, \\ 0 \cdot x + 0 \cdot y + 0 \cdot z = 1. \end{cases}$$

Since the solution set of $0 \cdot x + 0 \cdot y + 0 \cdot z = 1$ is $\varnothing$, we conclude that the solution set of the given system is $\varnothing$.

Exercise 11.3

Study Examples 1 and 2 carefully before doing Exercises 1–20. In each of these, solve the given linear system.

1. $\begin{cases} x + y + 2z = 1, \\ x + 2y + z = 1, \\ 2x + y + z = 1. \end{cases}$
   2. $\begin{cases} x + y - 2z = 0, \\ 3x + 4y + z = 1, \\ 3y - 2z = -1. \end{cases}$

3. $\begin{cases} 7u + 7v + 7w = 0, \\ -u + 2v + w = -7, \\ 2u - v - 4w = 15. \end{cases}$
   4. $\begin{cases} y + 2z = 0, \\ x + 3z = 1, \\ -2y - 2z = 2. \end{cases}$

5. $\begin{cases} 2x - 5y - z = 3, \\ 5x + 14y - z = -11, \\ 7x + 9y - 2z = -5. \end{cases}$
   6. $\begin{cases} 3x - 4y = 1, \\ 2x + 4y - z = -1, \\ x + y + z = 0. \end{cases}$

7. $\begin{cases} 3u + 5v + 6w = 2, \\ 2u + 3v + 2w = 7, \\ u + 2v + 4w = 13. \end{cases}$

8. $\begin{cases} 4x - y + 2z = 1, \\ 6x + y = -7, \\ x + 3z = 1. \end{cases}$

9. $\begin{cases} -x + 3y + z = -1, \\ x - 3z = 0, \\ 4x + 2y = 1. \end{cases}$

10. $\begin{cases} 4x + z = -10, \\ 3x + 2y = 0, \\ 6x - y - z = 2. \end{cases}$

11. $\begin{cases} r + s \phantom{- \frac{1}{2}t} = 2, \\ r - \frac{1}{2}t = \frac{1}{2}, \\ \frac{1}{3}s - \frac{1}{2}t = \frac{1}{6}. \end{cases}$

12. $\begin{cases} x + y + 2z = 1, \\ -x - y + 3z = 2, \\ 6x + y + z = 1. \end{cases}$

13. $\begin{cases} x + 2y - z = 0, \\ y + 3z = 1, \\ x + z = 0. \end{cases}$

14. $\begin{cases} x + 2y + 3z = 1, \\ 2x + 4y + 6z = 2, \\ x + y + z = 0. \end{cases}$

15. $\begin{cases} 2x - 4y + 6z = 2, \\ -3x + y - 3z = 3, \\ x - 2y + z = 3. \end{cases}$

16. $\begin{cases} r + s + 2t = 1, \\ 2r + 3s + t = 0, \\ r - s + t = 2. \end{cases}$

17. $\begin{cases} 2u + 3v - w = 5, \\ u + 2v + 4w = -1, \\ 4u - v + 2w = 0. \end{cases}$

18. $\begin{cases} 3r + 2s + 2t = 0, \\ 4r + 3s - t = 0, \\ 6r + 4s + 4t = 1. \end{cases}$

19. A collection of coins consists of nickels, dimes, and quarters. If all the coins were nickels, the total value would be $4.80. If all the nickels were replaced by dimes, the total value would be $10.50. If all the quarters were replaced by dimes, the total value would be $7.10. How many of each sort of coin are in the collection?

20. Determine a, b, and c if $(1, 0, -1)$, $(3, -1, -2)$, and $(0, 1, -2)$ are solutions of the equation $ax + by + cz = 1$.

Read Example 2 before working Exercises 21–28. In each of these, use the augmented matrix notation to solve the given system.

21.
$$\begin{cases} 3x + 2y + 4z = 5, \\ 2x + y + z = 1, \\ x - 2y - 3z = 1. \end{cases}$$

22.
$$\begin{cases} x - y + 2z = 1, \\ 2x + y - z = 2, \\ x - 3y + z = 3. \end{cases}$$

23.
$$\begin{cases} 2r + s + t = 1, \\ r + 2s + t = 0, \\ r + s + 2t = 0. \end{cases}$$

24.
$$\begin{cases} 3p + q - 2r = 4, \\ 7p + 9q + r = 7, \\ 10p + 10q - r = 12. \end{cases}$$

25.
$$\begin{cases} 3x + 4y + 2z = 0, \\ x + y + z = 2, \\ 4x + 5y + 3z = 7. \end{cases}$$

26.
$$\begin{cases} p + q - r = 0, \\ 7p - 2q + 2r = 1, \\ 3p + q = 0. \end{cases}$$

27.
$$\begin{cases} 4u - w = 1, \\ 4u + v + 2w = 0, \\ 3v - w = 0. \end{cases}$$

28.
$$\begin{cases} y - z = 2, \\ 3x + 4z = 1, \\ 2x - y = 1. \end{cases}$$

*29. Prove the theorem stated in Section 11.3.

11.4 SOME NONLINEAR SYSTEMS OF TWO EQUATIONS IN TWO VARIABLES

Equations such as $xy = 1$ or $x^2 + y = 0$ are said to be *nonlinear*. In the present section we wish to consider solutions of some systems of two equations in two variables, where at least one of the equations is quadratic in one of the variables. We illustrate the basic ideas and techniques with the aid of examples.

Example 1. Solve the system

$$\begin{cases} 4x + 3y = 0, \\ x^2 + y^2 = 25. \end{cases}$$

Solution: Note that the given system is the open sentence

$$4x + 3y = 0 \land x^2 + y^2 = 25,$$

whose solution set is

$$\{(x, y) \mid 4x + 3y = 0\} \cap \{(x, y) \mid x^2 + y^2 = 25\}.$$

We know that the graph of the equation $4x + 3y = 0$ is a straight line. We also recognize that the graph of the equation $x^2 + y^2 = 25$ is a circle with center at the origin and radius 5. Figure 11.3

shows the two graphs and indicates that the graph of the solution set of the given system consists of those points at which the straight line intersects the circle. Thus we know that the solution set will have two distinct ordered pairs.

Figure 11.3 suggests that these ordered pairs could be $(-3, 4)$ and $(3, -4)$. Since

$$4 \cdot (-3) + 3 \cdot 4 = 0$$

$$\wedge \ (-3)^2 + 4^2 = 25,$$

$$4 \cdot 3 + 3 \cdot (-4) = 0$$

$$\wedge \ 3^2 + (-4)^2 = 25$$

are true propositions, we conclude that the solution set is $\{(-3, 4), \ (3, -4)\}$.

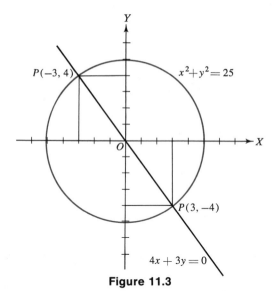

Figure 11.3

In Example 1 we solved the given nonlinear system using a graphical method. Very often this method will fail to produce accurate results. It is therefore preferable to solve such systems algebraically. The next example illustrates the algebraic method.

Example 2. Solve

$$\begin{cases} y + 2 = \dfrac{x^2}{4}, \\[2mm] 4x^2 + 9y^2 = 36. \end{cases}$$

Solution: Basically we wish to generate a chain of equivalent open sentences such as the following:

$$\begin{cases} y + 2 = \dfrac{x^2}{4}, \\ 4x^2 + 9y^2 = 36; \end{cases}$$

$$\begin{cases} 4 \cdot (y + 2) = x^2, \\ 4x^2 + 9y^2 = 36; \end{cases}$$

$$\begin{cases} 4 \cdot (y + 2) = x^2, \\ 4 \cdot 4 \cdot (y + 2) + 9y^2 = 36 \quad \text{(since we can substitute} \\ \qquad\qquad\qquad\qquad\qquad\quad 4 \cdot (y + 2) \text{ for } x^2); \end{cases}$$

$$\begin{cases} 4 \cdot (y + 2) = x^2, \\ 9y^2 + 16y - 4 = 0; \end{cases}$$

$$\begin{cases} 4 \cdot (y + 2) = x^2, \\ y = \dfrac{-16 \pm \sqrt{16^2 + 4 \cdot 4 \cdot 9}}{2 \cdot 9}; \end{cases}$$

$$\begin{cases} 4 \cdot (y + 2) = x^2, \\ y = -2 \ \lor \ y = \tfrac{2}{9}; \end{cases}$$

$$\begin{cases} x^2 = 4 \cdot (y + 2), \\ y = -2; \end{cases} \quad \lor \quad \begin{cases} x^2 = 4 \cdot (y + 2), \\ y = \tfrac{2}{9}; \end{cases}$$

$$\begin{cases} x^2 = 4 \cdot (-2 + 2), \\ y = -2; \end{cases} \quad \lor \quad \begin{cases} x^2 = 4(\tfrac{2}{9} + 2), \\ y = \tfrac{2}{9}; \end{cases}$$

$$\begin{cases} x^2 = 0, \\ y = -2; \end{cases} \quad \lor \quad \begin{cases} x = \pm\tfrac{4}{3}\sqrt{5}, \\ y = \tfrac{2}{9}. \end{cases}$$

Thus the solution set is $\{(0, -2), (-\tfrac{4}{3}\sqrt{5}, \tfrac{2}{9}), (\tfrac{4}{3}\sqrt{5}, \tfrac{2}{9})\}$.

We wish to comment on an important aspect of the technique displayed in the foregoing solution—namely, the fact that we indeed have a chain of equivalent open sentences. A simple way to see this is to take each pair of adjacent open sentences and examine their equivalence. For example, take the open sentences

$$(1) \quad \begin{cases} 4(y + 2) = x^2, \\ 4x^2 + 9y^2 = 36; \end{cases} \qquad (2) \quad \begin{cases} 4(y + 2) = x^2, \\ 4 \cdot 4(y + 2) + 9y^2 = 36. \end{cases}$$

To show that they are equivalent we may argue as follows:

If an ordered pair of real numbers (x_0, y_0) is a solution of (1), then the proposition

$$4(y_0 + 2) = x_0{}^2 \wedge 4x_0{}^2 + 9y_0{}^2 = 36$$

is true and therefore the proposition

$$4(y_0 + 2) = x_0{}^2 \wedge 4 \cdot 4(y_0 + 2) + 9y_0{}^2 = 36$$

is true. On the other hand, if an ordered pair of real numbers (x_0, y_0) is a solution of (2), then the proposition

$$4(y_0 + 2) = x_0{}^2 \wedge 4 \cdot 4(y_0 + 2) + 9y_0{}^2 = 36$$

is true and hence

$$4(y_0 + 2) = x_0{}^2 \wedge 4x_0{}^2 + 9y_0{}^2 = 36$$

is true. Thus every solution of system (1) is a solution of system (2) and conversely. We conclude that the two systems are equivalent.

The student should strive to develop his ability to perform simple equivalence arguments, such as the foregoing, mentally. This will greatly enhance his proficiency in solving systems of equations.

As a rough check on our solution we sketch the graphs of the equations $y + 2 = \dfrac{x^2}{4}$ (parabola) and $4x^2 + 9y^2 = 36$ (ellipse). (See Figure 11.4.) The graphs confirm the fact that the solution set of the system has three distinct members.

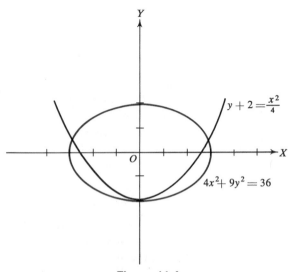

Figure 11.4

Example 3. If $a + bi$ is a complex number in standard form, solve the equation $z^2 = a + bi$ over C (the set of complex numbers).

Solution: If $b = 0$, then from Section 7.4 we know that the solution set is $\{-\sqrt{a}, \sqrt{a}\}$, where a is the principal square root of a.

We assume now that $b \neq 0$. Let $z = x + yi$, where x and y are variables with replacement set R. Now we have the following chain of equivalent statements:

$$z^2 = a + bi.$$

$$(x + yi)^2 = a + bi.$$

$$(x^2 - y^2) + 2xyi = a + bi.$$

$$\begin{cases} x^2 - y^2 = a. \\ 2xy = b. \end{cases}$$

Thus the problem is reduced to the task of determining all ordered pairs of real numbers that are solutions of the system

$$\begin{cases} x^2 - y^2 = a, \\ 2xy = b. \end{cases}$$

In Figure 11.5 we have sketched the graphs of the equations $x^2 - y^2 = a$, $2xy = b$ for the particular case $a = 1$, $b = 2$. Figure 11.5 suggests that it is reasonable to expect two distinct solutions of the system $x^2 - y^2 = a$, $2xy = b \, (b \neq 0)$.

$$\begin{cases} x^2 - y^2 = a, \\ 2xy = b; \end{cases}$$

$$\begin{cases} x^2 - y^2 = a, \\ y = \dfrac{b}{2x}; \end{cases}$$

$$\begin{cases} x^2 - \left(\dfrac{b}{2x}\right)^2 = a, \\ y = \dfrac{b}{2x}; \end{cases}$$

$$\begin{cases} x = \pm\sqrt{\tfrac{1}{2}a + \tfrac{1}{2}\sqrt{a^2 + b^2}}, \\ y = \dfrac{b}{2x}. \end{cases}$$

(We leave it as an exercise for the student to justify the equivalence of $x^2 - \left(\dfrac{b}{2x}\right)^2 = a$ and $x = \pm\sqrt{\tfrac{1}{2}a + \tfrac{1}{2}\sqrt{a^2 + b^2}}$ relative to the set $R - \{0\}$).

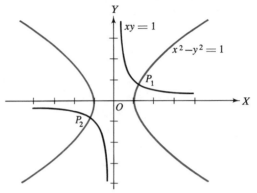

Figure 11.5

It is clear now that the solution set of $x^2 - a^2 = a \wedge 2xy = b$ is

$$\{(-a_0, -b_0), (a_0, b_0)\},$$

where

$$a_0 = \sqrt{\tfrac{1}{2}a + \tfrac{1}{2}\sqrt{a^2 + b^2}}, \; b_0 = \frac{b}{2a_0}.$$

Summarizing the foregoing discussion, we can now assert that the solution set of $z^2 = a + bi$ is

$$\{-\sqrt{a}, \sqrt{a}\}, \qquad\qquad \text{if } b = 0,$$

$$\{-a_0 - b_0 i, a_0 + b_0 i\}, \qquad \text{if } b \neq 0.$$

Example 4. Calculate the square roots of the complex number $1 + 2i$.

Solution: From Example 3, the solution set of $z^2 = 1 + 2i$ is the set

$$\{-a_0 - b_0 i, a_0 + b_0 i\}$$

where

$$a_0 = \sqrt{\tfrac{1}{2} \cdot 1 + \tfrac{1}{2}\sqrt{1^2 + 2^2}} = \sqrt{\tfrac{1}{2}(1 + \sqrt{5})}$$

and

$$b_0 = \frac{2}{2\sqrt{\tfrac{1}{2}(1 + \sqrt{5})}} = \frac{1}{\sqrt{\tfrac{1}{2}(1 + \sqrt{5})}}.$$

Thus the square roots of the number $1 + 2i$ are the numbers

$$-k_0 - \frac{1}{k_0}i, \quad k_0 + \frac{1}{k_0}i, \qquad \text{where } k_0 = \sqrt{\tfrac{1}{2}(1 + \sqrt{5})}.$$

Question: What are the coordinates of the points P_1 and P_2 in Figure 11.5?

Exercise 11.4

Study Example 1 before doing Exercises 1–4. In each of these, solve the given system graphically.

1. $\begin{cases} x + y = 0, \\ x^2 + y^2 = 8. \end{cases}$ 2. $\begin{cases} x - y = 0, \\ x^2 + 3y^2 = 4. \end{cases}$

3. $\begin{cases} x + y = 1, \\ x^2 + y^2 = 4. \end{cases}$ 4. $\begin{cases} 2x + y = 2, \\ -4x^2 + y^2 = 0. \end{cases}$

Study Example 2 carefully before working Exercises 5–14. In each of these, solve the given system algebraically and check your result graphically.

5. $\begin{cases} x + 7y = 7, \\ x + y^2 = -5. \end{cases}$ 6. $\begin{cases} x^2 + 3y^2 = 9, \\ x - y = 0. \end{cases}$

7. $\begin{cases} \frac{1}{4}(y - 1) = x^2, \\ 4x^2 + y^2 = 1. \end{cases}$ 8. $\begin{cases} x^2 - y = 2, \\ x + 2y = 1. \end{cases}$

9. $\begin{cases} x^2 - 4y = 4, \\ x^2 + y^2 = 1. \end{cases}$ 10. $\begin{cases} x^2 - y^2 = 4, \\ x + 2y = -3. \end{cases}$

11. $\begin{cases} x^2 + y^2 = 4, \\ (x - 2)^2 + y^2 = 4. \end{cases}$ 12. $\begin{cases} 4x^2 + 9y^2 = 36, \\ 2x - y = -4. \end{cases}$

13. $\begin{cases} 4x^2 + y^2 = 4, \\ x^2 + (y - 3)^2 = 9. \end{cases}$ 14. $\begin{cases} x^2 - 4y^2 = 9, \\ 3x + 4y^2 = -5. \end{cases}$

Study Examples 3 and 4 before doing Exercises 15–26. In each of these, find the two square roots of the given complex number.

15. $8 + 6i$.   16. $-5 + 12i$.
17. $4 - 3i$.   18. $2i$.
19. $1 - i$.   20. $8 - 6i$.
21. $2 + 3i$.   22. $-i/2$.
23. $a - ai$, with $a > 0$.   24. $5 + 12i$.
25. $(a + 1) + (a - 1)i$, with $a > 1$.
26. $-3 - 4i$.

27. Solve $(x + iy)(1 - 5i) = 2 + 5i$. [*Hint:* Two complex numbers are equal if and only if their real parts are equal and their imaginary parts are equal.]

28. The sum of the areas of two circles is 74π, while the sum of their circumferences is 24π. What are the radii of the circles?

29. The annual income from an investment is \$32. If the amount invested were \$200 more and the rate $\frac{1}{2}$ percent less, the annual income would be \$35. What are the amount and the rate of investment?

30. If a train were to increase its average speed between two towns by 12 mph, the time required would be 1 hr less. On the other hand, if it were to decrease its average speed by 3 mph, the time required would be $\frac{1}{3}$ hr more. What is its average speed, and how far apart are the towns?

31. The area of a rectangle is 144 sq in. and its perimeter is 50 in. What are its dimensions?

*32. At what points do tangent lines drawn from the point $(1, 2)$ intersect the circle $x^2 + y^2 = 1$? [*Hint:* A tangent line is perpendicular to the radius from the center to the point of tangency. Also see Example 1 of Section 10.8.]

**33. Prove that a line tangent to a circle is perpendicular to the radius at the point of tangency. [*Hint:* Let $x^2 + y^2 = a^2$ be an equation of the circle and consider the two cases: (1) the tangent line has an equation of the form $x = c$, or an equation of the form $y = d$; (2) the tangent line has an equation of the form $y = mx + b$, $m \neq 0$.]

*34. Prove that if $b \neq 0$, and $a, b \in R$,

$$R - \{0\} \quad \begin{array}{l} x^2 - \left(\dfrac{b}{2x}\right)^2 = a \\[2ex] x = \pm \sqrt{\tfrac{1}{2}a + \tfrac{1}{2}\sqrt{a^2 + b^2}}. \end{array}$$

35. Solve the equation $x^2 = -3$ over C.

11.5 GRAPHS OF INEQUALITIES

Suppose we are asked to sketch the graph of the inequality $3x + y < 1$. By definition, a point $P(x_0, y_0)$ is on the graph of $3x + y < 1$ if and only if $3x_0 + y_0 < 1$ —that is, if and only if (x_0, y_0) is a solution of the inequality $3x + y < 1$. Let L be the line with equation $3x + y = 1$ or, equivalently, $y = 1 - 3x$. Now we have the following equivalent statements:

$P(x_0, y_0)$ is on the graph of $3x + y < 1$.
$3x_0 + y_0 < 1$.
$y_0 < 1 - 3x_0$.

Since the point $(x_0, 1 - 3x_0)$ is on L, we conclude that $P(x_0, y_0)$ is on the graph of the given inequality if and only if $P(x_0, y_0)$ is below the line L. Thus the graph of $3x + y < 1$ consists of all those points which are below the line L (see Figure 11.6). The graph is pictured as a shaded region in the plane. That L is shown as a broken line indicates that points on L are not part of the graph of $3x + y < 1$.

Note that the line L divides the plane into three disjoint sets, namely:

(1) the line L, which is the graph of $3x + y = 1$,
(2) the region below L, which is the graph of $3x + y < 1$,
(3) the region above L, which is the graph of $3x + y > 1$.

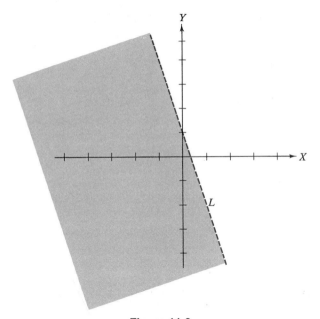

Figure 11.6

The student should realize that this observation ties in very closely with the law of trichotomy.

The next example depicts a more complicated situation.

Example 1. Sketch the graph of the system of inequalities.

$$\begin{cases} x > 1, \\ y < 4, \\ 2x - y \leqslant 4. \end{cases}$$

Solution: The given system is the open sentence $x > 1 \land y < 4$ $\land\ 2x - y \leqslant 4$, and its solution set is

$$\{(x, y) \,|\, x > 1\} \cap \{(x, y) \,|\, y < 4\} \cap \{(x, y) \,|\, 2x - y \leqslant 4\}.$$

Hence we obtain the desired graph by displaying those points which are common to the graphs of each of the inequalities $x > 1$, $y < 4$, $2x - y \leqslant 4$.

The graph of $x > 1$ consists of all those points which are located to the right of the line with equation $x = 1$. Further, the graph of $y < 4$ consists of all those points which are below the line with equation $y = 4$. Writing $2x - y \leqslant 4$ in the equivalent form $2x - 4 \leqslant y$, we observe that the graph of $2x - y \leqslant 4$ includes all those points which are on or above the line with equation $2x - 4 = y$. The graph of the given system is shown as a shaded region in Figure 11.7.

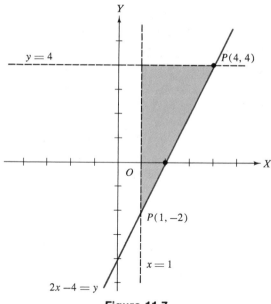

Figure 11.7

Question: Are the points $P(1, -2)$ and $P(4, 4)$ on the graph of the given system of inequalities?

So far in this section we have treated graphs whose boundaries are straight lines. That other types of curves may serve as boundaries of graphs is illustrated in the next example.

Example 2. Sketch the graph of the system of inequalities

$$\begin{cases} x^2 + y^2 < 16, \\ y > x^2 - 1. \end{cases}$$

Solution: The graph of $x^2 + y^2 = 16$ is a circle with center at the origin and radius 4. The following statements are equivalent:

(1) $\quad x_0^2 + y_0^2 < 16.$
(2) $\quad \sqrt{x_0^2 + y_0^2} < 4.$
(3) The distance of the point $P(x_0, y_0)$ from the origin is less than 4 units.

Hence the graph of $x^2 + y^2 < 16$ consists of all points that are interior to the circle with equation $x^2 + y^2 = 16$. (See Figure 11.8.) Note that the circumference of the circle is not part of the graph.

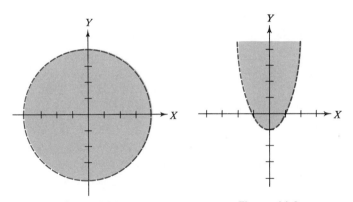

Figure 11.8   **Figure 11.9**

The graph of the equation $y = x^2 - 1$ is a parabola extending upward. If $P(x_0, y_0)$ is a point and $P(x_0, u_0)$ is a point on the parabola, then $u_0 = x_0^2 - 1$, hence $P(x_0, y_0)$ is on the graph of $y > x^2 - 1$ if and only if $y_0 > u_0$. Thus the graph of $y > x^2 - 1$ contains all those points which are interior to the parabola $y = x^2 - 1$. (See Figure 11.9.) Note that the parabola itself is not part of the graph.

The graph of

$$\begin{cases} x^2 + y^2 < 16, \\ y > x^2 - 1, \end{cases}$$

consists of those points which are common to the graphs of $x^2 + y^2 < 16$, and $y > x^2 - 1$. (See Figure 11.10.)

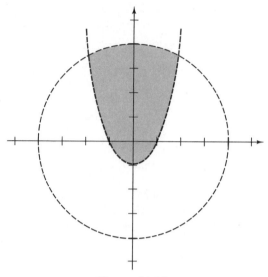

Figure 11.10

Exercise 11.5

Read Example 1 before doing Exercises 1–5. In each of these, sketch the graph of the system of inequalities.

1. $\begin{cases} x > 2, \\ y > 1, \\ 3x - 2y < 2. \end{cases}$

2. $\begin{cases} x \geqslant 3, \\ x + y < 2, \\ 3x - 2y > 1. \end{cases}$

3. $\begin{cases} x - y \geqslant -1, \\ x + 2y > -2, \\ x - 3y > -6, \\ x < 4. \end{cases}$

4. $\begin{cases} x < 3, \\ y > 2, \\ y < 4, \\ -x + y < 6. \end{cases}$

5. $\begin{cases} x - 3y < 2, \\ 5x + 6y > 1, \\ 6x - 4y \leqslant 6. \end{cases}$

Read Example 2 before working Exercises 6–11. In each of these, sketch the graph of the given system of inequalities.

6. $\begin{cases} x^2 + y^2 \le 25, \\ y + x > 2. \end{cases}$ **7.** $\begin{cases} 4x^2 + 9y^2 < 36, \\ x^2 - y < 1. \end{cases}$

8. $\begin{cases} y + x^2 < 2, \\ 2x^2 - y < 4. \end{cases}$ **9.** $\begin{cases} x^2 + y^2 < 36, \\ (x - 4)^2 + y^2 < 9, \\ x - 2y < 5. \end{cases}$

*10. $\begin{cases} y \le \sqrt{4 - x^2}, \\ y > -2. \end{cases}$ ***11.** $\begin{cases} x < \sqrt{9 - y^2}, \\ x < 1. \end{cases}$

In Exercises 12–15, sketch the graph of the given system.

12. $\begin{cases} x^2 + y^2 < 4, \\ x^2 + y^2 \ge 1. \end{cases}$ **13.** $\begin{cases} x^2 + y^2 \le 4, \\ x^2 + y^2 \ge 1. \end{cases}$

14. $\begin{cases} |x| \le 1, \\ |y| \le 1. \end{cases}$ **15.** $\begin{cases} |y| < |x|, \\ |x| < 1. \end{cases}$

11.6 LINEAR PROGRAMMING (OPTIONAL)

Many problems, whose solutions are of practical importance, are closely related to problems involving systems of linear inequalities. The following example serves as an illustration.

Example 1. A farmer has 200 acres on which he may raise peanuts and corn. He has accepted orders requiring 10 acres of peanuts and 5 acres of corn. Moreover, he must follow a regulation that the acreage for corn must be at least twice the acreage for peanuts. If the profit from corn and peanuts is $100 per acre and $200 per acre, respectively, how many acres of each crop will give him the greatest profit?

Solution: Let x and y denote the acreage allocated to peanuts and corn, respectively. We have the following system of inequalities to begin with:

$\begin{cases} x + y \le 200, \\ \\ 10 \le x, \\ 5 \le y, \\ \\ 2x \le y, \end{cases}$

since the farmer cannot use more than 200 acres for crops;

since at least 10 acres must be allowed for peanuts and at least 5 acres must be allowed for corn;

since the acreage for corn must be at least twice the acreage for peanuts.

The graph of this system of inequalities can be readily obtained by the methods of Section 11.5. It is shown in Figure 11.11 as a shaded area of a polygon. Solving the linear systems

$$\begin{cases} x = 10, \\ y = 2x, \end{cases} \qquad \begin{cases} y = 2x, \\ x + y = 200, \end{cases} \qquad \begin{cases} x = 10, \\ x + y = 200, \end{cases}$$

we find that the vertices of the polygon have the coordinates $(10, 20)$, $\left(\frac{200}{3}, \frac{400}{3}\right)$, and $(10, 190)$.

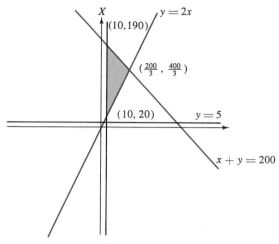

Figure 11.11

Let $z = 200x + 100y$ — that is, z is the farmer's profit on peanuts and corn. The problem will be solved if we can answer the question: which point $P(x_0, y_0)$ in the shaded polygonal region has the property that, given any other point $P(x, y)$ in the same region, $200x_0 + 100y_0 \geqslant 200x + 100y$?

At first the situation may appear rather hopeless. How are we to know that there is such a point — and, if it exists, how can we find it among the infinitely many points in the shaded region? This is where the theory of a branch of mathematics called *linear* *linear programming* comes to our rescue. Applied to our par-*programming* ticular problem, this theory asserts that z attains its maximum value at the coordinates of one of the vertices of the polygon. Since there are only three vertices to check, the rest is now routine. Choosing $(x, y) = (10, 190)$, we find that $z = 200 \cdot 10 + 100 \cdot 190 = 21{,}000$. Similarly, if $(x, y) = \left(\frac{200}{3}, \frac{400}{3}\right)$, $z = \frac{80{,}000}{3}$, and if $(x, y) = (10, 20)$, $z = 4000$. Since among the numbers $21{,}000$, $\frac{80{,}000}{3}$, and 4000, the number $\frac{80{,}000}{3}$ is the greatest, we conclude that the ordered pair $(\frac{200}{3}, \frac{400}{3})$ has the desired property, namely that

$$200 \cdot \tfrac{200}{3} + 100 \cdot \tfrac{400}{3} \geq 200x + 100y$$

where (x, y) are coordinates of any point in the shaded polygonal region. Therefore, to attain the maximum profit the farmer should plant $66\frac{2}{3}$ acres of peanuts and $133\frac{1}{3}$ acres of corn.

It now seems appropriate to discuss some basic concepts of linear programming. Let us start by inquiring what the characteristics were of the problem we just solved. We had a set of linear inequalities involving variables (such inequalities are some- *constraints* times called *constraints*), and we wished to maximize a linear expression, subject to the given constraints. The following concepts will help us understand the techniques and reasoning used in dealing with problems of this type.

convex **Definition.** If S is a set of points in a plane, S is said to be *convex* if, whenever, P and Q are points in S, then the line segment from P to Q is also in S.

Figure 11.12 displays two sets of points, where one is convex and the other is not.

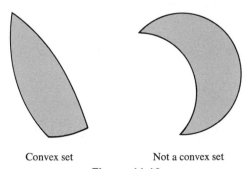

Convex set   Not a convex set

Figure 11.12

If L is a line in a plane, then L determines two plane regions *half planes* H_1 and H_2, called *half planes*, such that the plane is the union of H_1, H_2, and L, and H_1, H_2, L are three mutually disjoint sets *closed half* (see Figure 11.13). The sets $H_1 \cup L$, $H_2 \cup L$ are called *closed planes half planes*.

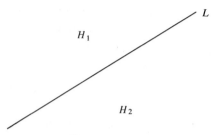

Figure 11.13

We state here without proof the fact that each of the sets H_1, H_2, $H_1 \cup L$, $H_2 \cup L$ is a convex set. It can be shown that the intersection of any number of convex sets is a convex set. Hence, in particular, the intersection of closed half planes is a convex set.

Question: Why is the shaded region in Figure 11.11 a convex set?
If S is a nonempty plane region such that
(1) S is the intersection of a finite number of closed half planes,
bounded   (2) S is *bounded* (that is, there is a positive number r such that whenever a point $P(x, y)$ is in S, then $x^2 + y^2 < r$),
convex polygonal then S is called a *convex polygonal region*. For example, the
region shaded region in Figure 11.11 is a convex polygonal region.
In a typical linear programming problem in two variables, we have a convex polygonal region S and a polynomial expression $f(x, y) = ax + by + c$ and we wish to determine the points
maximum $P(x_0, y_0)$ in S such that f attains its *maximum or minimum values*
minimum at (x_0, y_0) — that is, points $P(x_0, y_0)$ such that

$$ax_0 + by_0 + c \geq ax + by + c$$

for all points $P(x, y)$ in S or

$$ax_0 + by_0 + c \leq ax + by + c$$

for all points $P(x, y)$ in S. That some of the vertices of a convex polygonal region must have the desired property is a consequence of the following theorem.

Theorem. If $P(x_1, y_1)$ and $P(x_2, y_2)$ are distinct points in a plane and $ax + by + c$ is a linear expression such that $ax_2 + by_2 + c \geq ax_1 + by_1 + c$, then for every point $P(x, y)$ on the line segment joining $P(x_1, y_1)$ and $P(x_2, y_2)$,

$$ax_1 + by_1 + c \leq ax + by + c \leq ax_2 + by_2 + c.$$

Further questions related to linear programming will be taken up in the problem section. We conclude with the remark that the general linear programming problem does not require a geometric interpretation, nor is it limited to two or three variables.

Exercise 11.6

1. (a) Graph the convex polygonal region $S = \{(x, y) \,|\, x \leq y \land x \geq 0 \land y \leq 6\}$.

(b) If $Q(x, y) = 3x - 2y$, determine points in S at which Q attains its maximum or minimum values on S. What are the maximum and minimum values of Q on S?

2. (a) Graph the region $T = \{(x, y) \,|\, 0 \leq 2x + y < 4 \land 0 \leq y - x \leq 4\}$.

(b) If $P(x, y) = 3x - y + 5$, determine points in T at which P attains its maximum or minimum values on T. What are the maximum and minimum values of P on T?

Read carefully Example 1 in this section before doing Exercises 3–8.

3. A farmer has 200 acres available for planting with crops A and B. Because of existing contracts, he must plant at least 10 acres of crop A and 5 acres of crop B. Furthermore, the acreage for B must be at least twice the acreage for A. If the profits from crops A and B are \$40 and \$30 per acre, respectively, how many acres of each crop will give him the greatest profit? What is his greatest profit?

4. A manufacturer can produce two types of refrigerators, called models A and B. Model A can be sold for \$70 each and model B for \$60 each. After production the items are shipped in two stages, first by truck and then by train. Train rates are \$3 for each model A and \$2 for each model B refrigerator. Truck rates are \$5 for each model A and \$4 for each model B. The factory accountant insists that the total shipping charges by train must not exceed \$40. The same limit is imposed on total shipping charges by truck. How much of each item should be produced to maximize gross sales?

5. A farmer has 200 acres available for planting with crops A, B, and C. The profits from crops A and B are \$40 and \$30 per acre, respectively. He cannot plant more than 100 acres of A or 120 acres of C. How many acres should he assign to each crop in order to maximize his profit, if the profit per acre on C is \$25?

6. A canned-food manufacturer wishes to include three ingredients, say foods A, B, C, in his recipe. Foods A, B, C cost 80, 30, and 60 cents per pound, respectively. To make the final product tasty, at least 25 percent but not more than 60 percent should be A. For a balanced diet at least 10 percent should be food C and the amount of food C should not exceed "three times the amount of B plus the amount of A"; the amount of B must not exceed twice the amount of A. If no other foods are to be included in this recipe, find the amounts of each food to be used for producing 100 lb of canned food so that the manufacturer's cost is minimized.

7. An investor wishes to buy 100 shares of common stocks. His broker recommends stocks A and B and suggests that in order to reduce the risks of a fluctuating market at least 30 shares of stock A and not more than 70 shares of stock B should be bought. Stock A pays an annual dividend of 80 cents per share and stock B pays an annual dividend of $1 per share. If the investor wishes to follow his broker's advice, how many shares of each stock should he buy to maximize his income from stock dividends?

8. An owner of a gasoline station is committed to buy 800 gal of gasoline per day from his supplier. On a certain day he estimates that his regular customers will buy at least 300 gal of regular gasoline and at least 100 gal of premium gasoline. He also knows from experience that he can never sell more than four times as much regular gasoline than premium gasoline. If his profits from regular gasoline and premium gasoline are 3 cents and 1 cent per gallon, respectively, how many gallons of each kind of gasoline should he buy on that particular day to maximize his profit?

9. Give an example of a convex set that is the union of two nonconvex sets.

10. Give an example of a convex set that is the intersection of two nonconvex sets.

11. Show that the region $T = \{(x, y) \mid y \leqslant 2x \wedge y \geqslant -x + 1 \wedge x \leqslant 1\}$ is a convex polygonal region by displaying three convex sets H_1, H_2, H_3 such that $T = H_1 \cap H_2 \cap H_3$ and a positive real number r such that whenever $(x, y) \in T$, then $x^2 + y^2 < r$.

*12. Let $P_1(x_1, y_1)$, $P_2(x_2, y_2)$, $P_3(x_3, y_3)$ be vertices of a triangle. If $Q(x, y) = ax + by + c$ attains its maximum value on the boundary of the triangle at (x_1, y_1), show that for any point $P(x, y)$ interior to the triangle, $Q(x, y) \leqslant Q(x_1, y_1)$. [*Hint:* Use the theorem in Section 11.6.]

REVIEW EXERCISES

In Exercises 1–4 sketch the graph of each equation and solve each system algebraically.

1. $\begin{cases} 2x + y = -1, \\ x + y = 0. \end{cases}$ **2.** $\begin{cases} x + y = 2, \\ -2x - 2y = 1. \end{cases}$

3. $\begin{cases} 3x + y = 5, \\ x + \frac{1}{3}y = \frac{5}{3}. \end{cases}$ **4.** $\begin{cases} x - 5y = 0, \\ 3x + 2y = 34. \end{cases}$

5. (a) Evaluate the determinant $\begin{vmatrix} 2 & 2 \\ 1 & 1 \end{vmatrix}$.

(b) Explain why Cramer's rule cannot be used to solve the system

$$\begin{cases} 2x + 2y = 3, \\ x + y = 1. \end{cases}$$

In Exercises 6–9 solve each system algebraically.

6. $\begin{cases} x + y + z = 6, \\ x - y + 2z = 3, \\ 2x - y - z = 3. \end{cases}$ **7.** $\begin{cases} 2x + 3y + 3z = 3, \\ 4x + 6y + 3z = 5, \\ x - y - z = -\frac{1}{6}. \end{cases}$

8. $\begin{cases} 2x + y + z = 8, \\ x - y + z = 4, \\ x + 2y = 4. \end{cases}$ **9.** $\begin{cases} 2x + y + z = 7, \\ x - y + z = 4, \\ 4x - y + 3z = 11. \end{cases}$

In Exercises 10–13 sketch the graph of each equation and solve each system algebraically.

10. $\begin{cases} x^2 + y^2 = 4, \\ x - y = 0. \end{cases}$ **11.** $\begin{cases} 4x^2 + y^2 = 4, \\ x - 2y = 0. \end{cases}$

12. $\begin{cases} y^2 = x + 2, \\ x^2 + y^2 = 4. \end{cases}$ **13.** $\begin{cases} y - x^2 + 4 = 0, \\ y + x^2 - 4 = 0. \end{cases}$

In Exercises 14–19 sketch the graph of each system of inequalities.

14. $\begin{cases} y - x - 1 \geqslant 0, \\ y + x \geqslant 1. \end{cases}$ **15.** $\begin{cases} x^2 + y^2 < 4, \\ x - y < 0. \end{cases}$

16. $\begin{cases} y - x - 1 \geqslant 0, \\ y + x \geqslant 1, \\ y \leqslant 2. \end{cases}$ **17.** $\begin{cases} y + 2x > 0, \\ y + x - 2 < 0. \end{cases}$

18. $\begin{cases} y - x^2 + 4 > 0, \\ y + x^2 - 4 < 0. \end{cases}$ **19.** $\begin{cases} y^2 \leqslant x + 2, \\ x^2 + y^2 < 9. \end{cases}$

20. Solve $x^2 = -5 + 12i$ over C.

21. Solve $x^2 = 3 - 4i$ over C.

22. Show that the open sentences

$$x^2 - \left(\frac{b}{2x}\right)^2 = a, \qquad x = \pm \sqrt{\tfrac{1}{2}a + \tfrac{1}{2}\sqrt{a^2 + b^2}}$$

are equivalent relative to the replacement set $R - \{0\}$, if $\{a, b\} \subseteq R$ and $b \neq 0$.

23. Solve the equation $x^2 + (4 + 3i)x + (7 + i) = 0$ over C. [*Hint:* Find a square root of the discriminant.]

24. (a) Show that the region

$$T = \{(x, y) \mid y - x - 1 \leqslant 0 \wedge y + x \geqslant 1 \wedge y \leqslant 2\}$$

is a convex polygonal region by displaying three convex sets H_1, H_2, H_3 such that $T = H_1 \cap H_2 \cap H_3$ and a positive real number r such that whenever $(x, y) \in T$, then $x^2 + y^2 < r$.

(b) If $f(x, y) = -5x + 2y + 3$, find points in T at which f attains its maximum or minimum values on T. What are the maximum or minimum values of f on T?

FUNCTIONS
AND
GRAPHS

○

12.1 INTRODUCTION

Almost everyone is familiar with the meaning of the word function. For example, the expression "the amount of knowledge you will get from this course is a function of the amount of time you devote to it" would be clearly understood by most students. Engineers, physicists, and chemists often use the word function in this sense. That is, there are two variable quantities, so related that the way one of them varies depends on the way the other varies.

This concept is so common in the physical and social sciences that it has become necessary to make it precise. The best way to do this is to define it in terms of mathematical concepts that have already been introduced.

Let us recall that whenever we have a variable, we must have a replacement set for it. Hence, since we want to describe a correspondence between two variables, we should consider two replacement sets.

function   A *function* is a rule that assigns to each member of a set S one and only one member of a set T. The set S is called the
domain *domain* of the function.

In general we use letters such as f, h, g, and so on to denote a function. If f is the symbol that denotes a function as we just described, and $a \in S$, then $f(a)$ (read "f of a") denotes the unique element of T that corresponds to a. (See Figure 12.1.)

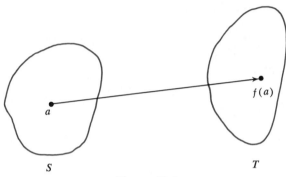

Figure 12.1

If f is a function and D_f is its domain, the set $\{f(x) \mid x \in D_f\}$
range is called the *range* of the function f and is often denoted by R_f.

There are several common ways of describing functions. The most natural way perhaps, but the least used, is to describe the two sets S and T and to state the rule that associates with each member of S one and only one member of T.

A second way is to use a table that at once describes both sets and the rule. For example:

x	1	2	3	4	5	6
$f(x)$	0	-2	π	3	0	-1

describes a function f with domain the set $\{1, 2, 3, 4, 5, 6\}$ and range the set $\{-2, -1, 0, 3, \pi\}$. For $a \in D_f$, $f(a)$ is the number written below a in the table. For example, $f(2)$ is -2, $f(3)$ is π, and so on.

A third way, which is in some sense closely related to the second way, is to describe a function f as a set of ordered pairs. Then the set that consists of all the first elements of these pairs is the domain, the set of all second elements is the range, and the rule is that if b is in the domain of the function f, then $f(b)$ is c if and only if (b, c) is a member of the given set.

If we analyze this situation carefully, we see that to find $f(b)$ we search for an ordered pair in the given set that has b as its first element. Then we conclude that the second element of that ordered pair is $f(b)$. Since we know that for b in the domain of f, $f(b)$ is unique, it means that we cannot find two distinct ordered pairs in the given set that have b as their first element.

For example, the set

$$\{(1,3),\ (5,2),\ (7,-1),\ (8,2)\}$$

describes a function, since no two distinct members of that set have the same first element. On the other hand, the set $\{(3,1),\ (6,2),\ (3,7)\}$ does not describe a function, since $(3,1)$ and $(3,7)$ are distinct and have the same first element.

Although we have described a function as a rule that associates with each member of one set a unique member of another set, we choose to use our third way of describing a function as the formal definition.

function **Definition.** A *function* is a set of ordered pairs, no two distinct members of which have the same first element.

Thus, if we are given a set of ordered pairs and we are asked to verify that it is a function, we consider (a, b) and (a, c) to be members of that set and we show that these two ordered pairs cannot be distinct (otherwise the definition would be violated). That is, we show that $b = c$.

On the other hand, if we want to show that a set of ordered pairs is not a function, it is sufficient to find two of its members that are distinct and have the same first element.

Example 1. Show that the set $\{(x, y) \mid x \in R \wedge y = x^3 + 1\}$ is a function.

Solution: Each member of the given set is an ordered pair. Thus we need only verify that if two of its members have the same first element, they cannot be distinct.

For convenience, let f denote our set and suppose that $(a, b) \in f \wedge (a, c) \in f$ is true. We must show that (a, b) and (a, c) are not distinct. That is, we must show that $b = c$.

We proceed as follows:

$$(a, b) \in f \quad \wedge \quad (a, c) \in f.$$

$$(a \in R \wedge b = a^3 + 1) \quad \wedge \quad (a \in R \wedge c = a^3 + 1).$$

Therefore, $b = c$, since b and c are both equal to $a^3 + 1$.

Example 2. Show that the set $\{(x, y) \mid x = y^2\}$ is not a function.

Solution: The given set is a set of ordered pairs. To show that it is not a function we need only verify that there are two distinct members of that set with the same first element. With a little practice the student will often be able to perform this task by inspection. For example, it is easy to see that $(1, -1)$ and $(1, 1)$ both belong to $\{(x, y) \mid x = y^2\}$, since $1 = (-1)^2 \land 1 = 1^2$ is true.

Example 3. Show that the set $\{(1, 2), (2, 4), 3\}$ is not a function.

Solution: The given set is not a set of ordered pairs, since 3 is one of its elements and is not an ordered pair. Hence, the given set is not a function.

Example 4. If f denotes the function described in Example 1, find $f(0), f(1), f(3), f(-2)$.

Solution: In general,

$$
\left. \begin{array}{l}
b = f(a). \\[2mm]
(a, b) \in f. \\[2mm]
a \in R \land b = a^3 + 1.
\end{array} \right.
$$

Therefore,

$$f(a) = a^3 + 1.$$

In particular,

$$f(0) = 0^3 + 1 = 0 + 1 = 1,$$

$$f(1) = 1^3 + 1 = 1 + 1 = 2,$$

$$f(3) = 3^3 + 1 = 27 + 1 = 28,$$

and

$$f(-2) = (-2)^3 + 1 = -8 + 1 = -7.$$

We now give an example to illustrate that the domain and range of a function need not be sets of real numbers.

Example 5. Let $F = \{(x, y) \mid x$ is a living American and y is x's father$\}$. Then F is a function, since each living American has one and only one father.

Exercise 12.1

Read the first three pages of Section 12.1 before doing Exercises 1–10. In each of Exercises 1–8, a table or a set of ordered pairs is given. Determine in each case whether or not a function is described.

1. $\{(1,5), (2,6), (3,-2)\}$.
2. $\{(0,5), (-4,1), (8,2), (0,2)\}$.
3. $\{(-5,2), (6,1), (8,-2), (6,7)\}$.
4. $\{(5,1), (6,8), (9,4), a, (-3,2)\}$.
5. $\{(9,1), (0,5), (6,7)\}$.

6.

x	1	2	3	4	5
$f(x)$	-1	3	4	1	2

7.

t	-1	-3	4	-1	2
$g(t)$	0	5	6	1	4

8.

z	i	$3+i$	$4-i$
$f(z)$	-1	-3	$3-4i$

9. For each of the sets described in Exercises 1–5 that is a function, find its domain and range.

10. Repeat Exercise 9 for each of the tables given in Exercises 6–8.

Read Example 1 carefully before doing Exercises 11–15. In each of these, show that the given set is a function.

11. $\{(x, y) \mid y = 3x - 1\}$. 12. $\{(x, y) \mid y = -4x + 2\}$.
13. $\{(x, y) \mid y = x^2 - 5x + 2\}$.
14. $\{(r, s) \mid s = r^3 - 3r^2 - 2r + 1\}$.
15. $\{(u, v) \mid v = -u^4 + 2u^2 + 5\}$.

Read Examples 2 and 3 before doing Exercises 16–22. In each of these show that the given set is not a function.

16. $\{(x, y) \mid x = y^4\}$. 17. $\{(x, y) \mid x^2 = y^2\}$.
18. $\{(u, v) \mid u = v^6\}$. 19. $\{(r, s) \mid r = s^2 - 1\}$.
20. $\{(z, u) \mid z = u^4 - 32\}$.
21. $\{(1,2), (3,5), (4,6), \pi, (1,7)\}$.
22. $\{(1,3), (-4,2), b, (1,6)\}$.

Read Example 4 before doing Exercises 23–27.

23. If $f = \{(x, y) \mid y = 3x + 2\}$, find $f(-2), f(3), f(0), f(\pi)$.
24. If $g = \{(u, v) \mid v = u^2 + 2u - 1\}$,
    find $g(0), g(-2), g(-4), g(a), g(a+2)$.
25. If $H = \{(x, y) \mid y = 3x + 2\}$,
    find $H(5), H(3), H(5) - H(3), [H(5) - H(3)]/2$.
26. If H is as in Exercise 25 above and $b \neq 0$, find

$$\frac{H(a+b) - H(a)}{b}.$$

27. Let f be the function $f(x) = 3x + 1$.

(a) Calculate $\dfrac{f(x+h) - f(x)}{h}$.

(b) Give a geometric interpretation of the result you obtained in part (a).

12.2 FUNCTIONS DESCRIBED BY EQUATIONS

In the preceding section we described a function as a rule that assigns to each element of some set a unique element of another set.

We indicated that a convenient way of describing such a relation between two sets was through the use of ordered pairs. In fact, a function was defined formally as a set of ordered pairs in which any two members having the same first element cannot be distinct.

An example of a function was the set

$$\{(x, y) \mid x \in R \wedge y = x^3 + 1\}.$$

Note that in the foregoing description of the function the equation $y = x^3 + 1$ is given.

It is often the case that, to save space, scientists and mathematicians will give only an equation when they wish to describe a function. For example, the equations

$$y = x^3 + 1 \qquad \text{or} \qquad f(x) = x^3 + 1$$

could have been used to describe the function

$$\{(x, y) \mid x \in R \wedge y = x^3 + 1\}.$$

The intended meaning is usually understood from context. Thus, we must make a few statements concerning the common practice of describing functions by equations.

(1) Unless otherwise specified, the domain of a function is always the largest set of real numbers that can be used so that the range is also a set of real numbers.

For example, if

$$f(x) = \frac{1}{x - 2}$$

and

$$g(x) = \sqrt{5 - x},$$

then the domain of f is $R - \{2\}$ and the domain of g is $\{x \mid 5 - x \geqslant 0\}$. In the first case we had to exclude 2 from the domain of f, since x cannot be replaced by 2 in $\dfrac{1}{x - 2}$ (division by zero is not defined). In the second case, we needed to require $5 - x \geqslant 0$, for otherwise $\sqrt{5 - x}$ does not represent a real number.

    (2) The equation $y = 2x + 3$ could describe the function $\{(x, y) \mid y = 2x + 3\}$ or the function $\{(y, x) \mid y = 2x + 3\}$. As we shall soon see, these two functions are related. However, they are different functions.

    To clarify the situation, we need to introduce two more terms. If an equation in two variables describes a function, the variable whose replacement set is the domain of the function is called the *independent variable* and the one whose replacement set contains the range of the function is called the *dependent variable*.

independent variable
dependent variable

    Now, we agree that unless otherwise specified, the variable that is isolated on one side of the equation is the dependent variable. For example, if

$$u = 2t + 3$$

defines a function, then it is the function $\{(t, u) \mid u = 2t + 3\}$ rather than $\{(u, t) \mid u = 2t + 3\}$, unless otherwise indicated. Note that the independent variable appears first and the dependent variable second in the ordered pair (t, u). This is natural, since the replacement set for t is the domain of the function, while the replacement set for u (the dependent variable) is the range of the function. In the case where no variable is isolated on one side of the equation, as in $2x + 3y = 5$, for example, we must be told which variable is the dependent one. However, it is customary (but not necessary) to use x for the independent variable and y for the dependent variable.

Example 1. If $y = \dfrac{1}{(x - 1)(x - 3)}$ describes a function f, then from context we conclude the following:

    (1) $D_f = R - \{1, 3\}$, since division by zero is not defined,
    (2) x is the independent variable and y is the dependent variable,
    (3) $f = \left\{(x, y) \,\middle|\, y = \dfrac{1}{(x - 1)(x - 3)} \wedge x \in R - \{1, 3\}\right\}$.

Example 2. If the function g is defined by the equation

$$t = \sqrt{5v + 3},$$

then from context we claim that

(1) v is the independent variable and t is the dependent variable,

(2) $D_g = \{v \mid 5v + 3 \geqslant 0 \wedge v \in R\}$,

(3) $g = \{(v, t) \mid t = \sqrt{5v + 3} \wedge v \in R \wedge 5v + 3 \geqslant 0\}$.

Example 3. If f is the function described in Example 1, find $f(2)$ and $f(0)$.

Solution: Since

$$f(x) = \frac{1}{(x - 1)(x - 3)},$$

then

$$f(2) = \frac{1}{(2 - 1)(2 - 3)} = -1,$$

and

$$f(0) = \frac{1}{(0 - 1)(0 - 3)} = \frac{1}{3}.$$

Exercise 12.2

Read Examples 1 and 2 before doing Exercises 1–10. In each of these, an equation that describes a function f is given. In each case do the following: (a) describe the domain of f, (b) name the independent and dependent variables, (c) describe f as a set of ordered pairs.

1. $y = \dfrac{1}{x - 1}$.

2. $y = \dfrac{x}{(x - 1)(x - 4)}$.

3. $y = \dfrac{x + 1}{(x^2 + 1)(x - 2)}$.

4. $y = \dfrac{x + 3}{(x - 4)(x + 6)(x + 8)}$.

5. $r = \dfrac{2s + 7}{(s + 1)(s - 2)}$.

6. $t = \dfrac{5u}{(u + 2)(u - 1)}$.

7. $z = \sqrt{5 - u}$.

8. $v = \sqrt{3 + u^2}$.

9. $t = \dfrac{1}{\sqrt{u - 7}}$.

10. $w = x^2 + \dfrac{1}{x}$.

Read Example 3 before doing Exercises 11–17.

11. If f is the function of Exercise 1, find $f(3)$, $f(-4)$.

12. If f is the function of Exercise 2, find $f(0)$, $f(6)$.

13. If f is the function of Exercise 3, find $f(4)$, $f(-1)$.

14. If f is the function of Exercise 4, find $f(6)$, $f(-2)$.

15. If f is the function of Exercise 5, find $f(a)$, $f(a+h)$.

16. If f is the function of Exercise 6, find $f(2) - f(0)$.

17. If f is the function of Exercise 7, find $f(1)f(4)$.

18. If $g(x) = x^2 + 1$, calculate $\dfrac{1}{h}g(x+h)$.

19. For the function g defined by $g(x) = x + \dfrac{1}{x}$ find a formula that corresponds to (a) $g\left(\dfrac{1}{x}\right)$, (b) $g\left(x - \dfrac{1}{x}\right)$, (c) $x^2 g(x) - g(1)$.

20. Express the volume of a cube as a function of its surface area.

21. The volume and surface area of a sphere are related to the radius by the equations $V = \frac{4}{3}\pi r^3$ and $S = 4\pi r^2$, respectively. Find a function f such that $V = f(S)$.

*22. A linear relationship exists between centigrade temperature C and Fahrenheit temperature F. Given that $F = 32°$ when $C = 0°$ and $F = 212°$ when $C = 100°$, derive a formula relating the two temperature scales. At what temperature would the Fahrenheit reading agree with the centigrade reading?

12.3 GRAPHS OF FUNCTIONS

In the preceding sections we defined a function as a set of ordered pairs, no two distinct members of which have the same first element.

Henceforth, unless otherwise specified, the domain and range of each function under consideration will be subsets of the set of real numbers.

Recall also that to each ordered pair of real numbers corresponds a unique point in a Cartesian coordinate plane. Hence, if f is a function and E is a Cartesian coordinate plane, then we may consider the following set of points in E:

$$\{P(a, b) \mid (a, b) \in f\}.$$

graph This set is called the *graph* of the function f.

Example 1. If $g = \{(1, 2), (3, -1), (5, 4)\}$, the graph of g consists of the three points plotted in Figure 12.2.

If a function f is defined by an equation (in two variables), then $(a, b) \in f$ if and only if the equation becomes a true proposition when the independent and dependent variables are replaced respectively by a and b. Thus, in this case, the graph of the function is identical with the graph of the equation.

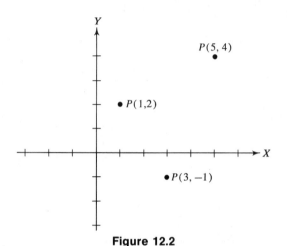

Figure 12.2

Example 2. Let $f = \{(x, y) \mid y = 2x + 1\}$. We have seen that the graph of an equation of the form $y = mx + b$ is a straight line. Note that $(0, 1) \in f$ and $(2, 5) \in f$, since $1 = 2 \cdot 0 + 1$ and $5 = 2 \cdot 2 + 1$ are both true propositions. Thus, $P(0, 1)$ and $P(2, 5)$ belong to the graph of f. Now we obtain the graph by drawing a straight line through these two points. (See Figure 12.3.)

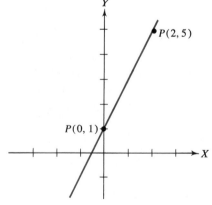

Figure 12.3

Example 3. The function h is defined by $h(x) = -x^3 + 4$. Sketch its graph.

Solution: We see that h has infinitely many members, since for each $x \in R$ there is an ordered pair $(x, h(x)) \in h$. In order to sketch an approximation to the graph of h, we replace x successively by several real numbers and we find the corresponding values for $h(x)$. We tabulate the results below.

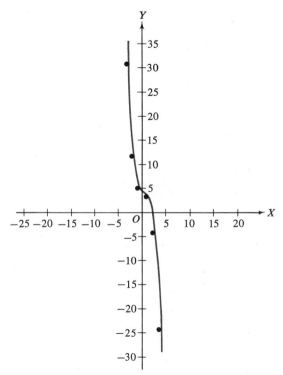

Figure 12.4

x	$h(x)$	$(x, h(x))$
-3	31	$(-3, 31)$
-2	12	$(-2, 12)$
-1	5	$(-1, 5)$
0	4	$(0, 4)$
1	3	$(1, 3)$
2	-4	$(2, -4)$
3	-23	$(3, -23)$

We now plot the points whose coordinates are the ordered pairs that appear in the third column of the table. Next we sketch a smooth curve passing through these points. (See Figure 12.4.)

Example 4. The function f is defined as follows:

$$f(x) = \begin{cases} -1 & \text{if } x < 0, \\ 1 & \text{if } x \geq 0. \end{cases}$$

Sketch its graph.

Solution: The graph is shown in Figure 12.5.

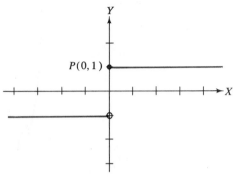

Figure 12.5

Example 5. Sketch the graph of the function g defined as follows:

$$g = \{(x, y) \mid xy = 1\}.$$

Solution: The graph of g is identical with the graph of the equation $xy = 1$. We have seen in Section 10.7 that the graph of this equation is symmetric with respect to the origin and also with respect to the graph of $y = x$. Hence we first consider the following table.

x	$y = \dfrac{1}{x}$	(x, y)
1	1	$(1, 1)$
2	$\frac{1}{2}$	$(2, \frac{1}{2})$
3	$\frac{1}{3}$	$(3, \frac{1}{3})$
4	$\frac{1}{4}$	$(4, \frac{1}{4})$

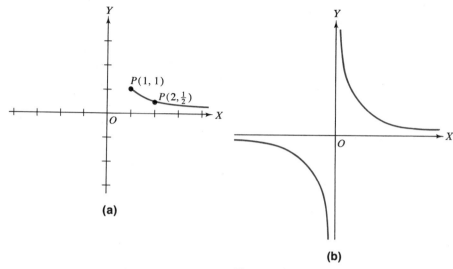

Figure 12.6

Then we plot the points whose coordinates appear in the third column of the table and we draw a smooth curve through these points. (See Figure 12.6(a).) We obtain the other part of the graph by using the symmetric properties of the graph. (See Figure 12.6(b).)

Exercise 12.3

Read Example 1 before doing Exercises 1–5. In each of these, a function is given. Sketch its graph.

1. $f = \{(1, 3),\ (2, 5),\ (7, 2)\}$.
2. $f = \{(-3, 2),\ (0, 5),\ (2, 7)\}$.
3. $g = \{(-5, 0),\ (-3, 1),\ (-2, 1),\ (0, 1)\}$.
4. $h = \{(-4, 1),\ (-3, 1),\ (-2, 1),\ (-1, 1),\ (0, 1),\ (1, 1)\}$.
5. $g = \{(1, 2)\}$.

Read Example 2 before doing Exercises 6–10. In each of these, a function is given. Sketch its graph.

6. $f = \{(x, y)\,|\,y = 3x + 2\}$.
7. $q = \{(x, y)\,|\,y = -5x + 1\}$.
8. $h = \{(x, y)\,|\,y = -x + 6\}$.
9. $g = \{(x, y)\,|\,y = 7 - 4x\}$.
10. $f = \{(u, v)\,|\,v = 2u - 6\}$.

Read Examples 3, 4, and 5 before doing Exercises 11–22. In each of these, a function is described. Sketch its graph.

11. $h(x) = x^2 + 5$.

12. $f(x) = -2x^3 + 2$.

13. $g(x) = -x^2 + 5x - 1$.

14. $H(x) = 3x^2 + 6x - 4$.

15. $h(x) = \dfrac{1}{x^2}$.

16. $g = \{(x, y) \mid y = x^3 - 4\}$.

17. $H = \{(x, y) \mid y = x^2 \wedge x \geqslant 0\}$.

18. $g = \{(x, y) \mid x^2 + y^2 = 5 \wedge y \geqslant 0\}$.

19. $f(x) = \begin{cases} -2, & \text{if } x \geqslant 1, \\ 1, & \text{if } x < 1. \end{cases}$

20. $f(x) = \begin{cases} x, & \text{if } x > 0, \\ 1, & \text{if } x \leqslant 0. \end{cases}$

21. $g(x) = \begin{cases} -3, & \text{if } x > 1, \\ 0, & \text{if } -1 \leqslant x \leqslant 1, \\ -x, & \text{if } x < -1. \end{cases}$

22. $g = \{(x, y) \mid x^4 y = 1\}$.

12.4 MORE ON FUNCTIONS

It is often possible to combine several functions to form new functions. A detailed study of the algebra of functions, however, is a topic for more advanced courses. In the present section we give a very short preview of this study through some examples.

Example 1. If f and g are functions defined by the equations

$$f(x) = x^2 + 1 \quad \text{and} \quad g(x) = 3x + 2,$$

describe a function h such that $h(x) = f(x) + g(x)$ for all $x \in R$.

Solution:

$$\begin{aligned} h(x) &= f(x) + g(x) \\ &= (x^2 + 1) + (3x + 2) \\ &= x^2 + 3x + 3. \end{aligned}$$

Example 2. If f and g are functions defined by

$$f(x) = \frac{2}{x} \quad \text{and} \quad g(x) = \frac{x}{x+1},$$

describe a function P such that $P(x) = f(x)g(x)$ for all real numbers $x \in D_f \cap D_g$.

Solution:

$$P(x) = f(x)g(x)$$

$$= \frac{2}{x} \cdot \frac{x}{x+1}$$

$$= \frac{2}{x+1}.$$

Note that we divided numerator and denominator of the fraction $\dfrac{2x}{x(x+1)}$ by x. The justification is that $x \neq 0$. To see this, note that $x \in D_f \cap D_g$ and therefore $x \in D_f$, which together with $0 \notin D_f$ implies $x \neq 0$. Thus we must have

$$P(x) = \frac{2}{x+1}, \qquad x \in D_f \cap D_g,$$

where $D_f \cap D_g = R - \{-1, 0\}$.

Example 3. If

$$f = \{(x, y) \mid y = x^2 + 1\} \qquad \text{and} \qquad g = \left\{(x, y) \mid y = \frac{x}{x+1}\right\},$$

find a function h such that $h(x) = \dfrac{f(x)}{g(x)}$ for all x for which the right side of the equality is defined.

Solution:

$$f(x) = x^2 + 1 \qquad (x \in R)$$

and

$$g(x) = \frac{x}{x+1} \qquad (x \in R - \{-1\}).$$

Therefore,

$$h(x) = \frac{f(x)}{g(x)}$$

$$= \frac{x^2 + 1}{\dfrac{x}{x+1}} \qquad \text{(provided } x \in \{-1, 0\}).$$

$$= \frac{(x^2 + 1)(x + 1)}{x}.$$

Therefore,

$$h(x) = \frac{(x^2 + 1)(x + 1)}{x}, \qquad \text{where } x \in R - \{-1, 0\}.$$

Example 4. If $f(x) = x^2 + 1$ and $g(x) = 3x + 2$, describe a function F such that $F(x) = f(g(x))$ for all x for which the expression $f(g(x))$ represents a real number.

Solution: Since $f(x) = x^2 + 1$, we find $f(g(x))$ by replacing each x by $g(x)$. Thus,

$$f(g(x)) = [g(x)]^2 + 1.$$

But

$$g(x) = 3x + 2.$$

Therefore,

$$f(g(x)) = (3x + 2)^2 + 1$$
$$= 9x^2 + 12x + 5.$$

Hence

$$F(x) = 9x^2 + 12x + 5$$

and

$$F = \{(x, y) \mid y = 9x^2 + 12x + 5\}.$$

Example 5. If $f(x) = 3x + 2$ and $g(x) = \frac{x - 2}{3}$, describe two functions F and G such that $F(x) = f(g(x))$ and $G(x) = g(f(x))$ for all x for which these expressions represent real numbers.

Solution: Since

$$f(x) = 3x + 2,$$
$$f(g(x)) = 3g(x) + 2.$$

But

$$g(x) = \frac{x - 2}{3}.$$

Therefore,

$$f(g(x)) = 3\left(\frac{x - 2}{3}\right) + 2$$
$$= (x - 2) + 2$$
$$= x.$$

Hence,

$$F(x) = x \qquad \text{for all } x \in R.$$

Similarly,

$$g(x) = \frac{x-2}{3}.$$

Therefore,

$$g(f(x)) = \frac{f(x) - 2}{3}.$$

But

$$f(x) = 3x + 2.$$

Hence,

$$g(f(x)) = \frac{(3x+2) - 2}{3}$$

$$= \frac{3x}{3}$$

$$= x.$$

It follows that

$$G(x) = x, \qquad \text{for all } x \in R.$$

identity / *functions* *Remark.* Since the functions F and G both have the same domain (R) and for each $x \in R$ the corresponding member is x itself, these functions are equal. They are called *identity functions*.

Exercise 12.4

Read Example 1 before doing Exercises 1–5. In each of these, two functions f and g are given. Describe a function h such that $h(x) = f(x) + g(x)$ for all $x \in D_f \cap D_g$.

1. $f(x) = x^2 + 3x - 1$; $g(x) = x^3 - 4x + 6$.
2. $f(x) = x^4 - 3x^2 + 2$; $g(x) = x^2 + 2$.

3. $f(x) = x^3 + 3x^2 + 4x + 6$; $g(x) = x^2 + \dfrac{1}{x}$.

4. $f(x) = \dfrac{1}{x+1}$; $g(x) = \dfrac{1}{x-1}$.

5. $f(x) = \dfrac{x}{x^2 - 4}$; $g(x) = \dfrac{x}{(x-2)(x+3)}$.

Read Example 2 before doing Exercises 6–10.

6. Describe a function P such that $P(x) = f(x) \cdot g(x)$ for all $x \in D_f \cap D_g$, where f and g are the functions of Exercise 1.

7. Repeat Exercise 6 with the functions of Exercise 2.

8. Repeat Exercise 6 with the functions of Exercise 3.

9. Repeat Exercise 6 with the functions of Exercise 4.

10. Repeat Exercise 6 with the functions of Exercise 5.

Read Example 3 before doing Exercises 11–15. In each of these, two functions f and g are given. Describe a function Q such that

$$Q(x) = \frac{f(x)}{g(x)} \qquad \text{for all } x \in D_f \cap D_g - \{x \mid g(x) = 0\}.$$

11. $f(x) = x^2 - 1$; $g(x) = x + 1$.

12. $f = \{(x, y) \mid y = x^3 - 1\}$; $g = \{(x, y) \mid y = x^2 + x + 1\}$.

13. $f(x) = \dfrac{1}{x - 1}$; $g(x) = \dfrac{x - 2}{x - 1}$.

14. $f(x) = x^2 + 5x - 6$; $g(x) = 2x^2 + x - 3$.

15. $f(x) = \dfrac{x^2 + 1}{x^3 - 1}$; $g(x) = \dfrac{x^2 + 1}{x^2 - 2x + 1}$.

Read Examples 4 and 5 before doing Exercises 16–20. In each of these, two functions f and g are given. Describe two functions F and G such that $F(x) = f(g(x))$ and $G(x) = g(f(x))$ for all x for which these expressions define real numbers.

16. $f(x) = x^2 + 1$; $g(x) = 5x - 2$.

17. $f(x) = x - 5$; $g(x) = x^2 + 2x - 6$.

18. $f = \{(x, y) \mid y = 3x + 2\}$; $g(x) = \frac{1}{3}(x - 2)$.

19. $f = \{(x, y) \mid xy = 1\}$; $g = \{(x, y) \mid xy = 1\}$.

20. $f(x) = \dfrac{1}{x - 1}$; $g(x) = \dfrac{1}{x + 1}$.

21. If $f = \{(1, 2), (2, 1), (3, 3)\}$ and $g = \{(1, 2), (2, 2), (3, 1)\}$, calculate $f[g(1)], f[g(2)], f[g(3)]$ and $g[f(1)], g[f(2)], g[f(3)]$.

12.5 INVERSE FUNCTIONS

In Example 5 of the preceding section we had two functions f and g for which the following two propositions were true:

$$f(g(x)) = x \qquad \text{for all } x \in D_g$$

and

$$g(f(x)) = x \qquad \text{for all } x \in D_f.$$

A situation such as this is pictured in Figure 12.7.

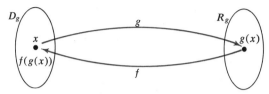

Figure 12.7

It can be shown that two functions f and g have the properties just described if and only if $g = \{(x, y) \mid (y, x) \in f\}$.

The first question that arises is: If f is a function, is the set $\{(x, y) \mid (y, x) \in f\}$ necessarily a function?

Let k denote the given set. Then k is a function if and only if no two of its members have the same first element and distinct second elements. Also $(x, y) \in k$ if and only if $(y, x) \in f$. We conclude that k is a function if and only if no two members of f have the same second element and distinct first elements.

Functions such as these are said to be one-to-one. Formally, we give the following:

one-to-one **Definition.** A *one-to-one function* is a function no two distinct
function members of which have the same second elements.

Symbolically, the function f is one-to-one if and only if $(a, b) \in f$ and $(c, b) \in f \rightarrow a = c$.

Example 1. Show that the function $f = \{(x, y) \mid y = 3x - 1\}$ is one-to-one.

Solution: We have the following:

$$(a, b) \in f \ \wedge \ (c, b) \in f.$$
$$b = 3a - 1 \ \wedge \ b = 3c - 1.$$
$$3a - 1 = 3c - 1.$$
$$3a = 3c.$$
$$a = c.$$

Thus, f is one-to-one.

It is easy to see that if the function f is one-to-one, the set
inverse $\{(x, y) \mid (y, x) \in f\}$ is a function. It is called the *inverse* of f and is denoted f^{-1}.

Example 2. Give a simple description of f^{-1} for the function f of Example 1.

Solution:

$(x, y) \in f^{-1}$.

$(y, x) \in f$.

$x = 3y - 1$.

$y = \frac{1}{3}(x + 1)$.

Therefore,

$$f^{-1} = \{(x, y) \mid y = \frac{1}{3}(x + 1)\}, \qquad \text{hence } f^{-1}(x) = \frac{1}{3}(x + 1).$$

In Section 10.7, we introduced the notion of symmetry. We now give one more definition on this subject.

symmetric **Definition.** Two sets of points S and T are *symmetric with respect to a line L* if and only if for each $P \in S$ there is a $Q \in T$ such that P and Q are symmetric with respect to L and also for each $Q' \in T$ there is a $P' \in S$ such that Q' and P' are symmetric with respect to L.

Recalling that $P(a, b)$ and $P(c, d)$ are symmetric with respect to the graph of $y = x$ if and only if $a = d$ and $b = c$, we can prove the following:

Theorem. Suppose that f and g are functions. Then $g = f^{-1}$ if and only if the graphs of f and g are symmetric with respect to the graph of $y = x$.

The proof is left as an exercise.

Example 3. Sketch the graphs of the functions of Examples 1 and 2.

Solution: The graphs are sketched in Figure 12.8. Note the symmetry.

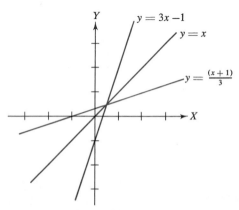

Figure 12.8

Exercise 12.5

Read Example 1 carefully before doing Exercises 1–5. In each of these, show that the given function is one-to-one.

    1. $f = \{(x, y) \mid y = 3x - 2\}$.
    2. $f = \{(x, y) \mid y = -x + 4\}$.
    3. $f = \{(x, y) \mid y = x^3\}$.
    4. $f = \{(x, y) \mid y = x^2 \land x \geqslant 0\}$.
    5. $f = \{(x, y) \mid y = x^2 \land x \leqslant 0\}$.

Read Example 2 before working Exercises 6–10.

    6. Give a simple description of f^{-1} for the function f of Exercise 1.
    7. Repeat Exercise 6 for the function f of Exercise 2.
    8. Repeat Exercise 6 for the function f of Exercise 3.
    9. Repeat Exercise 6 for the function f of Exercise 4.
    10. Repeat Exercise 6 for the function f of Exercise 5.

Read Example 3 before doing Exercises 11–15.

    11. On the same coordinate system, sketch the graphs of the functions f and f^{-1} of Exercises 1 and 6.
    12. Repeat Exercise 11 for f and f^{-1} of Exercises 2 and 7.
    13. Repeat Exercise 11 for f and f^{-1} of Exercises 3 and 8.
    14. Repeat Exercise 11 for f and f^{-1} of Exercises 4 and 9.
    15. Repeat Exercise 11 for f and f^{-1} of Exercises 5 and 10.
    *16. Let A be a set with n elements. How many one-to-one functions are there, such that their domain and range are A? [*Hint:* Recall the fundamental principle of counting.]
    *17. Let f be the function $f(x) = ax + b, a \neq 0$. If $f^{-1}(x) = f(x)$ for every real number x and $f(1) = 2$, calculate $f(2)$, $f(3)$, $f(4)$, and $f(5)$.
    **18. Prove that the graphs of f and f^{-1} are symmetric with respect to the graph of $y = x$.

REVIEW EXERCISES

In Exercises 1–6 some sets are given. In each of these **(a)** determine whether the given set is a function, **(b)** if it is a function, state its domain and range.

    1. $\{1, (3, 2)\}$.   **2.** $\{(1, 5), (2, 7), (1, 1)\}$.
    3. $\{(1, 1), (2, 1), (3, 1)\}$.   **4.** $\{(1, 3), (2, 7), (3, 5)\}$.
    5. $\{(x, y) \mid y \in R \land x = y^4\}$.
    6. $\{(x, y) \mid x \in R \land y = x^2 + 1\}$.

7. Show that $f = \{(x, y) \mid x + 2y = 1\}$ is a one-to-one function and describe f^{-1}. Compute $f^{-1}(2)$ and $f^{-1}(-1)$.

In Exercises 8–13

$$G = \left\{ (x, y) \mid y = \frac{1}{x+1}, \ x \in R - \{-1\} \right\}$$

and $F = \{(x, y) \mid y = x^2 + 1, \ x \in R\}$.

8. Describe the function H such that $H(x) = G(x) \cdot F(x)$ for every x in $D_F \cap D_G$. State the domain of H.

9. Describe the function H such that $H(x) = F(G(x))$ for every x in D_G.

10. Describe the function H such that $H(x) = G(F(x))$ for every x in D_F.

11. Show that G is a one-to-one function and describe its inverse.

12. Show that F is not a one-to-one function.

13. Sketch the graphs of G and G^{-1} on the same coordinate system.

In Exercises 14–19 an equation is used to define a function f. In each of these **(a)** describe the domain of f, **(b)** name the independent and dependent variables, **(c)** describe f as a set of ordered pairs.

14. $y = 3x + 5$.   **15.** $y = 2x^2 + 1$.

16. $y = \sqrt{(x-1)(x-2)}$.   **17.** $y = \dfrac{1}{(x-2)(x+3)}$.

18. $y = \sqrt{\dfrac{1}{(x-2)(x+3)}}$.   **19.** $y = \sqrt[4]{x} + |x|$.

In Exercises 20–25, $f = \{(2, 1), (1, 2), (3, 3)\}$, $g = \{(1, 1), (2, 3), (3, 2)\}$, and you are asked to describe a function h that has the stated property by explicitly listing the members of h in braces.

20. $h(x) = f(x) + g(x)$ for every x in $D_f \cap D_g$.

21. $h(x) = f(x) - g(x)$ for every x in $D_f \cap D_g$.

22. $h(x) = f(x) \cdot g(x)$ for every x in $D_f \cap D_g$.

23. $h(x) = \dfrac{f(x)}{g(x)}$ for every x in $D_f \cap D_g$.

24. $h(x) = f(g(x))$ for every x in D_g.

25. $h(x) = g(f(x))$ for every x in D_f.

26. Let f be a function such that $f(f(x)) = x$ for every x in the domain of f. Prove that f is one-to-one and $f = f^{-1}$.

27. State carefully the definition of a function.

28. State carefully the definition of a one-to-one function.

LOGARITHMS

13.1 INTRODUCTION

We already have studied certain basic laws of exponents for numbers of the form a^x, where x represents a rational number. (See Chapter 9.) To define a^x precisely when x represents an irrational number requires some concepts that we have not presented in this text. It is enough for our purposes to know that a definition can be given in such a way that for all real numbers x and y (rational or irrational) and positive real numbers a and b the following fundamental laws hold*:

$$a^x a^y = a^{x+y},$$

$$\frac{a^x}{a^y} = a^{x-y},$$

$$(ab)^x = a^x b^x,$$

$$\left(\frac{a}{b}\right)^x = \frac{a^x}{b^x},$$

$$(a^x)^y = a^{xy}.$$

Further, the following are also true:

$$a^x > 0.$$

If $a \neq 1$, $a^x = a^y \leftrightarrow x = y$.

If $a > 1$, $y < x \leftrightarrow a^y < a^x$.

If $a < 1$, $x < y \leftrightarrow a^y < a^x$.

* See A. L. Yandl, *The Non-Algebraic Elementary Functions, a Rigorous Approach* (Englewood Cliffs, N.J.: Prentice-Hall, Inc., 1964).

The foregoing laws of exponents provide a very useful device for computational purposes. This, of course, may not be as important today as it was a hundred years ago because of the availability of many electronic computing machines. However, the ideas are interesting enough and the theory is useful enough to deserve a brief introduction in a basic algebra course.

We begin with an example.

Example 1. Estimate the size of the number 2^{30}.

Solution:

$$2^1 = 2, \qquad 2^6 = 64,$$
$$2^2 = 4, \qquad 2^7 = 128,$$
$$2^3 = 8, \qquad 2^8 = 256,$$
$$2^4 = 16, \qquad 2^9 = 512,$$
$$2^5 = 32, \qquad 2^{10} = 1024.$$

Hence we see that 2^{10} is slightly larger than 1000, which is 10^3. Therefore, we argue that $2^{30} = (2^{10})^3$ is somewhat larger than $(10^3)^3 = 10^9 = 1,000,000,000$.

We see that if a gambler were losing at a certain game and betting double or nothing, he would not have to lose too many games in a row before going bankrupt.

Example 2. Sketch the graph of the equation $y = 2^x$, assuming that this graph is a smooth curve. Then use this sketch to estimate graphically the product $(2.5)(4.1)$.

Solution: We first perform a few simple calculations and obtain the following table.

x	-2	-1	0	2	3
2^x	$\frac{1}{4}$	$\frac{1}{2}$	1	4	8
$(x, 2^x)$	$(-2, \frac{1}{4})$	$(-1, \frac{1}{2})$	$(0, 1)$	$(2, 4)$	$(3, 8)$

We then plot the points whose coordinates are the ordered pairs that appear in the third row of the table. Finally we draw a smooth curve passing through these points. (See Figure 13.1.)

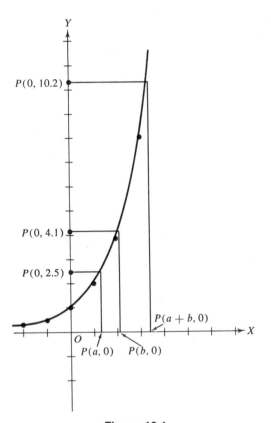

Figure 13.1

We now consider the points $P(0, 4.1)$ and $P(0, 2.5)$ on the Y-axis. The perpendiculars raised from these points to the Y-axis intersect the graph of $y = 2^x$ at points that we denote $P(a, 2.5)$ and $P(b, 4.1)$. Hence we have

$$2^a = 2.5 \quad \text{and} \quad 2^b = 4.1.$$

Therefore,

$$(2.5)(4.1) = 2^a 2^b = 2^{a+b}.$$

We next draw a perpendicular to the X-axis through the point $P(a + b, 0)$. This perpendicular intersects the graph of $y = 2^x$ at the point $P(a + b, c)$. Hence, $c = 2^{a+b}$. Since both c and $(2.5)(4.1)$ are equal to 2^{a+b}, we have $c = (2.5)(4.1)$. From the diagram we estimate that $c = 10.2$, which closely agrees with the value 10.25 of the product $(2.5)(4.1)$.

Careful analysis of the foregoing example shows that we estimated the product (2.5)(4.1) as follows:

First we found numbers a and b such that $2.5 \doteq 2^a$ and $4.1 \doteq 2^b$.*

Then we *added* a and b.

Finally, we found a number c such that $c = 2^{a+b}$. We then were able to conclude that $(2.5)(4.1) \doteq c$.

Therefore, we have changed a multiplication problem to a problem of addition. In lengthy calculations, this idea may be very useful. The principle involved here is also used in designing slide rules.

logarithm   If $y = 2^x$, we say that x is the *logarithm of y* (to the base 2).

Exercise 13.1

Read Example 1 before doing Exercises 1–6. In each of these, estimate the size of the given power.

1. 2^{20}.	**2.** 2^{40}.
3. 2^{50}.	**4.** 2^{60}.
5. 2^{90}.	**6.** 2^{100}.

Read Example 2 before doing Exercises 7 and 8.

7. Sketch the graph of the equation $y = 3^x$. Use this sketch to estimate the product (4.6)(3.2).

8. Sketch the graph of the equation $y = 5^x$. Use this sketch to estimate (2.3)(7.6).

Study the last sentence of Section 13.1 before doing Exercises 9–21.

9. Find the logarithm of 4 (to the base 2).
10. Find the logarithm of 8 (to the base 2).
11. Find the logarithm of 16 (to the base 2).
12. Find the logarithm of $\frac{1}{2}$ (to the base 2).
13. Find the logarithm of $\frac{1}{4}$ (to the base 2).
14. Find the logarithm of $\frac{1}{16}$ (to the base 2).
15. Find the logarithm of 2 (to the base 2).
16. Find the logarithm of $\sqrt[3]{2}$ (to the base 2).
17. Find the logarithm of 1 (to the base 2).

* The symbol $\doteq$ indicates an approximation.

18. Find the logarithm of 9 (to the base 3).
19. Find the logarithm of 81 (to the base 3).
20. Find the logarithm of $\frac{1}{9}$ (to the base 3).
21. Find the logarithm of 1 (to the base 3).
22. Solve $2^x = \frac{1}{4}$ over R.
23. Solve $(\frac{2}{3})^x = \frac{16}{81}$ over R.
24. Solve $\log_2 (x + 1) = 3$ over R.
25. Solve $\log_2 x^2 = 2$ over R.
26. Solve $\log_2 (x(x + 2)) = 3$ over R.

13.2 THE LAWS OF LOGARITHMS

In the preceding section we have seen that if $y = 2^x$, then x is called the logarithm of y to the base 2. We also illustrated that logarithms (which are essentially exponents) may be used for computational purposes.

It can be shown that if $b > 0$ and $b \neq 1$, then the function $E_b = \{(x, y) \mid y = b^x\}$ is one-to-one. Hence, E_b has an inverse, which is called the logarithm function to the base b. Formally we give the following

logarithm **Definition.** If $b > 0$ and $b \neq 1$, the *logarithm function to the*
function to *base b* is denoted $\log_b$ and is defined by the equation
the base b
$$\log_b = \{(x, y) \mid (y, x) \in E_b\}.$$

Equivalently,

$$\log_b = \{(x, y) \mid x = b^y\}.$$

Example 1. Find $\log_2(4)$ and $\log_3(\frac{1}{27})$.

Solution:

$$\log_2(4) = x.$$
$$(4, x) \in \log_2.$$
$$(x, 4) \in E_2.$$
$$2^x = 4.$$
$$2^x = 2^2.$$
$$x = 2.$$

Hence, $\log_2(4) = 2$.

Similarly,

$$\log_3 \left(\tfrac{1}{27}\right) = x.$$

$$\left(\tfrac{1}{27}, x\right) \in \log_3.$$

$$\left(x, \tfrac{1}{27}\right) \in E_3.$$

$$3^x = \tfrac{1}{27}.$$

$$3^x = \frac{1}{3^3}.$$

$$3^x = 3^{-3}.$$

$$x = -3.$$

Hence, $\log_3 \left(\tfrac{1}{27}\right) = -3$.

Since a logarithm function is the inverse of some exponential function, it is natural that we should have some *laws of logarithms* which correspond to the basic laws of exponents. We have in fact the following:

Theorem. If b is a positive number distinct from the number 1 and x, y are positive real numbers, then the following are true:

$$\log_b (xy) = \log_b (x) + \log_b (y).$$

$$\log_b \left(\frac{x}{y}\right) = \log_b (x) - \log_b (y).$$

$$\log_b (x^p) = p \cdot \log_b (x).$$

$$\log_b \sqrt[r]{x} = \frac{1}{r} \log_b (x).$$

Proof:

$$\log_b (x) = r \quad \wedge \quad \log_b (y) = s,$$

$$x = b^r \wedge y = b^s,$$

$$xy = b^r \cdot b^s,$$

$$xy = b^{r+s},$$

$$r + s = \log_b (xy),$$

$$\log_b (xy) = \log_b (x) + \log_b (y).$$

The remaining parts of the theorem are proved in a similar manner.

The most widely used logarithm function for purposes of common computation is the function $\log_{10}$, called the *common logarithm function*. For brevity in this text we write log instead of $\log_{10}$. Further, it is usual to write log x in lieu of log (x).

common logarithm function

Example 2. Given log 2 = 0.3010 and log 3 = 0.4771, find (a) log 6, (b) log 8, (c) log $\sqrt{3}$, (d) log $\frac{1}{4}$, (e) log 18, (f) log 5.

Solution:

(a)
$$\log 6 = \log (2 \cdot 3)$$
$$= \log 2 + \log 3$$
$$= 0.3010 + 0.4771$$
$$= 0.7781.$$

(b)
$$\log 8 = \log 2^3$$
$$= 3 \cdot \log 2$$
$$= 3(0.3010)$$
$$= 0.9030.$$

(c)
$$\log \sqrt{3} = \tfrac{1}{2} \log 3$$
$$= \tfrac{1}{2}(0.4771)$$
$$= 0.23855.$$

(d)
$$\log \tfrac{1}{4} = \log \frac{1}{2^2}$$
$$= \log 2^{-2}$$
$$= (-2)(\log 2)$$
$$= (-2)(0.3010)$$
$$= -0.6020.$$

(e)
$$\log 18 = \log (2 \cdot 9)$$
$$= (2 \cdot 3^2)$$
$$= \log 2 + \log 3^2$$
$$= \log 2 + 2 \log 3$$
$$= 0.3010 + 2(0.4771)$$
$$= 0.3010 + 0.9542$$
$$= 1.2552.$$

(f) To find log 5, first note that in general $\log_b b = 1$. Hence, log 10 = 1. Therefore,

$$\log 5 = \log \tfrac{10}{2}$$
$$= \log 10 - \log 2$$
$$= 1 - 0.3010$$
$$= 0.6990.$$

Remark. Since the logarithm function is the inverse of the exponential function, it is one-to-one. Hence, $(\log x = \log y) \leftrightarrow (x = y \wedge x > 0)$.

Example 3. Solve the following equation:

$$\log (x - 2) + \log (x - 17) = 2.$$

Solution: We first note that we must have $x - 2 > 0$ and $x - 17 > 0$, since the domain of the logarithm function is the set of positive real numbers. Thus, the replacement set is

$$S = \{x \in R \,|\, x > 17\}.$$

Also, we note that $2 = \log 100$. Hence, we obtain the following chain of equivalent equations:

$\uparrow$ $\log (x - 2) + \log (x - 17) = 2.$

$\log (x - 2) + \log (x - 17) = \log 100.$

$\log (x - 2)(x - 17) = \log 100$ (by the law of logarithms).

$(x - 2)(x - 17) = 100$ (since the logarithm function is one-to-one).

S $x^2 - 19x + 34 = 100.$

$x^2 - 19x - 66 = 0.$

$(x - 22)(x + 3) = 0.$

$x - 22 = 0 \quad \vee \quad x + 3 = 0.$

$x = 22 \quad \vee \quad x = -3.$

$x = 22 \quad$ (since $-3 \notin S$).

Exercise 13.2

Read Example 1 before doing Exercises 1–6.

    1. Find $\log_2 (8)$.   **2.** Find $\log_2 (32)$.
    3. Find $\log_2 (\frac{1}{16})$.   **4.** Find $\log_3 (81)$.
    5. Find $\log_5 (125)$.   **6.** Find $\log_6 (1)$.

Read Example 2 before doing Exercises 7–16.

    7. Find $\log 8$.   **8.** Find $\log 48$.
    9. Find $\log 54$.   **10.** Find $\log 25$.
    11. Find $\log \sqrt[4]{2}$.   **12.** Find $\log 24$.
    13. Find $\log \frac{20}{3}$.
    14. Given $\log 7 = 0.8451$, and $\log 2 = 0.3010$, find (a) $\log 49$, (b) $\log 28$, (c) $\log 3.5$.
    15. Find $\log 1.4$ (see Exercise 14).
    16. Find $\log 560$ (see Exercise 14).
    17. Show that for any real number $b > 0$ $(b \neq 1)$, $\log_b b = 1$.
    18. Recall that if x is a real number and a is a positive real number, then $a^x > 0$. Further, it can be shown that if $a > 0$ and $a \neq 1$, then for any positive real number y there is a real number x such that $a^x = y$.

Use these facts to show that the domain of a log function is the set of all positive real numbers and the range is the set of all real numbers.

    19. Prove the last three parts of the theorem of Section 13.2.
    *20. Prove that $(\log_b a)(\log_a b) = 1$.
    21. Sketch the graph of $\log_2$. [*Hint:* Recall the symmetry of the graphs of $\log_2$ and $y = 2^x$ with respect to the graph of $y = x$.]

Read Example 3 before doing Exercises 22–30. In each of these, find the largest possible replacement set and solve the given equation over that set.

    22. $\log (x - 3) + \log (x - 18) = 2$.
    23. $\log (x - 19) + \log (x - 4) = 2$.
    24. $\log (x - 5) + \log (x - 20) = 2$.
    25. $\log (x - 1) + \log (x - 16) = 2$.
    *26. $2 \log (x + 3) + \log (x + 2) = 2$.
    27. $\log (x - 3) + \log x = 1$.
    28. $\log (x - 8) - \log (x + 1) = 1$.
    29. $\log (3x + 35) - \log (2x - 5) = 1$.
    30. $\log (10x^2 + 40x + 40) + \log 4 = 3$.

13.3 TABLE OF COMMON LOGARITHMS

It is evident from our discussion that logarithms may be used to advantage in long and tedious arithmetic calculations. Therefore, the student must become proficient with the following two problems:

(1) Given a positive number N, find log N.

(2) Given log M, find M.

Tables are provided to enable us to deal with these two tasks. The present section offers a brief presentation of these tables.

First, since many logarithms are irrational numbers, the tables necessarily give only approximations. (Recall that it is physically impossible to write down an infinite nonperiodic decimal.) To arrive at such approximations, we use the tech-

rounding off nique of "*rounding off*" a decimal. The procedure we shall adopt is the following. Look at the first of all the digits dropped. If it is a member of the set $\{0, 1, 2, 3, 4\}$, the last digit that was not dropped is retained. On the other hand, if it is a member of the set $\{6, 7, 8, 9\}$, the last digit that was not dropped is increased by 1. If the first of the dropped digits is a 5, we keep the last of the digits that was not dropped if it is even or we increase it by 1 if it is odd.

Example 1. Round off the following numbers to four decimal places: (a) 0.333̲3, (b) 3.41625̲12, (c) 21.617351̲4, (d) 0.001267̲14, (e) 3.141592653589 (an approximation of π).

Solution: (a) 0.3333 is the four-decimal-place approximation of 0.333̲3, since the first of the digits dropped is a 3.

(b) 3.4162 is the correct approximation to the given number, since the first of the dropped digits is a 5 and the last digit that was not dropped is even.

(c) To get a four-decimal-place approximation of 21.617351̲4, note that the fifth decimal is a 5 and the fourth is 3, which is odd. Hence we increase the fourth decimal by 1 and take 21.6174 as our approximation.

(d) We take 0.0013 as the four-decimal-place approximation of 0.001267̲14, since the first of the dropped digits was a 6.

(e) The four-decimal-place approximation for π is 3.1416.

In order to understand how to use the tables of logarithms, it is important to note that any positive real number can be written as a product of the form $c \cdot 10^k$, where k is an integer and $1 \leq c < 10$. For example,

$$21.634 = (2.1634) \cdot 10^1,$$

$$34151.41 = (3.415141) 10^4,$$

$$0.00031 = (3.1)(10^{-4}).$$

scientific
notation

When a number has been expressed in this form, it is said to be written in *scientific notation*. The reason for this terminology is that in the sciences we frequently use very large or very small numbers, and the notation just introduced is very convenient in dealing with such numbers.

For example, the speed of light is $(2.99776)(10^{10})$ cm/sec, that of sound is $(3.3)(10^4)$ cm/sec, and the mass of an electron is $(9.107)(10^{-28})$ g.

The usefulness of the scientific notation in dealing with common logarithms stems from the fact that log 10 = 1. (See Exercise 17 of the preceding section.) Thus if x is a positive real number and in scientific notation $x = y \cdot 10^k$, then

$$\log x = \log (y \cdot 10^k)$$

$$= \log y + \log 10^k$$

$$= \log y + k \cdot \log 10$$

$$= \log y + k \cdot 1$$

$$= k + \log y.$$

Therefore, if we can find the common logarithms of all numbers between 1 and 10, then we can find the common logarithms of all positive numbers.

characteristic

mantissa

In the expression $\log x = k + \log y$ (where $1 \leqslant y < 10$), the integer k is called the *characteristic* of log x and the non-negative number log y is called the *mantissa* of log x. (See Exercise 25 below.)

When we evaluate the logarithm of a given positive number, we obtain the mantissa from the tables and the characteristic by inspection. Note that the characteristic is always an integer and the mantissa is always between 0 and 1, or 0.

Before proceeding to Example 2, we must make a few remarks concerning the tables shown on pp. 410–411. First note that all decimal points have been left out. If

$$\{a, b, c\} \subset \{0, 1, 2, 3, 4, 5, 6, 7, 8, 9\}$$

and we want to find the logarithm of the nonzero number $a + b \cdot 10^{-1} + c \cdot 10^{-2}$ (*a.bc* in decimal form), we find the number $10a + b$ in the column headed by N, and the number c in the row containing N. The number located at the intersection of the row and column containing the numbers $10a + b$ and c, respectively, is the logarithm we want, except for the decimal point. For example, if we want log 3.74, we find 37 in the column headed by N, then we find 4 in the row containing N. The number at the intersection of the row containing the 37 and the column headed by 4 is 5729. Hence log 3.74 = 0.5729.

Example 2. Find log 210 and log 0.00563.

Solution: We first write 210 in scientific notation:

$$210 = (2.1)(10^2).$$

Therefore,

$$\log 210 = 2 + \log 2.1.$$

From the table we see that log 2.10 = 0.3222. Hence,

$$\log 210 = 2 + 0.3222 = 2.3222.$$

Next we write 0.00563 in scientific notation. Thus

$$0.00563 = (5.63)(10^{-3}).$$

Therefore,

$$\log 0.00563 = -3 + \log 5.63.$$

To find log 5.63 in the table, refer to the number 56 in the left column. Then, on the same row as this number, find the number in the column headed by 3. This number is 7505. Therefore, log 5.63 = 0.7505 and

$$\log 0.00563 = -3 + 0.7505$$

$$= -2.2495.$$

Note that it is best in practice to leave a logarithm such as the one just obtained in a form from which the characteristic and mantissa are readily recognized. It is often useful to add and subtract the same positive integer k. For example, we could write log 0.00563 as

$$(-3 + 0.7505 + 4) - 4 = 1.7505 - 4$$

or as

$$(-3 + 0.7505 + 10) - 10 = 7.7505 - 10.$$

Exercise 13.3

Read Example 1 before doing Exercises 1–6. In each of these, round off the given numbers to (a) three decimal places, (b) four decimal places.

1. 51.5<u>31</u>.

2. 0.013<u>21</u>.

3. 21.<u>31</u>.

4. 61.210<u>164</u>.

5. 121.0<u>12</u>.

6. 3.15<u>61</u>.

In each of Exercises 7–12, write the given number in scientific notation.

7. 0.000001602.

8. 1,291,615.2.

9. 1,000,561,000.

10. 0.00000000012.

11. 1,561,000,000.

12. 7,100,310.

13. Find three quantities in a physics book that are expressed using scientific notation.

Read Example 2 before working Exercises 14–21. In each of these, use the tables to find the logarithm of the given number.

14. 213.

15. 0.000516.

16. 0.0000217.

17. 31.5.

18. 3,240,000.

19. 0.000000913.

20. 6,130,000,000.

21. 0.000812.

22. If $\log x = 0.8069$, find x. (Use the tables.)

23. If $\log x = 3.9101$, find x. [*Hint:* $3.9101 = 3 + 0.9101$.]

24. If $\log x = 6.5263 - 10$, find x. [*Hint:* $6.5263 - 10 = -4 + 0.5263$.]

25. Show that the following are true: (a) If a number is more than one, the characteristic of its logarithm is non-negative and is one less than the number of digits to the left of the decimal point of the number. (b) If a positive number is less than one, the characteristic of its logarithm is negative and its absolute value is one more than the number of consecutive zeros immediately to the right of the decimal point of the number.

13.4 LINEAR INTERPOLATION

The table of common logarithms on pages 410–411 may be used to determine the logarithm of a number x with four significant figures. If the last significant figure is zero, the logarithm can be found directly from the tables, as we saw in the preceding section. If the last significant figure is not zero, we must use a *linear* process called *linear interpolation.*

interpolation   We illustrate this technique with some examples.

Example 1. Find $\log 2.317$.

Solution: We first remark that

$$2.310 < 2.317 < 2.320.$$

Since it can be shown that

$$\log x < \log y \leftrightarrow (0 < x < y),$$

we conclude that

$$\log 2.310 < \log 2.317 < \log 2.320.$$

Using the tables, we find $\log 2.310 = 0.3636$ and $\log 2.320 = 0.3655$, and we write $0.3636 < \log 2.317 < 0.3655$.

The information we have tells us that the points with coordinates $(2.310, 0.3636)$, $(2.320, 0.3655)$ are on the graph of the logarithm function. If $z = \log 2.317$, then $P(2.317, z)$ is also on the graph of that function. In Figure 13.2 we show that z can be approximated by the length of the segment AB. But $|AB| = |AC| + |CB|$, and from the similar triangles we have

$$\frac{|FC|}{|FD|} = \frac{|CB|}{|DE|}.$$

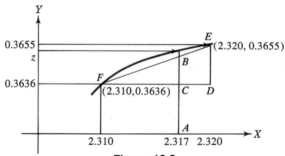

Figure 13.2

Now

$$\frac{|FC|}{|FD|} = \frac{2.317 - 2.310}{2.320 - 2.310} = \frac{0.007}{0.010} = \frac{7}{10}.$$

Also, $|DE| = 0.3655 - 0.3636 = 0.0019$. Therefore,

$$\frac{7}{10} = \frac{|CB|}{0.0019}$$

and

$$|CB| = \frac{7}{10}(0.0019) = 0.00133.$$

Rounding off this last result we get

$$|CB| \doteq 0.0013.$$

Therefore,

$$|AB| = |AC| + |CB| \doteq 0.3636 + 0.0013 = 0.3649.$$

Hence, $z \doteq 0.3649$ and we conclude that log $2.317 \doteq 0.3649$.

It is helpful to arrange the work as follows.

$$0.010 \left[0.007 \begin{bmatrix} 2.310 & \log 2.310 \doteq 0.3636 \\ 2.317 & \log 2.317 = ? \\ 2.320 & \log 2.320 \doteq 0.3655 \end{bmatrix} x \right] 0.0019.$$

Hence,

$$\frac{x}{0.0019} \doteq \frac{0.007}{0.010} = \frac{7}{10}.$$

Therefore,

$$x \doteq (0.0019)\frac{7}{10} = 0.0013 \qquad \text{(rounded off to four decimals).*}$$

It follows that

$$\log 2.317 \doteq 0.3636 + 0.0013$$

$$\doteq 0.3649.$$

Example 2. If log $N = 8.6142 - 10$, find N.

Solution: The characteristic is $8 - 10 = -2$ and the mantissa is 0.6142. Therefore, in scientific notation, $N = M \cdot 10^{-2}$, where log $M = 0.6142$. From the tables we obtain the following:

$$0.010 \left[x \begin{bmatrix} 4.110 & \log 4.110 = 0.6138 \\ M & \log M = 0.6142 \\ 4.120 & \log 4.120 = 0.6149 \end{bmatrix} 0.0004 \right] 0.0011.$$

Hence,

$$\frac{x}{0.010} = \frac{0.0004}{0.0011}$$

and $\qquad x = 0.00\underline{36} \qquad$ (rounded off, $x \doteq 0.004$).

Therefore,

$$M \doteq 4.110 + 0.004 = 4.114.$$

It follows that

$$N \doteq (4.114)(10^{-2}) = 0.04114.$$

* It is customary to perform the round-off before addition or subtraction.

antilogarithm   In general, if log $x = y$, then x is called the *antilogarithm* of
y. In the preceding example, 0.04114 approximates the anti-
logarithm of $8.6142 - 10$.

Exercise 13.4

Read Example 1 before doing Exercises 1–10. In each of these,
use linear interpolation to find an approximation of the given
logarithm.

1. log 2.342.	**2.** log 5.617.
3. log 3.102.	**4.** log 7.112.
5. log 9.665.	**6.** log 24.16.
7. log 415.1.	**8.** log 0.001342.
9. log 0.000006174.	**10.** log 215.4.

Read Example 2 before doing Exercises 11-20. In each of
these, find the antilogarithm of the given number. Recall that if
log $N = x$, you must first write x as $y + k$, where k is an integer
and $0 \leqslant y < 1$. Then y and k are, respectively, the mantissa and
characteristic of log N.

11. 2.5612.	**12.** 5.4321.
13. 6.0137.	**14.** 2.1561.
15. 3.0148.	**16.** $5.1623 - 10$.
17. $3.1749 - 5$.	**18.** $8.0016 - 10$.
19. $2.9817 - 3$.	**20.** $8.1926 - 10$.

21. Explain why the following is true. If N is a real number
and in scientific notation

$$N = M \cdot 10^k,$$

then the decimal point of N has to be moved $|k|$ places to obtain
M. Further, $k > 0$ if the point has to be moved to the left and
$k < 0$ if the point has to be moved to the right.

13.5 CALCULATIONS USING LOGARITHMS

We are now ready to illustrate how logarithms may be used to
efficiently approximate the results of lengthy computational
problems. It is best to present the ideas through examples.
    First recall the following important facts.

    (1) log N is defined if and only if N is a positive real number.
    (2) If in scientific notation $N = M \cdot 10^k$ (where $1 \leqslant M$

< 10), then the mantissa of log N is log M and the characteristic is k. Conversely, if the characteristic of the logarithm of N is k, then in scientific notation $N = M \cdot 10^k$, where M is the antilogarithm of the mantissa of log N.

Example 1. Approximate the product $(1.01)(30.2)(80)$.

Solution:

log $[(1.01)(30.2)(80)]$

$$= \log 1.01 + \log 30.2 + \log 80$$

$$= \log 1.01 + \log [(3.02)(10)] + \log [8 \cdot 10]$$

$$\doteq 0.0043 + 1.4800 + 1.9031$$

$$= 3.3874$$

$$= 3 + 0.3874.$$

Therefore, in scientific notation,

$$(1.01)(30.2)(80) = x \cdot 10^3, \qquad \text{where } \log x \doteq 0.3874.$$

From the tables we see that $x \doteq 2.44$. Hence,

$$(1.01)(30.2)(80) \doteq (2.44)10^3 = 2440.$$

Example 2. Find the quotient $\dfrac{-7.77}{8.94}$.

Solution. Since we are dividing a negative number by a positive number, the quotient will be negative. We note that

$$\frac{-7.77}{8.94} = -\frac{7.77}{8.94}.$$

We first find $\dfrac{7.77}{8.94}$ using logarithms.

$$\log \frac{7.77}{8.94} = \log 7.77 - \log 8.94$$

$$\doteq 0.8904 - 0.9513.$$

Before subtracting these two numbers, we note that the first is smaller than the second and therefore the difference is a negative number. Since the mantissa of the logarithm of a number is always non-negative, and since we wish to recognize the mantissa of $\log \dfrac{7.77}{8.94}$ by inspection, we add a positive integer n large enough so that $n.8904 - 0.9513$ is positive and subtract that integer as follows. (In this case we may take n to be 1.)

$$0.8904 - 0.9513 = 1.8904 - 0.9513 - 1$$

$$= 0.9391 - 1.$$

Hence,

$$\log \frac{7.77}{8.94} \doteq 0.9391 - 1.$$

We see that the mantissa is 0.9391 and the characteristic is -1. Therefore, in scientific notation,

$$\frac{7.77}{8.94} = M \cdot 10^{-1}$$

where $\log M \doteq 0.9391$. From the tables, we find that

$$\log 8.690 \doteq 0.9390 \quad \text{and} \quad \log 8.700 \doteq 0.9395.$$

Since $0.9390 < 0.9391 < 0.9395$, we use interpolation as follows:

$$0.010 \left[x \begin{bmatrix} 8.690 & \log 8.690 \doteq 0.9390 \\ M & \log M = 0.9391 \\ 8.700 & \log 8.700 = 0.9395 \end{bmatrix} \begin{matrix} 0.0001 \\ \end{matrix} \right] 0.0005.$$

Hence,

$$\frac{x}{0.010} \doteq \frac{0.0001}{0.0005}$$

— that is,

$$x \doteq 0.002 \quad \text{and} \quad M \doteq 8.692.$$

Therefore,

$$\frac{7.77}{8.94} \doteq (8.692) \, 10^{-1}$$

$$= 0.8692,$$

and

$$\frac{-7.77}{8.94} \doteq -0.8692.$$

Example 3. Approximate $\sqrt[5]{8.31}$.

Solution:

$$\log \sqrt[5]{8.31} = \tfrac{1}{5} \log 8.31$$

$$\doteq \tfrac{1}{5}(0.9196)$$

$$= 0.18392$$

$$\doteq 0.1839 \qquad \text{(rounded off to four decimal places)}.$$

We must use linear interpolation.

$$0.010 \left[x \begin{bmatrix} 1.520 & \log 1.520 \doteq 0.1818 \\ \sqrt[5]{8.31} & \log \sqrt[5]{8.31} \doteq 0.1839 \\ 1.530 & \log 1.530 \doteq 0.1847 \end{bmatrix} 0.0021 \right] 0.0029.$$

Therefore,

$$\frac{x}{0.010} \doteq \frac{21}{29} \qquad \text{and} \qquad x \doteq 0.007 \quad \text{(rounded off)}.$$

Hence,

$$\sqrt[5]{8.31} \doteq 1.520 + 0.007 = 1.527.$$

Example 4. Calculate $\dfrac{(7.21)^2(189)^{2/3}}{\sqrt{46.7}}$.

Solution:

$$\log \frac{(7.21)^2(189)^{2/3}}{\sqrt{46.7}} = \log \left[(7.21)^2(189)^{2/3}\right] - \log (46.7)^{1/2}$$

$$= \log (7.21)^2 + \log (189)^{2/3} - \log (46.7)^{1/2}$$

$$= 2 \log 7.21 + \tfrac{2}{3} \log 189 - \tfrac{1}{2} \log 46.7$$

$$\doteq 2(0.8579) + \tfrac{2}{3}(2.2765) - \tfrac{1}{2}(1.6693)$$

$$\doteq 1.7158 + 1.5177 - 0.8346.$$

(Observe that the last two numbers were rounded off.) Hence,

$$\log \frac{(7.21)^2(189)^{2/3}}{\sqrt{46.7}} \doteq 2.3989 = 2 + 0.3989.$$

Therefore, the characteristic is 2 and the mantissa is 0.3989. It follows that, in scientific notation,

$$\frac{(7.21)^2(189)^{2/3}}{\sqrt{46.7}} = M \cdot 10^2,$$

where $\log M \doteq 0.3989$. Using linear interpolation,

$$0.010 \left[x \begin{bmatrix} 2.500 & \log 2.5 \doteq 0.3979 \\ M & \log M \doteq 0.3989 \\ 2.510 & \log 2.510 \doteq 0.3997 \end{bmatrix} 0.0010 \right] 0.0018$$

$$\frac{x}{0.010} \doteq \frac{10}{18}.$$

Hence, $x \doteq 0.006$ (rounded off). Therefore, $M \doteq 2.500 + 0.006 = 2.506$. We conclude that

$$\frac{(7.21)^2(189)^{2/3}}{\sqrt{46.7}} \doteq (2.506)10^2 = 250.6.$$

Example 5. Approximate $\sqrt[4]{0.00152}$.

Solution: $\log \sqrt[4]{0.00152} = \frac{1}{4} \log 0.00152$. In scientific notation, $0.00152 = (1.52)10^{-3}$. Therefore,

$$\log 0.00152 = -3 + \log 1.52 \qquad \text{(why?)}$$

$$\doteq -3 + 0.1818.$$

Since we need to divide $\log 0.00152$ by 4, and -3 is not divisible by 4, we find it convenient to write

$$\log 0.00152 = -3 - 1 + 1 + 0.1818$$

$$= -4 + 1.1818.$$

Hence,

$$\tfrac{1}{4} \log 0.00152 = -1 + 0.2954 \qquad \text{(rounded off)}.$$

It follows that

$$\sqrt[4]{0.00152} \doteq M \cdot 10^{-1}, \qquad \text{where } \log M = 0.2954.$$

Again we must use interpolation.

$$0.010 \left[\begin{array}{c} x \left[\begin{array}{ll} 1.970 & \log 1.970 \doteq 0.2945 \\ M & \log M \doteq 0.2954 \end{array} \right] 0.0009 \\ 1.980 \quad \log 1.980 = 0.2967 \end{array} \right] 0.0022$$

$$\frac{x}{0.010} \doteq \frac{9}{22}.$$

Hence, $x \doteq 0.004$ and $M \doteq 1.970 + 0.004 = 1.974$. Therefore,

$$\sqrt[4]{0.00152} \doteq (1.974)10^{-1} = 0.1974.$$

Exercise 13.5

Use logarithms to evaluate each of the following.

1. $(41.6)(0.00257)(316)$.   **2.** $(31.6)^3(2.51)^2(0.0127)$.

3. $(-81.2)(31.2)(-127)$. **4.** $(-61.2)^3(0.169)$.

5. $\sqrt[5]{12.6}$.

6. $\dfrac{(31.2)^3(12.6)^4}{\sqrt[3]{135}}$.

7. $\dfrac{(0.0031)^4(73.1)^3}{\sqrt{51.6}}$.

8. $\dfrac{(61.1)^5(0.0071)^3}{\sqrt[3]{12.6}}$.

9. $(\sqrt[3]{0.0129})(31.2)^{2/5}$. **10.** $\dfrac{(0.0016)^4(132)^5}{(0.071)^2\sqrt[4]{0.091}}$.

11. $(1.05)^{-17}$.

12. $\left(\dfrac{950}{43.4}\right)^{1/2}$.

13. $(0.471)^{0.26}$. **14.** $10^{2/5}$.

15. $\sqrt{7} + \sqrt{28}$. **16.** $\log(\log 2)$.

17. $10^{\log(\log 3)}$. **18.** $(\log 2)^{10}$.

19. $\dfrac{(8.91)^3\sqrt{7.22}}{36.58}$. **20.** $\sqrt[3]{0.0091}$.

21. If 100 g of radium is allowed to disintegrate freely, t years from now the amount A of radium present in grams is given by the formula

$$A = 100 \cdot \left(\tfrac{1}{2}\right)^{t/1600}.$$

After how many years will the amount present be 50 g? (The time required for radium to decrease to one half of the initial amount is called the *half-life* of radium.)

*22. Show that the half-life of radium is independent of the initial amount.

*23. Using tables, calculate $\log_3 10$. [*Hint:* Use Exercise 20, Section 13.2.]

REVIEW EXERCISES

In Exercises 1–10 find the numerical value of each logarithm without using tables.

1. $\log_2 16$. **2.** $\log_{1/2} 4$.

3. $\log_2 \tfrac{1}{16}$. **4.** $\log_{3/2} \tfrac{243}{32}$.

5. $\log_5 1$. **6.** $\log_b 1$, $b > 0$, $b \neq 1$.

7. $\log_{27} 3$. **8.** $\log_{10} 0.001$.

9. $\log_{10} 100$. **10.** $\log_5 \sqrt{5}$.

In Exercises 11–16 all letter symbols denote positive real numbers. In each of these, represent the given expression as a logarithm of a single expression.

11. $2 \log a + \frac{1}{2} \log b - \frac{1}{3} \log c$.

12. $\frac{1}{3} (\log a - \log b - \log c)$.

13. $5 \log b - 3 \log c$.   **14.** $\log_b a + \log_b c + 1$.

15. $\frac{1}{4} (\log c^2 + 2 \log c)$.   **16.** $\frac{1}{2} \log_{10} x + \frac{2}{3} \log_{10} y$.

In Exercises 17–22 find both the logarithm and the antilogarithm of each of the given numbers.

17. 2.312.   **18.** 100.2.

19. 1.715.   **20.** 3.412.

21. 1.005.   **22.** 3.712.

In Exercises 23–28 use logarithms to obtain a decimal approximation for each of the given numbers.

23. $\sqrt{0.0212}$.   **24.** $\dfrac{(8.13)(8.751)}{3.19}$.

25. $\sqrt[5]{\dfrac{2.86}{1.381}}$.   **26.** $\dfrac{\sqrt{31.8}\ \sqrt[3]{51.9}}{\sqrt[5]{81.23}}$.

27. $(2.8)^5 \cdot (3.1)^7$.   **28.** $(0.0213)^{10}$.

29. Obtain a decimal approximation of $\log_3 5.12$. [*Hint:* $\log_{10} 3 \cdot \log_3 x = \log_{10} x$ if $x > 0$.]

30. Solve $\log x + \log (x - 3) = 1$ over R.

31. Solve $\log (2x) + \log 3 = 2$ over R.

32. Solve $\log (x - 1) + \log (x + 1) = \log 8$ over R.

MISCELLANEOUS TOPICS

○

14.1 ARITHMETIC PROGRESSIONS

Suppose you were offered a job starting at $500 a month with a guaranteed increase of $25 a month each six months. You might be interested in calculating what your income would be after a certain period of time. You could proceed as follows.

Since your salary will change every six months, let A_1 dollars denote the amount you will earn the first six months, A_2 dollars the amount you will earn the second six months, and so on. Then

$$A_1 = 500 \times 6 = 3000,$$

$$A_2 = (500 + 25)6 = 3000 + 150 = A_1 + 150,$$

$$A_3 = (500 + 25 + 25)6 = 3000 + 150 + 150 = A_2 + 150,$$

$$A_4 = A_3 + 150, \qquad \text{and so on.}$$

Hence, we have a "sequence" of numbers $A_1, A_2, A_3, \cdots$, where each (except A_1) is obtained from the previous one by adding 150. A "sequence" such as this is called an arithmetic progression. We now introduce these terms formally.

sequence **Definition.** A *sequence* is a function whose domain is a set of consecutive positive integers.

If s is a sequence and $i \in D_s$, it is customary to write s_i in *ith term* lieu of $s(i)$ and to call s_i the *ith term* of the sequence s.

[331]

arithmetic **Definition.** An *arithmetic progression* is a sequence that has the
progression property that each term except the first can be obtained by adding
a fixed number to the previous term.

For example, the table below defines an arithmetic progression.

i	1	2	3	4	5
s_i	2	5	8	11	14

Note that

$$s_2 - s_1 = 5 - 2 = 3,$$

$$s_3 - s_2 = 8 - 5 = 3,$$

$$s_4 - s_3 = 11 - 8 = 3,$$

$$s_5 - s_4 = 14 - 11 = 3.$$

In general, if s is an arithmetic progression, the number $s_2 - s_1$
common is called the *common difference*.
difference An interesting problem is the following. Suppose you were
given an arithmetic progression and you were asked to find the
sum of its first k terms. Is there an efficient way of performing
this task?

We illustrate the idea with an example.

Example 1. A junior high school teacher who wanted to keep her
class occupied for a while asked her pupils to find the sum of
the first hundred odd integers. Almost immediately one of her
students came up to her and presented a small piece of paper
with the number 10,000 written on it. The teacher was astonished
and asked how he obtained his result. The pupil explained. To
add $1 + 3 + 5 \cdots + 195 + 197 + 199$, note that the sum of the
first and last terms $(1 + 199)$ is 200, so is the sum of the second
and next-to-last terms $(3 + 197)$, and also $(5 + 195)$. Observing
this pattern and noting that 50 such sums can be obtained, one
finds the required sum is equal to $(200)(50) = 10,000.$*

Note that in the foregoing example the pupil added the first
and last terms and multiplied this sum by half the number of
terms. We generalize this result as follows:

Theorem. If s is an arithmetic progression, then

$$s_1 + s_2 + \cdots + s_n = \frac{n}{2}(s_1 + s_n).$$

Proof: Let $d = s_2 - s_1$.

---

* The famous German mathematician C. F. Gauss was involved in a similar
situation in his early schooling. See James R. Newman, *The World of Mathematics* (New York: Simon and Schuster, 1956), p. 298.

$$A = s_1 + s_2 + \cdots + s_n = s_1 + (s_1 + d) + (s_1 + 2d) + \cdots$$
$$+ s_1 + (n-1)d,$$
$$A = s_n + s_{n-1} + \cdots + s_1 = s_n + (s_n - d) + (s_n - 2d) + \cdots$$
$$+ s_n - (n-1)d.$$

Adding the equalities, we get

$$2A = (s_1 + s_n) + (s_1 + s_n) + \cdots + (s_1 + s_n)$$
$$= n(s_1 + s_n) \qquad \text{(since there are } n \text{ equal terms } s_1 + s_n\text{).}$$

Therefore,

$$A = \frac{n}{2}(s_1 + s_n).$$

We can also write

$$A = \frac{n}{2}(s_1 + s_1 + (n-1)d) = \frac{n}{2}(2s_1 + (n-1)d).$$

Example 2. Suppose s is an arithmetic progression, $s_1 = 6$, and the common difference is 3. Find the sum of the first fifteen terms.

Solution: The fifteenth term is $6 + (15 - 1)3 = 6 + 14 \cdot 3 = 48$. Hence the sum of the first fifteen terms is $\frac{15}{2}(6 + 48) = \frac{15}{2}(54) = 405$.

Exercise 14.1

In Exercises 1–20, the first two terms of an arithmetic progression are given. (a) Find the common difference; (b) find the next three terms; (c) find the sum of the first 40 terms.

1. $2, 6$.
2. $-1, 6$.
3. $3, \frac{7}{2}$.
4. $4, \frac{13}{3}$.
5. $-6, -1$.
6. $-6, -3$.
7. $4, 12$.
8. $8, \frac{5}{2}$.
9. $-7, 1$.
10. $4, 2$.
11. $6, -2$.
12. $-3, 6$.
13. $-\pi, 2$.
14. $-51, -54$.
15. $\frac{3}{2}, \frac{5}{2}$.
16. $\frac{7}{3}, \frac{4}{3}$.
17. $8.5, -\frac{3}{10}$.
18. $-\frac{4}{3}, \frac{12}{5}$.
19. $\frac{6}{7}, \frac{3}{4}$.
20. $\pi/3, 2\pi/3$.

21. A man started a job at a salary of \$400 a month with a guaranteed raise of \$20 a month every six months. How much

will his total earnings have been in (a) four years? (b) six years?

22. Find the sum of the first 50 even positive integers.

23. If s is an arithmetic progression, $s_3 = 5$ and $s_5 = 9$, find (a) the common difference, (b) s_1, (c) s_8, (d) the sum of the first eight terms.

14.2 GEOMETRIC PROGRESSIONS

Suppose that a man buys a piece of property under the following conditions. He pays 1 cent the first day, 2 cents the second day, 4 cents the third day, and so on, each day paying twice the amount he paid the previous day. How much will he have paid in thirty days? The total amount in cents will be

$$A = 1 + 2 + 2^2 + 2^3 + \cdots + 2^{28} + 2^{29}.$$

Multiplying both sides by 2, we get

$$2A = 2 + 2^2 + 2^3 + \cdots + 2^{29} + 2^{30}.$$

Subtracting the first equality from the second, we obtain

$$2A - A = 2^{30} - 1.$$

Hence,

$$A = 2^{30} - 1.$$

As illustrated in Example 1, Section 13.1, 2^{30} is somewhat larger than 1,000,000,000. Hence, in thirty days the man will have paid over 1,000,000,000 cents ($10,000,000)!

We now formalize the ideas illustrated above.

geometric progression **Definition.** A *geometric progression* is a sequence that has the property that each term except the first can be obtained by multiplying the previous term by a nonzero fixed number. This fixed *common ratio* number is called the *common ratio*.

Example 1. The numbers 3, 6 are the first two terms of a geometric progression. Find the common ratio and the next three terms.

Solution: $s_1 = 3$ and $s_2 = 6$. Therefore the common ratio $r = s_2/s_1 = \frac{6}{3} = 2$. Hence,

$$s_3 = 6 \cdot 2 = 12,$$

$$s_4 = 12 \cdot 2 = 24,$$

$$s_5 = 24 \cdot 2 = 48.$$

Example 2. Suppose that s is a geometric progression and suppose that $s_2 = 4$ and $s_4 = 1$. If it is known that not all terms of this progression are positive, find the common ratio and the terms s_1, s_3, s_8.

Solution: Let the common ratio be r. Then $s_3 = s_2 r$ and $s_4 = s_3 r$ $= s_2 r^2$. Hence, $1 = 4r^2$ and $r^2 = \frac{1}{4}$. It follows that $r = \pm\frac{1}{2}$. Since s_2 and s_4 are positive, then if r were positive all terms of this geometric progression would be positive. But it was given that some terms must be negative. Thus, we choose $r = -\frac{1}{2}$.

Since $s_2 = s_1 r$, we have $4 = s_1 (-\frac{1}{2})$. Hence, $s_1 = -8$. Similarly, $s_3 = s_2 r = 4(-\frac{1}{2}) = -2$. In general,

$$s_2 = s_1 r, \qquad s_3 = s_2 r = s_1 r^2, \qquad s_4 = s_3 r = s_1 r^2 r = s_1 r^3,$$

$$\cdots, \qquad s_n = s_1 r^{n-1}.$$

Hence,

$$s_8 = s_1 r^7 = (-8)\left(-\frac{1}{2}\right)^7 = \frac{2^3}{2^7} = \frac{1}{2^4} = \frac{1}{16}.$$

Theorem. If s is a geometric progression with common ratio $r \neq 1$, then

$$s_1 + s_2 + \cdots + s_n = s_1 \left(\frac{1 - r^n}{1 - r}\right).$$

Proof: Let $A = s_1 + s_2 + s_3 + \cdots + s_n$. Then

$$A = s_1 + s_1 r + s_1 r^2 + \cdots + s_1 r^{n-1}.$$

Multiplying both sides of the foregoing equality by r, we obtain

$$Ar = s_1 r + s_1 r^2 + \cdots + s_1 r^n.$$

Subtracting the last equality from the previous one, we get

$$A - Ar = s_1 - s_1 r^n.$$

Hence,

$$A(1 - r) = s_1 (1 - r^n).$$

Since $r \neq 1$, we know that $1 - r \neq 0$, and we obtain

$$A = s_1 \left(\frac{1 - r^n}{1 - r}\right).$$

Example 3. Find the sum of the first ten terms of the geometric progression of Example 2.

Solution: For that progression, we found $s_1 = -8$ and $r = -\frac{1}{2}$. Hence, the sum of the first ten terms is

$$(-8)\frac{1 - (-\frac{1}{2})^{10}}{1 - (-\frac{1}{2})} = (-8)\frac{1 - (\frac{1}{2})^{10}}{1 + (\frac{1}{2})}$$

$$= (-2)^3 \frac{(2^{10} - 1)/2^{10}}{\frac{3}{2}}$$

$$= (-2)^3 \frac{(2^{10} - 1) \cdot 2}{2^{10} \cdot 3}$$

$$= -\frac{2^{10} - 1}{2^6 \cdot 3}$$

$$= -\frac{1023}{2^6 \cdot 3}$$

$$= -\frac{341}{2^6}$$

$$= -\frac{341}{64}.$$

Exercise 14.2

Read Example 1 before doing Exercises 1–6. In each of the following, the first two terms of a geometric progression are given. Find the common ratio and the next three terms.

1. 3, 12.   **2.** $-5, 5$.
3. $-1, 2$.   **4.** $-\frac{1}{2}, 2$.
5. $-\pi, 3$.   **6.** $\frac{1}{8}, -\frac{1}{16}$.

Read Example 2 before doing Exercises 7–11. In each of these, two terms of a geometric progression s are given. Find the common ratio, s_1, s_2, and s_3 in each case.

7. $s_4 = 8$, $s_5 = -4$.
8. $s_6 = 27$, $s_7 = 9$.
9. $s_4 = -32$, $s_6 = -8$ (all terms are negative).
10. $s_4 = -32$, $s_6 = -8$ (some terms are positive).
11. $s_5 = \frac{1}{2}$, $s_7 = \frac{1}{32}$.

Read Example 3 before doing Exercises 12–23.

12. Find the sum of the first 10 terms of the geometric progression of Exercise 1.

13. Find the sum of the first 12 terms of the geometric progression of Exercise 2.

14. Find the sum of the first 8 terms of the geometric progression of Exercise 3.

15. Find the sum of the first 6 terms of the geometric progression of Exercise 4.

16. Find the sum of the first 15 terms of the geometric progression of Exercise 5.

17. Find the sum of the first 9 terms of the geometric progression of Exercise 6.

18. Find the sum of the first 8 terms of the geometric progression of Exercise 7.

19. Find the sum of the first 7 terms of the geometric progression of Exercise 8.

20. Find the sum of the first 10 terms of the geometric progression of Exercise 9.

21. Find the sum of the first 12 terms of the geometric progression of Exercise 10.

22. Find the sum of the first 25 terms of the geometric progression of Exercise 11.

23. If a ping-pong ball is dropped vertically on a table from a height h in. above the table, it bounces back $\frac{2h}{3}$ in. If that ball is dropped from a point 24 in. above a table, find the distance it will have traveled at the instant it hits the table for the fifteenth time.

14.3 COMPOUND INTEREST

Suppose that a man deposits $500 in a bank that pays 6 percent interest compounded semiannually. How much will he have if he withdraws his money five years after he deposited the $500?

Each six months the bank pays 3 percent (half of the annual interest rate) of the capital the man has in his account. Thus, at the end of the first six months, $\frac{3(\$500)}{100} = \$15$ is added to the man's capital as interest. Thus, at the end of six months he has $500 + $15 = $515.

In the accompanying table we list the results of our computations for each conversion period.

Period	Capital	Interest	Total at the end of each Conversion Period
First six months	500	15	515
Second six months	515	15.45	530.45
Third six months	530.45	15.91	546.36
Fourth six months	546.36	16.39	562.75
Fifth six months	562.75	16.88	579.63
Sixth six months	579.63	17.39	597.02
Seventh six months	597.02	17.91	614.93
Eighth six months	614.93	18.45	633.38
Ninth six months	633.38	19.00	652.38
Tenth six months	652.38	19.57	671.95

We would like to find a more convenient way to perform such calculations. We offer the following theorem.

Theorem. Suppose that a sum P is deposited in a bank that pays compound interest. If the interest rate is r for each conversion period and if P_n denotes the compound amount at the end of n conversion periods, then

$$P_n = P(1 + r)^n.$$

Proof: The interest for the first period is Pr. Therefore, at the end of this first period,

$$P_1 = P + Pr = P(1 + r).$$

During the second period, interest is paid on $P(1 + r)$. Hence, that interest is $P(1 + r)r$. Therefore, at the end of two periods,

$$P_2 = P(1 + r) + P(1 + r)r = P(1 + r)(1 + r) = P(1 + r)^2.$$

Proceeding in exactly the same manner, after n steps we would obtain the desired formula:

$$P_n = P(1 + r)^n.$$

Now, referring to the discussion of the beginning of the section, there were ten conversion periods, with $r = \frac{3}{100} = 0.03$ and $P = 500$. Hence,

$$P_{10} = 500(1 + 0.03)^{10} = 500(1.03)^{10}.$$

Using logarithms, we get

$$\log P_{10} = \log [(500)(1.03)^{10}]$$
$$= \log 500 + \log (1.03)^{10}$$
$$= \log 500 + 10 \log 1.03$$
$$= 2.6990 + 10(0.0128)$$
$$= 2.6990 + 0.128$$
$$= 2.827.$$

Hence,

$$P_{10} \doteq 671.4.$$

Note that the foregoing result agrees very closely with the result obtained by direct calculations and listed in the table ($671.95).

Example 1. John wants to buy a car that costs $2,100. The dealer tells him he can have it with no down payment and a monthly payment of $100 for two years. The dealer tells John that at the end of two years he will have paid $2400, the extra $300 being interest and carrying charges on $2100 for two years. He tells John that he is charging $7\frac{1}{7}$ percent interest and carrying charge. Find a fallacy in the dealer's argument.

Solution: The fallacy in the dealer's argument lies in the fact that not all of the $2100 was borrowed for 24 months (since John pays part of it each month). To illustrate, imagine that John has 24 friends, each willing to lend him any amount of money for any length of time at a rate of 1 percent per month compounded monthly. Let $a_1, a_2, \cdots, a_{24}$ be these friends. John decides to borrow $\$\frac{2100}{24}$ from each, and he tells each a_i that he is borrowing this amount for i months. He then pays cash for his car.

At the end of one month he pays a_1 the amount

$$\$\tfrac{2100}{24} + \$\tfrac{2100}{24}(0.01) = \$\tfrac{2100}{24}(1.01).$$

At the end of two months he pays a_2 the amount

$$\$\tfrac{2100}{24}(1.01)^2.$$

$$\cdots$$

At the end of 24 months he pays a_{24} the amount

$$\$\tfrac{2100}{24}(1.01)^{24}.$$

He has paid a total of

$$\left[\tfrac{2100}{24}(1.01) + \tfrac{2100}{24}(1.01)^2 + \tfrac{2100}{24}(1.01)^3 + \cdots + \tfrac{2100}{24}(1.01)^{24}\right]$$

dollars.

The sum in brackets is the sum of the first 24 terms of geometric progression with first term $\tfrac{2100}{24}(1.01)$ and common ratio 1.01. Hence, this sum is

$$\tfrac{2100}{24}(1.01)\frac{1 - (1.01)^{24}}{1 - 1.01} = \tfrac{2100}{24}(1.01)\frac{(1.01)^{24} - 1}{1.01 - 1}$$

$$= \tfrac{2100}{24}(101)[(1.01)^{24} - 1].$$

Using logarithms, we find that $(1.01)^{24}$ is approximately 1.27. Hence,

$$\tfrac{2100}{24}(101)[(1.01)^{24} - 1] \doteq \tfrac{2100}{24}(101)(1.27 - 1)$$

$$= \tfrac{2100}{24}(101)(0.27)$$

$$\doteq 2386.$$

Thus we see that John could borrow from his friends at an annual rate of 12 percent compounded monthly and yet save approximately 14 dollars. Thus, the dealer is charging slightly over 12 percent interest and carrying charges.

Exercise 14.3

Use the theorem of Section 14.3 to do Exercises 1–4. Also read the discussion preceding and immediately following the theorem before attempting these exercises.

1. A man deposited $1000 in a bank that pays 4 percent interest compounded quarterly. How much will he have after (a) 5 years, (b) 10 years, (c) 15 years?

2. A man deposited $5000 in a bank that pays 5 percent interest compounded semiannually. How much will he have after (a) 5 years, (b) 10 years, (c) 15 years? (Use logarithms.)

3. A man deposited $7000 in a bank that pays 6 percent interest compounded quarterly. How much will he have after (a) 5 years, (b) 10 years, (c) 15 years? (Use logarithms.)

4. A man deposited $500 in a bank that pays 6 percent interest compounded quarterly. How long will it take for his capital to (a) double, (b) triple?

Read Example 1 carefully before doing Exercises 5–12.

5. Henry bought a motorcycle for $480 and paid $16 a month for three years. Show that he paid approximately 12 percent interest compounded monthly.

6. Elmar bought a car for $2800 and paid $100 a month for three years. Show that he paid approximately 15 percent interest compounded monthly.

7. Cecilia bought a coat for $110 and paid $10 a month for one year. Show that she paid approximately 18 percent interest compounded monthly.

8. Shirley bought a watch for $130 and paid $10 a month for 14 months. Show that she paid approximately 12 percent interest compounded monthly.

9. Jenny bought a sofa for $288 and paid $10 a month for 36 months. Show that she paid approximately 15 percent interest compounded monthly.

10. Michael bought a desk for $164 and paid $10 a month for 18 months. Show that he paid approximately 12 percent interest compounded monthly.

11. Steven bought an electric guitar for $425 and paid $20 a month for 24 months. Show that he paid approximately 12 percent interest compounded monthly.

12. Kris bought a car for $2000 and paid $100 a month for two years. Show that he paid approximately 18 percent interest compounded monthly.

14.4 ANNUITIES

In the preceding section we discussed problems involving a set of equal periodic payments on a debt, and we considered the interest paid on the money borrowed. It is often the case that equal amounts are invested at equal intervals at certain stated interest rates. These are examples of annuities.

annuity **Definition.** An *annuity* is a sequence of equal payments of P made at equal intervals of time between consecutive payments.

For simplicity we shall restrict our discussion to annuities for which the conversion period of the interest is the same as the time interval between consecutive payments.

amount of an If n is a positive integer, the *amount of an annuity* at the *annuity* end of n payment periods is the sum to which all payments and interest have accumulated during that time, and is denoted A_n.

Example 1. Find the amount of an annuity in which $500 is invested annually for ten years at 5 percent interest compounded annually.

Solution: The first payment of $500 is invested for 9 years. Thus it will amount to $500(1 + 0.05)^9$. The second payment of $500 is invested for 8 years; hence it will amount to $500(1 + 0.05)^8$. Continuing in this way, we see that the ninth payment of $500 earns interest for one period and will amount to $500(1 + 0.05)$. The last payment does not earn any interest. Thus, the total amount of the annuity is

$$[500 + 500(1 + 0.05) + 500(1 + 0.05)^2 + \cdots + 500(1 + 0.05)^8$$

$$+ 500(1 + 0.05)^9] \quad \text{dollars.}$$

The sum in brackets is the sum of the first ten terms of a geometric progression whose first term is 500 and common ratio 1.05. Thus, this sum is

$$500\frac{1 - (1.05)^{10}}{1 - 1.05} = 500\frac{(1.05)^{10} - 1}{1.05 - 1} = 500\frac{(1.05)^{10} - 1}{0.05}$$

$$= 10{,}000((1.05)^{10} - 1).$$

Using logarithms we find $(1.05)^{10} \doteq 1.629$. Hence, the amount of the annuity is approximately \$6290. In general, if \$$P$ is paid at the end of each period for n periods and the interest is r compounded at the end of each period, and if A_n represents the amount of the annuity, then

$$A_n = P \frac{(1+r)^n - 1}{r}.$$

In calculations, it is convenient to use logarithms to approximate $(1+r)^n$.

Let us look at a similar type of problem connected with annuities. If a payment of \$$P$ is to be made at equal intervals of time for n periods at an interest rate r, what is the present value of the annuity? Let \$$P_1$, \$$P_2, \cdots,$ \$$P_n$ be the present values of n payments such that if \$$P_i$ is invested for i conversion periods, it will amount to \$$P$.

Thus, we must have

$$P_1(1+r) = P,$$

$$P_2(1+r)^2 = P,$$

$$\cdot$$
$$\cdot$$
$$\cdot$$

$$P_n(1+r)^n = P.$$

Hence,

$$P_1 = P(1+r)^{-1},$$

$$P_2 = P(1+r)^{-2},$$

$$\cdot$$
$$\cdot$$
$$\cdot$$

$$P_n = P(1+r)^{-n}.$$

Hence, the present value of the annuity is

$$\$[P_1 + P_2 + \cdots + P_n]$$
$$= \$[P(1+r)^{-1} + P(1+r)^{-2} + \cdots + P(1+r)^{-n}].$$

Let $V_n = P(1+r)^{-1} + P(1+r)^{-2} + \cdots + P(1+r)^{-n}$. Multiplying both sides by $(1+r)^n$, we get

$$(1+r)^n V_n = P(1+r)^{n-1} + P(1+r)^{n-2} + \cdots + P.$$

In a way completely analogous to the solution of Example 1, we can show that the right side of the foregoing equality is A_n where

A_n would be the amount of the annuity after n equal payments. Thus,

$$V_n = \frac{A_n}{(1 + r)^n}.$$

Example 2. Find the present value of the annuity described in Example 1.

Solution: In Example 1 we found that the amount A_{10} of the annuity is approximately \$6290. Hence, the present value of this annuity is

$$\frac{\$6290}{(1 + 0.05)^{10}} = \frac{\$6290}{(1.05)^{10}} \doteq \frac{\$6290}{1.629} \doteq \$3861.$$

We remark that the foregoing result can be interpreted as follows: If \$3861 were deposited in a bank paying 5 percent interest compounded annually, at the end of ten years the capital and interest would have accumulated to \$6290.

Or, if \$3861 were deposited under the same conditions, one could withdraw \$500 at the end of each year for ten years and close his account at the end of ten years.

In other words, we say that \$3861 is the present value of the annuity, since one would be willing to pay that price to purchase a contract that would provide him with a \$500 payment at the end of each year for ten years.

Exercise 14.4

Read Example 1 carefully before doing Exercises 1–5. In each of these, find the amount of the annuity that is described.

1. Five hundred dollars is paid annually for 10 years and the interest is 4 percent compounded annually.
2. Five hundred dollars is paid annually for 15 years and the interest is 5 percent compounded annually.
3. One thousand dollars is paid annually for 10 years and the interest is 6 percent compounded annually.
4. One thousand dollars is paid annually for 15 years and the interest rate is 4 percent compounded annually.
5. One hundred fifty dollars is paid annually for 20 years and the interest rate is 5 percent compounded annually.

Read Example 2 carefully before doing Exercises 6–10.

6. Find the present value of the annuity described in Exercise 1.

7. Repeat Exercise 6 for the annuity of Exercise 2.

8. Repeat Exercise 6 for the annuity of Exercise 3.

9. Repeat Exercise 6 for the annuity of Exercise 4.

10. Repeat Exercise 6 for the annuity of Exercise 5.

11. A man buys a house by making a down payment of $5000 and agreeing to pay $1500 a year at the end of each year for the next 20 years. If the rate of interest is 6 percent compounded annually, what is the equivalent cash price of the house?

12. Mr. and Mrs. Smith want to provide an educational fund of $10,000 for their daughter payable 15 years hence. How much should they pay at the end of each month if they can purchase an annuity paying an interest rate of 0.5 percent per month compounded monthly?

*﹡**13.** Suppose that $2000 is deposited now in a bank that pays 5 percent interest compounded annually. How much can be withdrawn starting a year hence if we wish the balance to be $0 after ten equal withdrawals?

14.5 THE PIGEONHOLE PRINCIPLE

The pigeonhole principle* plays a very important role in mathematics. However, it is so clear intuitively that many people use it without thinking of it as a basic principle. The purpose of the present section is to describe this fundamental idea and to show how it can be used.

One way of stating the pigeonhole principle is the following: *If more than n pigeons are placed in n pigeonholes, then at least one hole will contain more than one pigeon.* We shall illustrate how to use this fundamental idea with some examples.

Example 1. Explain why if seven dice are thrown, at least two of them must come up with the same number.

Solution: Since a die has six faces, only one of the numbers 1, 2, 3, 4, 5, 6 can come up with each die. Think of these as representing six pigeonholes and imagine seven pigeons, each corresponding to one of the dice that are thrown. A pigeon is to be put in hole j if and only if the number j comes up on the corresponding die. Since there will be more than one pigeon in at least one of the holes, the same number must come up more than once.

---

* This principle was first discussed formally by a famous German mathematician, Peter Gustav Lejeune Dirichlet (1805–1859).

Question. Why is it that at a gathering of more than 366 people, at least two of them must celebrate their birthday the same day?

Example 2. In Example 2, p. 76, the same remainder (4) occurred twice in the division process. Explain why this fact was to be expected.

Solution: In dividing 11 by 7 using the familiar long-division process, we obtain a succession of remainders: 4, 5, 1, and so on. Since we are dividing by 7, the possible remainders are 0, 1, 2, 3, 4, 5, 6. If for any step the remainder is 0, then the division process stops. Otherwise, there are only six possible remainders (1, 2, 3, 4, 5, or 6). Think of each of these numbers as a pigeonhole, each step in the division representing a pigeon that is placed in a hole i if and only if the remainder i is obtained when performing that step. After more than six steps we must have a repetition in the set of remainders.

Thus, it was not surprising that the same remainder occurred twice in the division of 7 by 4.

Exercise 14.5

Read Examples 1 and 2 before doing Exercises 1–15.

1. Explain why if a coin is tossed three times the following proposition is true: "Heads come up at least twice or tails come up at least twice."

2. Explain why at least two people in Seattle have the same number of hairs on their heads (exclude the trivial case!).

3. Betty will go out on dates only with John, Roy, and Steven. She wants to have four dates this month. Explain why she must go out with the same boy (pigeon!) at least twice.

4. Given an arrangement of objects, a *switch* is the act of exchanging the positions of two of these objects. For example, we can go from the arrangement 5, 1, 3, 2, 4 to the arrangement 5, 1, 4, 2, 3 with one switch (exchanging the positions of 3 and 4). Explain how we can go from any arrangement of n objects to any other arrangement of these objects in not more than $(n - 1)$ switches.

5. Recall that if m and n are natural numbers, there exist unique non-negative integers q and r with $0 \leqslant r < n$ such that $m = n \cdot q + r$. For example, $7 = 4 \cdot 1 + 3$ and $13 = 3 \cdot 4 + 1$. We know that q is the quotient and r the remainder. Prove that if six natural numbers are divided by 5, we can find among them two that will yield the same remainder.

6. Repeat Exercise 5 for 14 natural numbers that are divided by 13.

7. Show that there exist distinct natural numbers m and n such that $2^m - 2^n$ is divisible by 5. [*Hint:* Consider the numbers 2^1, 2^2, 2^3, 2^4, 2^5, 2^6 and use exercise 5.]

8. Show that there exist distinct natural numbers m and n such that $3^m - 3^n$ is divisible by 13. [*Hint:* Consider the numbers 3^1, 3^2, 3^3, $\cdots$, 3^{14} and use the idea of Exercise 7.]

9. Recall that if S is a set with n elements, an operation on S is a rule that assigns to each ordered pair (a, b) of elements of S a unique element of S. Explain why if $n \geqslant 2$ there must be at least two distinct ordered pairs to which the same element of S is assigned.

10. Show that $\frac{7}{13}$ can be expressed as a periodic decimal. (See p. 75.)

11. Repeat Exercise 10 for $\frac{20}{17}$.

*__12.__ Prove that every rational number can be expressed as a periodic decimal.

*__13.__ Prove that if a, b, c are natural numbers such that a divides b and b divides c, then a divides c.

*__14.__ Prove that if a is a natural number, then any natural number that divides a is less than or equal to a. [*Hint:* Use the remarks on p. 124.]

*__15.__ Prove that if m is a natural number greater than 1, then there is a prime number that divides m. [*Hint:* Use Exercises 13 and 14 and the pigeonhole principle.]

REVIEW EXERCISES

1. The first term of an arithmetic progression is 2 and the third term is 8. Find the twentieth term and the sum of the first ten terms.

2. Show that the sum of the first n positive odd integers is equal to n^2.

3. Find the value or values of K for which the sequence $K - 3$, $K + 5$, $2K - 1$ is an arithmetic progression.

4. If S is an arithmetic progression and $S_2 = 3$, $S_{10} = 49$, find

    (a) the common difference,

    (b) S_{15},

    (c) the sum of the first ten terms.

5. A man accepted a job at a salary of $300 a month with a guaranteed raise of $60 a year. How much will his total earnings be in **(a)** five years, **(b)** eight years?

6. Suppose that S is a geometric progression of real numbers and $S_3 = 6$, $S_5 = 54$. If S has some negative terms, find the common ratio and the terms S_2 and S_8.

7. A ball is dropped from a height of 6 ft. When it hits the floor for the nth time, it rebounds to the height $6 \cdot (\frac{3}{4})^n$ ft. What will be the distance traveled by the ball when it hits the ground for the fifteenth time?

*8. The sequence a, b, c is an arithmetic progression with sum 18. If $a + 4$, $b + 4$, $c + 36$ is a geometric progression, find a, b, and c.

9. Water is removed from a tank by a pump that removes one-tenth of the remaining water at each stroke. What fractional part of the original amount remains after 5 strokes? After how many strokes will less than half of the original amount be present?

10. A sum of $100 is invested at 4 percent, compounded semiannually, for 8 years. Find the compound amount at the end of 8 years.

11. What amount of money should be invested at 4 percent interest, compounded semiannually, to have the compound amount of $1000 at the end of 9 years?

12. A house is offered for sale on the following terms: $4000 cash and $800 at the end of each 6 months for 8 years. What is the corresponding cash price for the house, if the seller wants to earn 6 percent compounded semiannually?

13. A man buys a house for $15,000. He pays $5000 on the date of purchase and is to pay the balance in 10 equal installments at intervals of 6 months over the next 5 years. What should be the amount of these payments, if money is worth 4 percent compounded semiannually?

14. The beneficiary of a life insurance policy is to receive 10 annual payments of $1200 each, the first payment to be made at the time of the death of the insured. What would be the equivalent cash settlement at the time of death of the insured, if money is worth 3 percent compounded annually?

15. A man who has no savings would like to retire at the end of 8 years with $15,000 in cash. How much should he deposit in a savings bank at the end of each six-month period if the bank pays 6 percent compounded semiannually?

16. A piece of land is offered for sale at $1500. An investor estimates that 6 years from now the land will be worth $2000. If he purchased the land now at $1500 and sold it 6 years later at $2000, what interest rate, compounded semiannually, would he be earning?

17. Explain why a function f with domain the set of the first 100 consecutive positive integers and range the first 95 consecutive positive integers cannot be one-to-one.

18. A teacher has a class of more than five students. If the possible grades are A, B, C, D, E, explain why at least two students in the class will have the same grade for the course.

19. Is it possible to have a function whose domain has n elements $(n \in N)$ and whose range has more than n elements? Justify your answer.

ANSWERS TO ODD-NUMBERED EXERCISES

Exercise 1.2

   1. Proposition.
   3. Statement, but not a proposition.
   5. Statement, but not a proposition.
   7. Statement, but not a proposition.
   9. Not a statement (hence not a proposition).
   11. Proposition.
   13. Statement, but not a proposition.
   15. Not a statement.
   17. Statement, but not a proposition.
   19. Statement, but not a proposition.

Exercise 1.3

   1. False. Some apple is a fruit. (True.)
   3. True. $2 + 2$ is not equal to 4. (False.)
   5. True. All squares are round. (False.)
   13. False.
   15. True.
   17. True.
   19. Some man is not good.
   21. No lion is hungry.
   23. Some bird does not fly.

Exercise 1.4

   1. The conclusion is true.
   3. Hypothesis is false.

5. Hypothesis is true, while the conclusion is false.

7. True.

9. True.

11. True.

13. True.

15. True.

17. True.

23. If $p \to q$ is true, it does not follow that p is true.

25. $q \to \sim p$.

27. If $2 + b$ is not an even integer, then b is not an even integer.

29. If two lines coincide, then they have a point in common.

Exercise 1.5

1. The given proposition is true. Converse: If $3 = 3$, then $1 = 2$. (F.) Contrapositive: If $3 \neq 3$, then $1 \neq 2$. (T.)

3. The given proposition is false. Converse: If $1 = 2$, then $2 + 2 \neq 5$. (T.) Contrapositive: If $1 \neq 2$, then $2 + 2 = 5$. (F.)

5. The given proposition is true. Converse: If p, then $p \vee (\sim p)$. (T.) Contrapositive: If $\sim p$, then $\sim (p \vee \sim p)$. (T.)

7. The given proposition is true. Converse: $\sim q \to p$. (T.) Contrapositive: $\sim (\sim q) \to \sim p$. (T.)

9. The given proposition is false. Converse: $5 + 1 = 7$ is sufficient for $2 + 1 = 3$. (T.) Contrapositive: $5 + 1 \neq 7$ is sufficient for $2 + 1 \neq 3$. (F.)

11. [*Hint:* Choose two propositions with the same truth value.]

13. [*Hint:* Choose two propositions with opposite truth values.]

REVIEW EXERCISES FOR CHAPTER 1

1. (a) Some automobile does not have four wheels.

(b) Mars is not a planet.

(c) Some proposition is not a statement.

(d) All counting numbers are even.

(e) Mt. Rainier is at least as high as Mt. Everest.

5. Converse: If all automobiles have four wheels, then Mars is a planet. Contrapositive: If some automobile does not have four wheels, then Mars is not a planet.

7. (a) False.

(b) False.

(c) True.

(d) True.

(e) True.

9. False.

11. True.

13. True. You should be able to justify this conclusion.

15. False.

17. True.

19. False.

21. True.

23. True.

25. True.

Exercise 2.1

1. $\{1, 2, 3, 4, 5, 6, 7, 8\}$.

3. $\{w, x, y, z\}$.

5. $\{11, 12, 13, 14, 15, 16, 17, 18, 19\}$.

7. $\varnothing$.

9. $\{$L. B. Johnson$\}$.

11. C is one of the following sets: $\varnothing$, $\{b\}$, $\{c\}$, $\{d\}$, $\{b, c\}$, $\{b, d\}$, $\{c, d\}$, $\{b, c, d\}$.

Exercise 2.2

1. **(a)** $\{2\}$, **(b)** $\{2\}$.

3. **(a)** $\varnothing$, **(b)** $\{\frac{4}{7}\}$.

Exercise 2.3

1. $\{2, 3\}$.

3. $\{2\}$.

5. $\varnothing$.

7. **(a)** $\{1\}$, **(b)** $\{-3, 1\}$, **(c)** $\{-3, \frac{1}{3}, 1\}$.

9. **(a)** The set of counting numbers, **(b)** the set of all nonzero integers, **(c)** the set of all ratios of integers of the form $\frac{a}{b}$, $a \neq 0$, $b \neq 0$.

11. **(a)** $\{7\}$, **(b)** $\{-4, -2, 7\}$, **(c)** $\{-4, -2, 7\}$.

13. **(a)** $\varnothing$, **(b)** $\{0\}$, **(c)** $\{-\frac{1}{6}, -\frac{1}{9}, 0\}$.

15. **(a)** $\varnothing$, **(b)** $\varnothing$, **(c)** $\{-\frac{1}{2}, -\frac{1}{3}, -\frac{1}{4}\}$.

17. $\{a, c\}$.

19. $\varnothing$.

21. The open sentence $(x + 1)(x - 3)(x - 4) = 0$ with the replacement set the set of all integers is one example. Many other examples are possible.

Exercise 2.4

1. $A \subset B$; also $A \subseteq B$.

3. $A \subseteq B$, but $A \not\subset B$.

5. $A \subseteq B$, but $A \not\subset B$.

7. $A \not\subseteq B$, and $A \not\subset B$.

9. $\varnothing$, $\{1\}$, $\{4\}$, $\{1, 4\}$.

11. $\varnothing$, $\{1\}$, $\{2\}$, $\{3\}$, $\{4\}$, $\{1, 2\}$, $\{1, 3\}$, $\{1, 4\}$, $\{2, 3\}$, $\{2, 4\}$, $\{3, 4\}$, $\{1, 2, 3\}$, $\{1, 2, 4\}$, $\{1, 3, 4\}$, $\{2, 3, 4\}$, $\{1, 2, 3, 4\}$.

13. $\varnothing$, $\{0\}$.

15. $\varnothing$, $\{a\}$, $\{b\}$, $\{c\}$, $\{d\}$, $\{a, b\}$, $\{a, c\}$, $\{a, d\}$, $\{b, c\}$, $\{b, d\}$, $\{c, d\}$, $\{a, b, c\}$, $\{a, b, d\}$, $\{a, c, d\}$, $\{b, c, d\}$, $\{a, b, c, d\}$.

17. $\varnothing$, $\{a\}$, $\{\{a\}\}$, $\{a, \{a\}\}$.

19. $\varnothing$, $\{\{0\}\}$, $\{\{\{0\}\}\}$, $\{\{0\}, \{\{0\}\}\}$.

23. $\{a\}$, $\{a, b\}$, $\{a, c\}$.

Exercise 2.5

1. (a) $\sim p(x)$: "x is not a college teacher." "For all $x \in A, p(x)$" is false. "For some $x \in A, \sim p(x)$" is true.

(b) $\sim p(x)$: "x is not a carpenter." "For all $x \in A, p(x)$" is false. "For some $x \in A, \sim p(x)$" is true.

(c) $\sim p(x)$: "x is not a woman." "For all $x \in A, \sim p(x)$" is false. "For some $x \in A, \sim p(x)$" is true.

(d) $\sim p(x)$: "x is not an artist." "For all $x \in A, p(x)$" is false. "For some $x \in A, \sim p(x)$" is true.

(e) $\sim p(x)$: "x is not an alcoholic." "For all $x \in A, p(x)$" is false. "For some $x \in A, \sim p(x)$" is true.

3. $\{x \in B | p(x)\} = \{9, 10, 11\}$. $\{x \in B | \sim p(x)\} = \{2, 3, 4, 5, 6, 7, 8\}$.

5. $\{x \in B | p(x)\} = \{7, 9, 11, 13\}$. $\{x \in B | \sim p(x)\} = \{x \in B | x$ is even or x is not between 6 and $14\} = \{1, 2, 3, 4, 5, 6, 8, 10, 12, 14, 15, 16, \cdots\}$.

7. $\{x \in B | p(x)\} = $ set of odd counting numbers. $\{x \in B | \sim p(x)\} = $ set of even counting numbers.

Exercise 2.6

1. $B = \{2, 12\}$. $A - B = \{4, 6, 8, 10, 14\}$.

3. $B = \{2\}$. $A - B = \{3, 5\}$.
5. $B = \{1, 2\}$. $A - B = \{3\}$.
7. $A - B = \{$A. L. Yandl$\}$.
9. $A - B = A$.
11. $A = B$.
13. $A - B = \{-1, 7, 11\}$, $B - A = \{36\}$.
15. $A - B = A$, $B - A = \varnothing$.
17. $A - B = A$ if and only if A and B have no elements in common.

Exercise 2.7

1. (a) $\{3, 6\}$, (b) $\{1, 2, 3, 4, 5, 6, 10, 12\}$, (c) $\{1, 2, 4, 5\}$, (d) $\{10, 12\}$.
3. (a) $\varnothing$, (b) set of counting numbers, (c) A, (d) B.
5. (a) $\varnothing$, (b) B; (c) $\varnothing$, (d) B.
7. (a) $\{2\}$, (b) $\{2, 5\}$, (c) $\varnothing$, (d) $\{5\}$.
9. (a) $\{2\}$, (b) B; (c) $\varnothing$, (d) $\{3, 4\}$.
11.

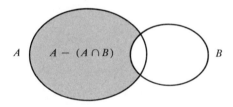

19. $A = \{a, b, c\}$, $B = \{a, c, d\}$.
21. (a) 37, (b) 12, (c) 27.
23. $A = B$.

REVIEW EXERCISES FOR CHAPTER 2

1. {Washington, Wisconsin, West Virginia, Wyoming}.
3. $\varnothing$.
5. (a) $\varnothing$, (b) $\varnothing$, (c) $\{2\}$.
7. (a) $S = \{\varnothing, \{1\}, \{2\}, \{1, 2\}\}$.
   (b) $S \subseteq S$ (true), $S \subset S$ (false), $S \in S$ (false), $\varnothing \in S$ (true), $\{\varnothing\} \in S$ (false), $1 \in S$ (false), $\{1\} \in S$ (true), $\{\{1, 2\}\} \subset S$ (true), $\{1, 2\} \subset S$ (false).
9. $\{1, 2, 3, 4, 5\}$.
11. $\{2, 3, 4, 5\}$.
13. $\{2\}$.
15. $\{3, 5\}$.

17. $\{2, 3, 5\}$.

19. $\{3, 5\}$.

21. $A = \{a\}$, $B = \{b, c\}$ will do. Other choices are possible.

Exercise 3.1

1. Yes.

3. Yes.

5. Yes.

7. Yes.

9. No. Note that $b \oplus a = c \notin A$.

11. (a) 4, (b) no. Note that $\{3, 5\} \subset A$, yet $3 \odot 5 = 4 \notin A$.

Exercise 3.2

1. (a) Yes, (b) yes, (c) b is the identity.

3. (a) No, (b) yes, (c) $\oplus$ has no identity.

5. x is the inverse of x; z has no inverse.

Exercise 3.3

1. 1.

3. 4.

5. $^-1$.

7. 112.

9. 0.

11. Commutative law of addition.

13. Commutative law of addition.

15. Distributive law.

17. $^-8$ is the additive inverse of 8.

19. 1 is the multiplicative identity.

Exercise 3.4

1. $(x + 3)^2$.

3. $(2x + 3)^2$.

5. $(t + 5)^2$.

7. $(4x + 1)^2$.

9. $(x + y) \cdot (x + 3y)$.

11. $(5x + y) \cdot (x + y)$.

13. $(2x + t) \cdot (x + t)$.

15. $4 \cdot (2r + t) \cdot (r + t)$.

Exercise 3.5

1. $^-15$.

3. $^-15$.

5. $^-156$.

7. $^-15$.

9. $^-61$.

11. $^-3$.

13. $^-1$.

15. 21.

17. $^-14$.

19. $^-4$.

21. 0.

23. $^-13$.

25. $^-7$.

27. $^-45$.

29. 90.

Exercise 3.6

1. $5^4 \cdot 8^7 = \underbrace{2 \cdot 2 \cdots \cdots 2}_{21 \text{ factors}} \cdot 5 \cdot 5 \cdot 5 \cdot 5 \quad 2^{20} \cdot 5^6 = \underbrace{2 \cdot 2 \cdots \cdots 2}_{20 \text{ factors}} \cdot$

$5 \cdot 5 \cdot 5 \cdot 5 \cdot 5 \cdot 5$. The numbers are not equal.

3. $10^5 \cdot 6^4 = 2 \cdot 2 \cdot 2 \cdot 2 \cdot 2 \cdot 2 \cdot 2 \cdot 2 \cdot 2 \cdot 3 \cdot 3 \cdot 3 \cdot 3 \cdot 5 \cdot 5 \cdot 5 \cdot 5 \cdot 5$.
$15^4 \cdot 2^{11} = 2 \cdot 2 \cdot 2 \cdot 2 \cdot 2 \cdot 2 \cdot 2 \cdot 2 \cdot 2 \cdot 2 \cdot 2 \cdot 3 \cdot 3 \cdot 3 \cdot 3 \cdot 5 \cdot 5 \cdot 5 \cdot 5$. The numbers are not equal.

5. No. Use the fundamental theorem of arithmetic to justify this conclusion.

7. No. Justify this conclusion!

9. No. Justify this conclusion!

11. 17, 37, 101.

13. 5, 17, 37, 101, 197.

15. No. Example: 6 divides $12 = 4 \cdot 3$, yet 6 does not divide 4 or 3.

17. (a) $12 = 7 + 5$, $14 = 7 + 7$, $16 = 13 + 3$, $18 = 11 + 7$, $20 = 17 + 3$.

(b) This is known as *Goldbach's conjecture*. No one so far has been able to prove or disprove it.

21. A natural number is divisible by 8, if the number formed by the three right-hand digits is divisible by 8. Prove this statement!

Exercise 3.7

1. $\{1, 2_1, 2_2, 2_3, 11_1\}$.
3. $\{1, 2_1, 2_2, 2_3, 3_1, 3_2, 5_1\}$.
5. $\{1, 5_1, 5_2, 5_3, 13_1, 13_2\}$.
7. 14,625.
9. 3920.
11. 4.
13. 1.
15. 18.
17. 1.
19. 1.

REVIEW EXERCISES FOR CHAPTER 3

3. Example 1, Section 3.1.
5. Example 4, Section 3.1.
7. $(2x - 1)^2$.
9. $(x + 2) \cdot (5x + 3)$.
11. $(x + 1)^2$.
15. 792.
17. 3276.

Exercise 4.1

1. $\frac{12}{5}$.
3. $\frac{18}{13}$.
5. $\frac{3}{25}$.
7. $\{\frac{1}{4}, 2, 3\}$.
9. $\{-3, -\frac{1}{2}, 3, 6\}$.
11. $a = b$.
13. $\{a, b\} \subseteq Q \wedge a \neq 0 \wedge b \neq 0$.
15. [*Hint:* $1 \cdot 1 = 1$.]

Exercise 4.2

9. [*Hint:* Show that $11 \cdot 25 \neq 13 \cdot 22$.]

Exercise 4.3

1. $\dfrac{79}{105}$.

3. $\dfrac{181}{525}$.

5. $-\dfrac{211}{420}$.

7. $\dfrac{8383}{19500}$.

9. $\dfrac{292}{319}$.

11. $\dfrac{41}{60}$.

13. $\dfrac{1337}{4896}$.

Exercise 4.4

1. $0.\underline{285714}$.
3. $0.\underline{923076}$.
5. 13.0
7. $2.3\underline{3} = \frac{7}{3}$.

9. $63.\underline{412} = \dfrac{63,349}{990}$.

11. $151.0\underline{13} = \dfrac{149,503}{990}$.

13. $71.513\underline{256} = \dfrac{71441743}{999000}$.

15. $3.\underline{9} = 4$.
17. $16.\underline{9} = 17$.
19. $\frac{1}{2} = 0.5\underline{0} = 0.4\underline{9}$. Can you give another example?

Exercise 4.5

1. $\varnothing$.
3. $\varnothing$.
5. $\varnothing$.
7. $\varnothing$.
9. $\{1\}$.
11. $\varnothing$.
23. (c) 2.

Exercise 4.6

3. [*Hint:* Suppose a is rational, $a \neq 0$, b is irrational, and $c = a \cdot b$. Show that c must be irrational.]

5. $\dfrac{(c + 1)(c + 4)}{c + 5}$.

7. x.

9. -2.

11. $\dfrac{a - 3}{a + 3}$.

13. $\dfrac{a - 3}{a - 2}$.

Exercise 4.7

1. $5 - i$.

3. 0.

5. $17 - i$.

7. $25 + 8i$.

9. $-i$.

11. i.

13. $6 + 8i$.

15. $\frac{11}{4} - \frac{13}{10}i$.

17. $\frac{111}{65} - \frac{53}{130}i$.

19. $\frac{88}{13} + \frac{15}{13}i$.

23. $\frac{18}{13} + \frac{12}{13}i$.

25. $-\frac{4}{13} + \frac{7}{13}i$.

REVIEW EXERCISES FOR CHAPTER 4

1. $\{-\frac{1}{2}\}$.

3. $\frac{78}{55}$.

5. $\frac{5}{16}$.

7. 2.75.

11. xy.

13. $1 + i$.

15. $\frac{5}{2} + \frac{34}{5}i$.

17. $\frac{3}{2} - \frac{11}{2}i$.

19. $-1 - 3i$.

21. True.
23. True.
25. False. Example: $(1 + i) \cdot (1 - i) = 2 \in R$.

Exercise 5.1

1. $3^8 \cdot 3^2 = 3^{10}$. First basic law of exponents.
3. $(25^6)^9 = 25^{54}$. Second basic law of exponents.
5. $[(a + b)^2]^3 = (a + b)^6$. Second basic law of exponents.
7. $-a^4$.

9. $\left(\dfrac{c}{d}\right)^5$.

11. $\left(\dfrac{ab}{b - a}\right)^3$.

13. $(ab)^{3n + m}$.
15. $(cd)^{3m - 5n}$.

17. $\left(\dfrac{a^n + b^n}{b^n - a^n}\right)^n$.

23. If n is greater than 1, $(1 + 2)^n \neq 1^n + 2^n$. Equality holds, if $a = 0$ or $b = 0$.

Exercise 5.2

1. (a) Polynomial, (b) polynomial, (c) not a polynomial, (d) not a polynomial.
3. $T(x) = -2x + 3$. (a) 1, (b) -2, (c) $-2x, 3$, (d) $-2, 3$.
5. $T(x) = 5$. (a) 0, (b) 5, (c) 5, (d) 5.
7. $T(x) = -5x^4 + (2 + i)x^2 + 3x - (2 + 4i)$. (a) 4, (b) -5, (c) $-5x^4, (2 + i)x^2, 3x, -(2 + 4i)$, (d) $-5, (2 + i), 3, -(2 + 4i)$.
9. $3y - 2$.
11. $-5x^2 + 8x + 3$.
13. $8u^2 + 2u + 6$.
15. $33V^{12} + 9V^{11} - 13V^{10} + 6V^9 + 10V^8 - V^7 - 6V^6 + 7V^5$.
17. $w^8 + w^6 + w^4 + w^2$.
19. $-2bi \cdot z^4 + (a + c)i \cdot z^3 + (a + e)i \cdot z^2 + bi \cdot z - bi$.
21. $10x^3 - 10x^2 + 10x + 10$.
23. $10u^8 - 4u^7 - u^6 + u^5 - 8u^4 + 9u^3 + u^2 - 3u + 1$.
25. $(6 + i)u^5 - (3 + 4i)u^4 + (3 + i)u^3 + (6 - 2i)u^2 - (7 + 2i)u - (3 + i)$.
27. $-dy^3 - diy^2 - by + bi$.

29. $13y - 9.$

31. $(3 + 7a)u + ab(a - 1).$

33. $(2d - a - c)x + d.$

Exercise 5.3

1. $6x^5 + 3x^4 - 4x^3 + x^2 - 2.$

3. $25x^4 - x^2 + 2x - 1.$

5. $8y^{13} - 19y^{11} + 15y^9 - 20y^7 + 4y^5.$

7. $y^2 + 3y - 1.$

9. $(u + 3)^4 - 2(u + 3)^2 + 1.$

15. $2x - \dfrac{x - 2}{2x^2 - 1}.$

17. $u^4 - 2u^2 + 4 - \dfrac{9}{u^2 + 2}.$

19. $2x - \dfrac{3x + 2}{3x^2 + 2}.$

21. $x^3 + x^2 + x + 1.$

23. $x^2 - \dfrac{7}{3}x + \dfrac{38}{9} - \dfrac{193x + 25}{9(3x^2 + 5x - 1)}.$

25. $6x^2 - 18x + 84 - \dfrac{349x^2 - 438x + 82}{x^3 + 3x^2 - 5x + 1}.$

Exercise 5.4

1. (a) 1, (b) -1, (c) $-1 - 4i$, (d) $3a^3 - a - 1.$

3. (a) 0, (b) 0, (c) -1, (d) $c^4 - 1.$

5. (a) Divisible by 9, (b) divisible by 9.

7. (a) Divisible by 9, (b) not divisible by 9.

9. Yes. Use the factor theorem to prove this assertion.

11. Yes. Use the factor theorem to prove this assertion.

13. No. Use the factor theorem to prove this assertion.

17. (a) [*Hint:* Study Example 2. Write $F(x) = G(x) \cdot (x + 1) + F(-1).$] (b) The only number divisible by 11 in Exercises 5–8 is 198.

Exercise 5.5

1. $(x - 7) \cdot (x + 7).$

3. $(2x + 5)^2$.

5. $(z + 2) \cdot (z^2 - 2z + 4)$.

7. $(x + 1) \cdot (x^4 - x^3 + x^2 - x + 1)$.

9. $(4u + 3b)^2$.

11. $(2y - 3c) \cdot (4y^2 + 6cy + 9c^2)$.

13. $2bx \cdot (4bx + 3a - 6)$.

15. $abx^2 \cdot ((a - b)x + c - d)$.

17. $(a - b)^2 u^2 ((a - b) \cdot u - 2)((a - b)u - 1)$.

19. $(3x + a) \cdot (1 - 2x)$.

21. $(b^2 + 1) \cdot (y - 1) \cdot (y^2 + y + 1)$.

23. $(V + 2) \cdot (V + 2a + 5)$.

25. $(x + 5) \cdot (x + 6)$.

27. $(y - 5) \cdot (y + 4)$.

29. $(x - 7)(x + 5)$.

31. $(3x + 2) \cdot (x + 1)$.

33. $(3r + 4) \cdot (r - 3)$.

35. $(3x + 2) \cdot (2x - 1)$.

37. $(2x - a + 1) \cdot (2x - a + 2)$.

39. $(4y - 3k - 1)^2$.

41. $cx(3x + ab - 5)(3x + ab + 2)$.

43. $(y^2 + ay + a^2)(y^2 - ay + a^2)$.

45. $(x + 3) \cdot (x - 2) \cdot (2x - 1) \cdot (x + 2)$.

REVIEW EXERCISES FOR CHAPTER 5

3. $2a^4 + 4a^2b^3 + b^6$.

5. $\dfrac{1}{a^2c^5}$.

7. $\dfrac{3y}{16a}$.

9. $2x^5 - 2x^2 + 10x + 3$.

11. $\dfrac{1}{3}x^2 + \dfrac{1}{9}x + \dfrac{10}{27} + \dfrac{37}{27(3x - 1)}$.

13. $4x^2 - 3x - 1 - \dfrac{2x}{x^2 + x + 1}$.

17. $F(0) = -1$, $F(1) = 3$, $F(a) = 3a^3 + a^2 - 1$.

21. $2x(x - 2)(x + 2)$.

23. $(5x - 2)(2x + 3)$.

25. $(3s + b + 2)(3s + b - 1)$.

27. $(x + 3i)(x - 3i)$.

Exercise 6.1

1. True, according to Example 1.
3. False. Example: $-2 < -1$, yet $-2 + (-1) \notin \mathscr{P}$.
5. False. Example: $1 < 2$, yet $(-1) \cdot 1 \nless (-1) \cdot 2$.
9. $R - \{1\}$. Solution set $= \{x \mid x > 1\}$.
11. $R - \{-\frac{1}{4}\}$. Solution set $= \{x \mid x > -\frac{1}{4}\}$.
13. $R - \{0\}$. Solution set $= \{x \mid x > 0\}$.

15. $R - \left\{\dfrac{bn}{c}\right\}$. Solution set $= \left\{y \mid y < \dfrac{bn}{c}\right\}$.

Exercise 6.2

1. $R - \{9\}$. Solution set $= \{x \mid x > 9\}$.
3. $R - \{\frac{1}{7}\}$. Solution set $= \{y \mid y < \frac{1}{7}\}$.
5. $R - \{-\frac{7}{2}\}$. Solution set $= \{y \mid y > -\frac{7}{2}\}$.
7. $R - \{-6\}$. Solution set $= \{u \mid u < -6\}$.

9. $R - \left\{\dfrac{d}{c}\right\}$. Solution set $= \left\{v \mid v > \dfrac{d}{c}\right\}$.

11. $R - \left\{\dfrac{a+b}{m+n}\right\}$. Solution set $= \left\{x \mid x < \dfrac{a+b}{m+n}\right\}$.

13. $R - \{-a\}$. Solution set $= \varnothing$.
15. The student's conclusion is incorrect. The two last inequalities are not equivalent. The solution set of the given inequality is $\{x \mid x > -1\}$.

Exercise 6.3

1. $\{x \mid x \geqslant \frac{1}{4}\}$.
3. $\{x \mid x \geqslant \frac{1}{6}\}$.

5. $\left\{x \mid x \leqslant -\dfrac{b}{a}\right\}$.

7. $\{x \mid -4 < x < -3\}$.
9. $\varnothing$.

11. $\left\{x \mid \dfrac{m-n}{a} \leqslant x < \dfrac{p-n}{a}\right\}$.

13. $\{x \mid -\frac{4}{5} \leqslant x < \frac{1}{3}\}$.
15. $\varnothing$.

17. $\left\{v \left| \dfrac{1}{3-a} \leqslant v < \dfrac{1-b}{2}\right.\right\}$.

19. $\{x|-3 < x < 1\}$.

21. $\{x|-1 \leqslant z \leqslant 1\}$.

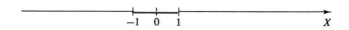

Exercise 6.4

1. (a) $\frac{1}{10}$, (b) 5, (c) -3.

5. 2.

7. 6.

9. $|k-1|$.

13. (a) $|x^2 + 1| \cdot |x^2 - 1|$, (b) $|x^2y^2 - 2| \cdot |x^2y^2 - 1|$, (c) $|uv + ab|$ $\cdot |u^2v^2 - abuv + a^2b^2|$.

15. $\{x|x \geqslant 0\}$.

17. $\{u|u < \frac{5}{6}\}$.

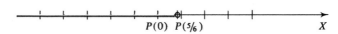

19. $\left\{x \left| x \leqslant \dfrac{d-b}{2c}\right.\right\}$.

Exercise 6.5

1. $\{x| -1 < x < 1\}$.

3. $\{u|-\frac{1}{2} < u < 5\}$.

5. $\{t|t \geqslant b \lor t \leqslant a\}$.

7. $\{x|-4 < x < 2\}$.

9. $\left\{V\left|-0.00001 + \dfrac{a}{10} < V < 0.00001 + \dfrac{a}{10}\right.\right\}$.

11. $\left\{ y \left| -\dfrac{e}{|a|} < y < \dfrac{2b+e}{|a|} \right. \right\}$.

13. $\{x \mid x > 10 \lor x < 4\}$.

15. $\{w \mid w > \frac{9}{7} \lor w < -\frac{1}{7}\}$.

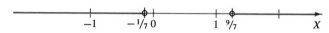

17. $\left\{ x \left| x > \dfrac{p}{m} \lor x < \dfrac{-p-2q}{m} \right. \right\}$.

19. $\{x \mid -3 \leqslant x < -1\}$.

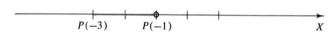

21. $\{r \mid 0 \leqslant r < \frac{1}{4}\}$.

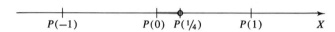

23. $\{y \mid y > -\frac{1}{4} \lor y \leqslant -4\}$.

27. $\{x \mid 1 < x < 2\}$.
29. $\sqrt{15} - \sqrt{14} > \sqrt{32} - \sqrt{31}$.
31. $\varnothing$.

REVIEW EXERCISES FOR CHAPTER 6

3. R. The graph of the solution set is the entire real number line.
5. $\{x \mid x > -1\}$.

7. $\{x \mid x \geqslant \frac{1}{2}\}$.

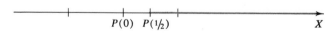

9. $\{y \mid y < 0\}$.

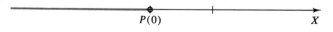

11. $\{x | -2 < x < 8\}$.

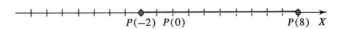

$P(-2)\ P(0)$ $P(8)$ X

13. $\{x | 2 < x < 5\}$.

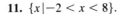

$P(0)$ $P(2)$ $P(5)$ X

15. $\{x | 1 < x < 5 \wedge x \neq 2\}$.

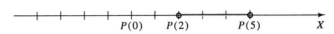

$P(0)\ P(1)\ P(2)$ $P(5)$ X

17. $\{x | -1 < x < 2\}$.

19. $\{x | -2 \leqslant x \leqslant 0 \vee x \geqslant 1\}$.

21. $\{5\}$.

23. (b) $\{x | x > 1 - c\}$.

25. $\{x | x > -\frac{1}{3} \wedge x \neq 2\}$.

$P(-\frac{1}{3})\ P(0)$ $P(2)$ X

Exercise 7.1

1. $\{1\}$.

3. $\left\{\dfrac{1}{a}\right\}$.

5. $\left\{-\dfrac{2}{b}\right\}$.

7.

$x - 2 = 0$ $(x - 2)(x - 1) = 0$

is true, and   is false.

$(x - 2)(x - 1) = 0$ $x - 2 = 0$

9.

$y = 0$ $y(y + 3) = 0$

is true;   is false.

$y(y + 3) = 0$ $y = 0$

11.

$(z - a)(z - b) = 0$ $(b - z)(a - z) = 0$

is true;   is true.

$(b - z)(a - z) = 0$ $(z - a)(z - b) = 0$

13. $R - \{0, 3\}$.

15. $R - \{0, 1\}$.

17. $R - \{-1, 1\}$.

19. $\varnothing$.

21. $c \geqslant 0$.

23. $\varnothing$.

25. No. $|3 - 2| \neq 2 - 3$.

27. Yes.

29. No. Justify this answer!

Exercise 7.2

1. Symmetric property.

3. Transitive property.

5. Transitive property.

7. $\{-1\}$.

9. $\{a - b\}$.

11. $\left\{ \dfrac{r_1^2(1 + r_2^2)}{(4r_1 - 3)r_2^2} \right\}$.

13. The third equation is redundant relative to the given equation. Note that 0 and -3 are solutions of the third equation that do not belong to the solution set of the given equation.

15. The third equation is defective relative to the given equation. Justify this statement!

17. (a) $K_1 = C - \{1\}$, (b) $K_2 = C - \{1\}$.

19. (a) $K_1 = C$, (b) $K_2 = C - \{-3\}$.

Exercise 7.3

1. $\{0, 2\}$.

3. If $b \neq 0$, solution set is $\varnothing$; if $b = 0$, solution set is $\{0\}$.

5. $\{-3, -2, -\frac{1}{2}\}$.

7. $(a + 3m + n)(b + 2)$.

9. $3 \cdot (t - 1) \cdot (t + 8)$.

11. (a) $\varnothing$, (b) $\{-ki, ki\}$.

13. (a) $\left\{ -\dfrac{c}{b}, \dfrac{c}{b} \right\}$, (b) $\left\{ -\dfrac{c}{b}, \dfrac{c}{b}, -\dfrac{c}{b}i, \dfrac{c}{b}i \right\}$.

15. (a) $\left\{ -\dfrac{ab}{c}, \dfrac{ab}{c} \right\}$, (b) $\left\{ -\dfrac{ab}{c}, \dfrac{ab}{c}, -\dfrac{ab}{c}i, \dfrac{ab}{c}i \right\}$.

17. $\left\{-\dfrac{b}{c}, \dfrac{b}{c}\right\}$.

19. $\{-bn, bn\}$.

Exercise 7.4

1. 10.

3. $1 + \dfrac{1}{\sqrt{3}}$.

5. $|x - 1|$.

7. $(y + \tfrac{5}{2})^2 - \tfrac{21}{4}$.

9. $3 \cdot (x - \tfrac{1}{6})^2 + \tfrac{35}{12}$.

11. $a\left(x - \dfrac{b}{2a}\right)^2 + \left(1 - \dfrac{b^2}{4a}\right)$.

13. $\left\{\dfrac{-5 + \sqrt{21}}{2}, \dfrac{-5 - \sqrt{21}}{2}\right\}$.

15. $\left\{\dfrac{1 + \sqrt{35}\,i}{6}, \dfrac{1 - \sqrt{35}\,i}{6}\right\}$.

17. $\left\{\dfrac{b + \sqrt{b^2 - 4a}}{2a}, \dfrac{b - \sqrt{b^2 - 4a}}{2a}\right\}$.

19. $\left\{\dfrac{-1 + 3i}{4}, \dfrac{-1 - 3i}{4}\right\}$.

21. $\left\{\dfrac{-1 + \sqrt{11}}{4}, \dfrac{-1 - \sqrt{11}}{4}\right\}$.

25. $\{-\tfrac{1}{3}, \tfrac{7}{3}\}$.

27. $r_1 = \dfrac{-1 + \sqrt{17}}{4}, r_2 = \dfrac{-1 - \sqrt{17}}{4}$.

31. $4i$.

33. i.

35. 10.

Exercise 7.5

1. $\left\{\dfrac{1 + i}{2}, \dfrac{1 - i}{2}\right\}$.

3. $\{0, \frac{3}{7}\}$.

5. $\left\{-\dfrac{n}{m}, -1\right\}$.

7. $\{-2y, y\}$.

9. (a) $\{k \mid k \leqslant \frac{9}{4}\}$; (b) $\{k \mid k > \frac{9}{4}\}$.

11. (a) $\{k \mid k \leqslant 0 \vee k \geqslant \frac{8}{9}\}$; (b) $\{k \mid 0 < k < \frac{8}{9}\}$.

13. (a) $\{k \mid k \leqslant -1 \vee k \geqslant 0\}$; (b) $\{k \mid -1 < k < 0\}$.

15. (a) $\{k \mid k \leqslant 0 \vee k \geqslant 12\}$; (b) $\{k \mid 0 < k < 12\}$.

17. $(3x - 2 - \sqrt{7})\left(x - \dfrac{2 - \sqrt{7}}{3}\right)$.

19. $\left(7x - \dfrac{\sqrt{2} - \sqrt{30}}{2}\right)\left(x - \dfrac{\sqrt{2} + \sqrt{30}}{14}\right)$.

21. $(y + 1) \cdot \left(y - \dfrac{1}{2} - \dfrac{i}{2}\right) \cdot \left(y - \dfrac{1}{2} + \dfrac{i}{2}\right)$.

23. $(2x + 1) \cdot (x - 1)$.

25. Not possible.

31. (a) $\{1 + i, 2 + i\}$, (b) $\left\{\dfrac{-4 - \sqrt{3} - 3i}{2}, \dfrac{-4 + \sqrt{3} - 3i}{2}\right\}$.

Exercise 7.6

1. $\{-7 \pm \sqrt{3}, -7 \pm \sqrt{2}\}$.

3. $\{2 \pm \sqrt{2}, 2 \pm \sqrt{3}\,i\}$.

5. $\{a \pm \sqrt{b}, a \pm \sqrt{b}\,i\}$.

7. $\{-1, 3\}$.

9. $\varnothing$.

11. $\left\{\dfrac{c - b}{a}, \dfrac{-b - c}{a}\right\}$.

13. $\{-1, 0\}$.

15. $\{3\}$.

17. $\varnothing$.

19. $\{\frac{5}{4}\}$.

21. $\{1, 3\}$.

23. $\{-n^2, m^2\}$.

25. $\{2\}$.

Exercise 7.7

1. $\{-\frac{7}{5}, 0\}$.

3. $\{3, 5\}$.

5. $\{-6\}$.

7. 55 miles per hour.

9. 95 games.

11. $\varnothing$.

13. $\{8\}$.

15. $\{\frac{9}{2}\}$.

17. $\left\{\dfrac{a(a+b)}{b}\right\}$.

19. $\{\frac{15}{7}\}$.

REVIEW EXERCISES FOR CHAPTER 7

1. (a) Symmetric property,
   (b) transitive property,
   (c) reflexive property.

3. The equations are not equivalent relative to R. (Why?)

5. The equations are not equivalent relative to C. (Why?)

7. $\{1 + i, -1 - i\}$.

9. (a) $(x + \frac{5}{2})^2 - \frac{9}{4}$,
   (b) $3(x + \frac{1}{3})^2 + \frac{2}{3}$.

11. $\{-\frac{3}{4}, 2\}$.

13. $\left\{\dfrac{1 + \sqrt{5}}{2}, \dfrac{-1 + \sqrt{5}}{2}\right\}$.

15. $\left\{\dfrac{-1 + \sqrt{7}i}{4}, \dfrac{-1 - \sqrt{7}i}{4}\right\}$.

17. $(4x + 3)(x - 2)$.

19. (a) $K = 0$, $K = 4$,
    (b) $\{K \mid K < 0 \vee K > 4\}$,
    (c) $\{K \mid 0 < K < 4\}$.

21. $\varnothing$.

23. $\{-\frac{1}{4}, 3\}$.

Exercise 8.2

1. The boy walked for 2 hr.

3. 20 and 30 mph, respectively.

5. 500 and 550 mph, respectively.

7. 3:45 P.M.

9. 900 and 880 miles, respectively.

Exercise 8.3

1. 15 lb of 70-cents-a-pound coffee and 45 lb of 90-cents-a-pound coffee.
3. $3\frac{1}{3}$ qt.
5. $45\frac{3}{5}$ oz.
7. $66\frac{2}{3}$ lb.
9. $18,000 invested in bonds and $6,000 invested in stocks.
11. $28,000 invested at 7 percent and $14,000 invested at 4 percent.

Exercise 8.4

1. $2\frac{2}{9}$ hr.
3. $1\frac{10}{23}$ hr.
5. $\frac{322}{405}$ hr.
7. $\frac{30}{31}$ hr.
9. 120 and 80 min.
11. $\frac{1}{2}$ hr.

Exercise 8.5

1. 5 yr and 30 yr.
3. 4 yr, 8 yr, and 32 yr.
5. 12 yr.
7. The twins are 4 yr old; the woman is 34.

Exercise 8.6

1. 22 and 24.
3. 22, 24, 26, 28.
5. 32.
7. 27.
9. 414.
13. (a) 3025, (b) 4225, (c) 5625.

Exercise 8.7

1. 3 quarters and 9 dimes.
3. 20 nickels, 30 dimes, 50 quarters.

5. 500 cows.

7. 250 10-cent cones and 400 15-cent cones.

9. 11,500 bleacher tickets and 10,010 grandstand tickets.

REVIEW EXERCISES FOR CHAPTER 8

1. 7:30 P.M.

3. 86 and 129.

5. 24.15 ft.

7. 100 and 105 ft.

9. 12 and 6 min.

11. 2.5 gal.

Exercise 9.1

1. 2.

3. 2.

5. 4.

7. 11.

9. $\frac{3}{5}$.

11. -0.1.

13. a.

15. $\frac{1}{2}$.

17. $|x + 1|$.

19. 2.

21. 4.

23. $\dfrac{6}{\sqrt[3]{516}}$.

25. 3.

Exercise 9.2

1. 16 (in both cases).

3. 9 (in both cases).

5. $(x + y)^3$ (in both cases).

7. $\sqrt{(-16)^3} = 64i$, $(\sqrt{-16})^3 = -64i$.

9. $\frac{25}{49}$.

11. 5.

13. 0.4.

15. $8\sqrt[3]{\dfrac{y}{x}}$.

17. $\frac{9}{4}$.

Exercise 9.3

1. Choosing $x = 0$, we get $1 = -1$. $\sqrt{x^2 - 2x + 1} = |x - 1|$ is correct.

3. Choosing $t = -2$, we get $-1 = -3$. $\sqrt{(t+1)^2} - \sqrt{4} = |t+1|$ -2 is correct.

5. $2\sqrt[3]{4}$.

7. $\frac{2}{3}\sqrt[6]{243}$.

9. $\sqrt[4]{2}$.

11. $\dfrac{3x^2}{4y^3}$.

13. $3\sqrt[4]{3}$.

15. $\sqrt[5]{5}$.

17. $5\sqrt[3]{5}\,x^2$.

19. $\frac{1}{3}\sqrt[3]{126}$.

21. $\frac{2}{3}\sqrt[5]{18}$.

23. 2.

25. $\dfrac{a}{b}\sqrt{3abc}$.

27. $\dfrac{ax^2}{b^2c}\sqrt[5]{(5abc^2)^2}$.

29. $\dfrac{mn}{p}\sqrt[12]{5nm^5p^5}$.

31. $-2\sqrt{3}$.

33. $8\sqrt{6} - 5\sqrt{2} + 3\sqrt{5}$.

35. $6\sqrt[3]{a}$.

37. $2\sqrt{x+y} - 3\sqrt[3]{r+t}$.

Exercise 9.4

1. $3ab^3\sqrt{2ab}$.

3. $2(a+b)\sqrt[3]{3(a+b)}$.

5. $\dfrac{2\sqrt{10}\,|a-b|}{5b^4}$.

7. $\dfrac{\sqrt{10}}{4|b|}$.

9. $\frac{3}{5}$.

11. $2\sqrt[4]{2} \cdot a \cdot \sqrt[12]{a^5 b^{11}}$.

13. $\sqrt[mn]{n^n m^m}$.

15. $\frac{1}{2}\sqrt[6]{72}$.

17. $-(2\sqrt{6}+5)$.

19. $\sqrt[3]{\frac{4}{3}}\sqrt[6]{a}$.

21. $4\sqrt{6}-9$.

23. 4.

Exercise 9.5

1. $\{\frac{1}{4}\}$.

3. $\varnothing$.

5. $\{0\}$.

7. $\{15\}$.

9. $\{7\}$.

11. $\varnothing$.

13. $\{-6\}$.

15. $\{-\frac{3}{2}\}$.

17. $\{-8, 64\}$.

REVIEW EXERCISES FOR CHAPTER 9

1. 2.

3. 0.1.

5. $\sqrt[6]{a^5}$.

7. $\dfrac{a}{2b}$.

9. $-5\sqrt{2}$.

11. $\sqrt{8+2\sqrt{13}}$.

13. $\sqrt{3}+\sqrt{13}$.

15. $\sqrt[3]{a^3-b^3}$.

17. 0.

19. $-\frac{13}{2}\sqrt{2}$.

21. $\{4\}$.

23. $\{-3, 1\}$.

25. $\{15\}$.

Exercise 10.1

1.

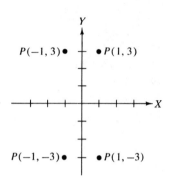

3.

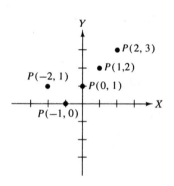

5.

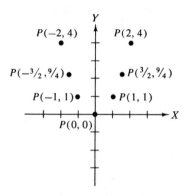

7. (a) Second quadrant, (b) third quadrant, (c) fourth quadrant.
9. (a) Third quadrant, (b) second quadrant, (c) first quadrant.
11. Second quadrant.

13. Third quadrant.
15. Fourth quadrant.

Exercise 10.2

1. 5.
3. $\sqrt{61}$.
5. 5.
7. $3\sqrt{5}$.
9. 8.
11. $2\sqrt{41}$.
13. $\sqrt{2}\,|b-a|$.
15. $|y-b|$.

Exercise 10.3

1. $(x-3)^2 + (y-3)^2 = 9$.
3. $(x+5)^2 + (y+3)^2 = 1$.
5. $x^2 + y^2 = 1$,
7. $(x-2)^2 + (y-1)^2 = 1$.
9. The graph of the second equation contains the origin.
11. Center at $(-2, 2)$, radius 4.
13. Center at $(2h, 2h)$, radius $2h$.
15. Center at $(ab, -ab)$, radius ab.
17. Center at $(-6, 4)$, radius $3\sqrt{6}$.

Exercise 10.4

1. $\frac{1}{5}$.
3. $-\frac{1}{3}$.
5. Slope is not defined, since the line is parallel to the Y-axis.
7. 0.
9. -1.
11. $y - 6 = 2(x - 4)$.
13. $y - 2 = -\frac{4}{3}x$.
15. $y - 6 = 2(x - 1)$.
17. $y + 6 = \frac{1}{2}(x + 1)$.
19. $y = x$.
21. L_2: $5y + 3x - 8 = 0$; L_3: $3y - 5x + 2 = 0$.
23. L_2: $y + x + 4 = 0$; L_3: $y = x$.
25. $c = 2$.

27.

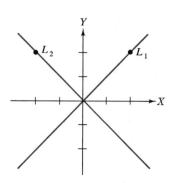

Graph of $x^2 = y^2$ is $L_1 \cup L_2$, where L_1 and L_2 are lines with equations $y = x$, $y = -x$, respectively.

Exercise 10.5

1. $x^2 + 16y = 0$.
3. $y = \frac{1}{6}\,[(x-2)^2 + 21]$.
5. $4x = (y-1)^2$.
7. $8 \cdot (x-3) + (y-3)^2 = 0$.
9. $16 \cdot (x+1) + (y-6)^2 = 0$.

Exercise 10.6

1. $\dfrac{x^2}{9} + \dfrac{y^2}{4} = 1$.

3. $\dfrac{x^2}{100} + \dfrac{y^2}{36} = 1$.

5. $\dfrac{x^2}{12} + \dfrac{y^2}{16} = 1$.

Exercise 10.7

1. (a) The points are symmetric with respect to the X-axis.
(b) The points are symmetric with respect to the Y-axis.
(c) The points are symmetric with respect to the origin.
(d) The points are symmetric with respect to the graph of $y = x$.
3. (a) The points are symmetric with respect to the graph of $y = x$.
(b) The points are symmetric with respect to the origin.
(c) The points are symmetric with respect to the origin and also with respect to the X-axis.

5. (a) The points are symmetric with respect to the X-axis.

   (b) The points are symmetric with respect to the Y-axis.

   (c) The points are symmetric with respect to the origin.

   (d) The points are symmetric with respect to the graph of $y = x$.

13. The graph is symmetric with respect to the Y-axis.

15. The graph is symmetric with respect to the origin, the X-axis, and the Y-axis.

17.

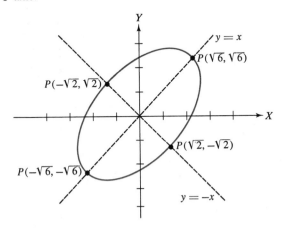

23. (a) $x = 1$, (b) $y = 1$, (c) $y = -x$.

Exercise 10.8

1. $\left(x - \dfrac{b-a}{2}\right)^2 + y^2 = \left(\dfrac{b+a}{2}\right)^2$. [*Hint:* To complete the proof, set $x = 0$.]

9. [*Hint:* Place the circle in the coordinate plane so that its center coincides with the origin and choose the chord to be parallel to the X-axis.]

11. (a) [*Hint:* Suppose that the line segment is determined by the real numbers a and b, $b > a$, and x is a real number such that $|x - a| = |x - b|$.]

REVIEW EXERCISES FOR CHAPTER 10

1. (a) First quadrant,

   (b) second quadrant,

   (c) fourth quadrant,

   (d) third quadrant.

3. $10 + 5\sqrt{2}$.

5. $(3, 2)$.

7. $(x-1)^2 + (y+1)^2 = 9$.

9. $2y - x - 5 = 0$.

11. $y^2 + 4x = 0$.

13.

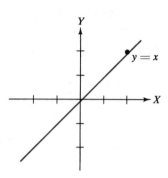

The graph is symmetric with respect to the origin and also with respect to the line $y = x$.

15.

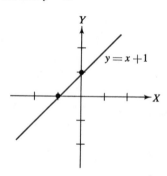

The graph lacks symmetry with respect to the origin, or the Y-axis, or the X-axis, or the line $y = x$.

17.

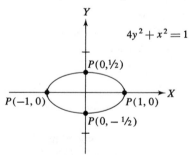

The graph is symmetric with respect to the origin, the Y-axis, and the X-axis.

25. $(x-2)^2 + (y+1)^2 = 25$.

Exercise 11.1

1. (a) $\begin{cases} y+1=1\cdot(x-0), \\ y-2=(-1)(x-0). \end{cases}$

   (b) $1; -1.$

   (c)

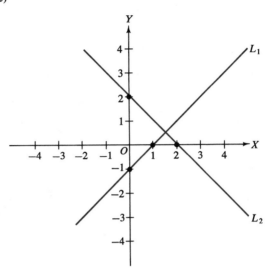

   (d) $\{(\tfrac{3}{2}, \tfrac{1}{2})\}.$

3. (a) $\begin{cases} y+1=3(x-0), \\ y-1=-\tfrac{1}{2}(x-0). \end{cases}$

   (b) $3; -\tfrac{1}{2}.$

   (c)

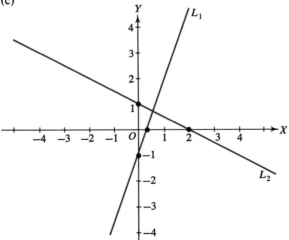

(d) $\{(\tfrac{4}{7}, \tfrac{5}{7})\}$.

5. (a) $\begin{cases} y + \tfrac{2}{3} = -\tfrac{4}{3}(x - 0), \\ y + \tfrac{3}{2} = \tfrac{5}{2}(x - 0). \end{cases}$

(b) $-\tfrac{4}{3}; \tfrac{5}{2}$.

(c)

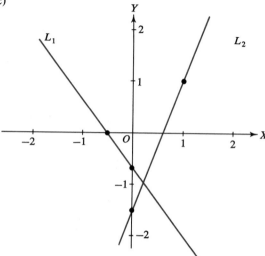

(d) $\{(\tfrac{5}{23}, -\tfrac{22}{23})\}$.

7. $\{(\tfrac{1}{3}, -\tfrac{2}{3})\}$.

9. $\{(\tfrac{4}{11}, \tfrac{5}{11})\}$.

11. $\{(x, y) \mid x - \tfrac{1}{3}y = -1\}$.

13. $\{(x, y) \mid 2x - \tfrac{5}{3}y = \tfrac{1}{6}\}$.

15. $\varnothing$.

17. $\left\{\left(-\dfrac{3}{5a}, \dfrac{7}{5}\right)\right\}$.

19. $\left\{\left(\dfrac{c}{a - b}, \dfrac{c}{a + b}\right)\right\}$.

21. $\left\{\left(\dfrac{1}{a - b}, 0\right)\right\}$.

23. (a) $a \neq b \wedge a \neq -b$, (b) $a = b$, (c) $a \neq b \wedge a = -b$.

25. $\{(2, 3)\}$.

27. $\left\{\left(\dfrac{a + c}{c}, \dfrac{b}{3}\right)\right\}$.

29.

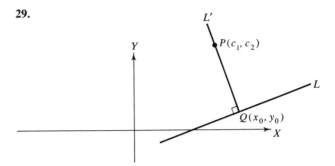

It will suffice to give the proof for the case where L is not parallel to the X-axis and $P(c_1, c_2) \notin L$. Let $Q(x_0, y_0)$ be the perpendicular projection of $P(c_1, c_2)$ on L. Since L has slope m, line L' through P and Q has slope $-\dfrac{1}{m}$. Then

$$-\frac{1}{m} = \frac{y_0 - c_2}{x_0 - c_1} \wedge y_0 = mx_0 + b, \qquad m(x_0 - c_1) \neq 0. \qquad (1)$$

From (1),

$$x_0 = \frac{mc_2 + c_1 - mb}{m^2 + 1}.$$

The distance d from P to Q is

$$d = \sqrt{(x_0 - c_1)^2 + (y_0 - c_2)^2} = \sqrt{(x_0 - c_1)^2} \cdot \sqrt{1 + \left(\frac{y_0 - c_2}{x_0 - c_1}\right)^2}$$

$$= \left| x_0 - c_1 \right| \cdot \sqrt{1 + \frac{1}{m^2}} \qquad = \left| \frac{mc_2 + c_1 - mb}{m^2 + 1} - c_1 \right| \cdot \frac{\sqrt{m^2 + 1}}{|m|}.$$

Hence,

$$d = \frac{|c_2 - mc_1 - b|}{\sqrt{m^2 + 1}}.$$

Exercise 11.2

1. 7.
3. 6.
5. 0.
7. 0.
9. $a_1 b_2 - a_2 b_1$.
11. 0.
13. $abx - cd$.
15. $\{(\frac{3}{2}, -\frac{1}{2})\}$.
17. $\{(\frac{27}{47}, \frac{12}{47})\}$.
19. $\{(\frac{6}{25}, \frac{6}{25})\}$.

21. $\left\{\left(\dfrac{c}{a+b}, \dfrac{c}{a+b}\right)\right\}.$

23. $\left\{\left(\dfrac{b-c}{b-a}, \dfrac{m(c-a)}{n(b-a)}\right)\right\}.$

Exercise 11.3

1. $\{(\frac{1}{4}, \frac{1}{4}, \frac{1}{4})\}.$

3. $\{(3, -1, -2)\}.$

5. $\varnothing.$

7. $\varnothing.$

9. $\{(\frac{3}{8}, -\frac{1}{4}, \frac{1}{8})\}.$

11. $\{(\frac{3}{4}, \frac{5}{4}, \frac{1}{2})\}.$

13. $\{(-\frac{1}{4}, \frac{1}{4}, \frac{1}{4})\}.$

15. $\{(-\frac{4}{5}, -\frac{12}{5}, -1)\}.$

17. $\{(\frac{45}{67}, \frac{62}{67}, -\frac{59}{67})\}.$

19. 50 nickels, 40 dimes, 6 quarters.

21. $\{(1, -3, 2)\}.$

23. $\{(\frac{3}{4}, -\frac{1}{4}, -\frac{1}{4})\}.$

25. $\varnothing.$

27. $\{(\frac{7}{40}, -\frac{1}{10}, -\frac{3}{10})\}.$

29. [*Hint:* Show that (a, b, c) is a solution of a system if and only if it is a solution of the other system.]

Exercise 11.4

1. $\{(-2, 2), (2, -2)\}.$ **3.**

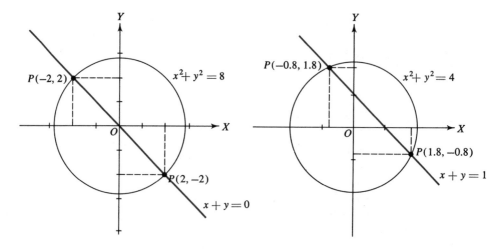

3. Thus $(-0.8, 1.8)$, $(1.8, -0.8)$ are approximate solutions.

$$\left(\text{Actually, the solution set is } \left\{\left(\frac{1+\sqrt{7}}{2}, \frac{1-\sqrt{7}}{2}\right), \left(\frac{1-\sqrt{7}}{2}, \frac{1+\sqrt{7}}{2}\right)\right\}.\right)$$

5. $\{(-21, 4), (-14, 3)\}$.

7. $\{(0, 1)\}$.

9. $\{(0, -1)\}$.

11. $\{(1, -\sqrt{3}), (1, \sqrt{3})\}$.

13. $\{(-\frac{2}{3}\sqrt{3} \cdot [\sqrt{33} - 5], \frac{2}{3}[6 - \sqrt{33}]),$

$$(\tfrac{2}{3}\sqrt{3}[\sqrt{33} - 5], \tfrac{2}{3}[6 - \sqrt{33}])\}.$$

15. $-3 - i,\ 3 + i$.

17. $-\dfrac{3\sqrt{2}}{2} + \dfrac{\sqrt{2}}{2}i,\ \dfrac{3\sqrt{2}}{2} - \dfrac{\sqrt{2}}{2}i$.

19. $-\sqrt{\tfrac{1}{2}(1 + \sqrt{2})} + \dfrac{1}{2\sqrt{\tfrac{1}{2}(1 + \sqrt{2})}}\,i,$

$$\sqrt{\tfrac{1}{2}(1 + \sqrt{2})} - \dfrac{1}{2\sqrt{\tfrac{1}{2}(1 + \sqrt{2})}}i.$$

21. $-\sqrt{1 + \dfrac{\sqrt{13}}{2}} - \dfrac{3}{2\sqrt{1 + \dfrac{\sqrt{13}}{2}}}i,\ \sqrt{1 + \dfrac{\sqrt{13}}{2}} + \dfrac{3}{2\sqrt{1 + \dfrac{\sqrt{13}}{2}}}i.$

23. $-\sqrt{\dfrac{a}{2}(1 + \sqrt{2})} + \dfrac{a}{2\sqrt{\dfrac{a}{2}(1 + \sqrt{2})}}i,$

$$\sqrt{\dfrac{a}{2}(1 + \sqrt{2})} - \dfrac{a}{2\sqrt{\dfrac{a}{2}(1 + \sqrt{2})}}i.$$

25. $-\sqrt{\dfrac{a + 1 + \sqrt{2a^2 + 2}}{2}} - \dfrac{a - 1}{2\sqrt{\dfrac{a + 1 + \sqrt{2a^2 + 2}}{2}}}i,$

$$\sqrt{\dfrac{a + 1 + \sqrt{2a^2 + 2}}{2}} + \dfrac{a - 1}{2\sqrt{\dfrac{a + 1 + \sqrt{2a^2 + 2}}{2}}}i.$$

27. $\left\{\left(-\dfrac{23}{26}, \dfrac{15}{26}\right)\right\}$.

29. \$800, 4 percent.

31. Length is 16 in., width is 9 in.

35. $\{-i\sqrt{3},\ i\sqrt{3}\}$.

Exercise 11.5

1.

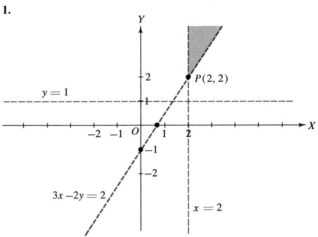

3.

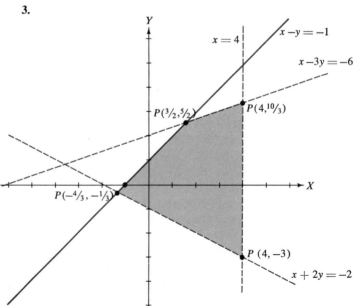

5.

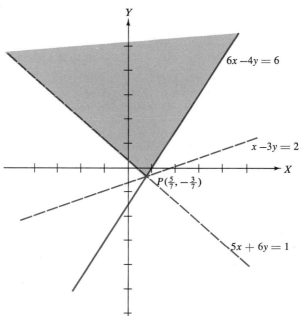

$6x - 4y = 6$

$x - 3y = 2$

$P(\frac{5}{7}, -\frac{3}{7})$

$5x + 6y = 1$

7.

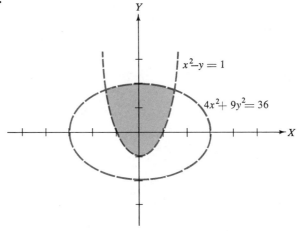

$x^2 - y = 1$

$4x^2 + 9y^2 = 36$

9.

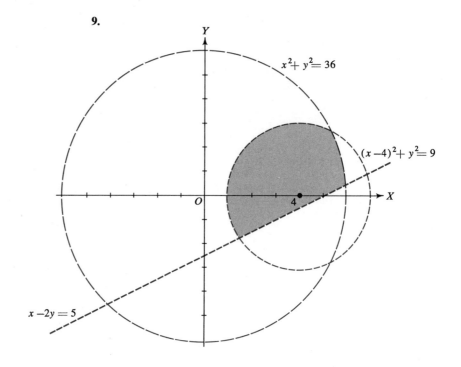

11.

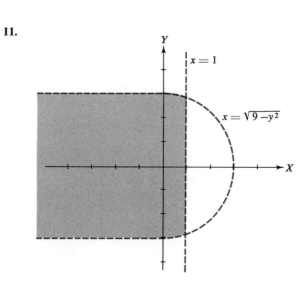

13.

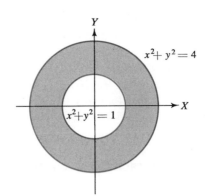

15.

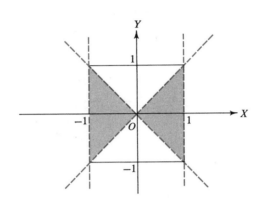

Exercise 11.6

1. (a)

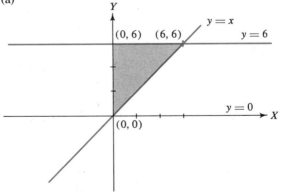

(b) Maximum value 6 at $(6, 6)$; minimum value -12 at $(0, 6)$.

3. $66\frac{2}{3}$ acres of crop A and $133\frac{1}{3}$ acres of crop B; $6666.67.

5. The farmer should assign 100 acres to crop A and 100 acres to crop B. Crop C should not be planted.

7. 30 shares of stock A and 70 shares of stock B.

9. Consider the regions $S = \{(x, y) \mid y \leqslant x^2\}$ and $T = \{(x, y) \mid y \geqslant -x^2 - 1\}$ in the plane. Both S and T are nonconvex. However, $S \cup T$ is convex.

11. Let $H_1 = \{(x, y) \mid y \leqslant 2x\}$, $H_2 = \{(x, y) \mid y \geqslant -x + 1\}$. $H_3 = \{(x, y) \mid x \leqslant 1\}$. H_1, H_2, H_3 are convex sets and $T = H_1 \cap H_2 \cap H_3$. If $K > 5$, then for every $(x, y) \in T$, $x^2 + y^2 < K$. Thus one may choose r to be any positive real number that is greater than 5. For example, the choice $r = 6$ will do.

REVIEW EXERCISES FOR CHAPTER 11

1. $\{(-1, 1)\}$

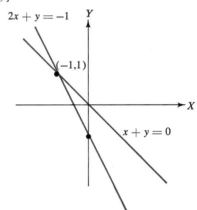

3.

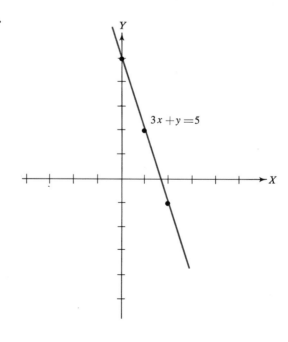

Note that the graph of the equation $x + \frac{1}{3}y = \frac{5}{3}$ is identical with the graph of the equation $3x + y = 5$.

5. (a) 0. (b) The system is inconsistent and $D = 0$.

7. $\{(\frac{1}{2}, \frac{1}{3}, \frac{1}{3})\}$.

9. $\varnothing$.

11. $\left\{\left(-\frac{4\sqrt{17}}{17}, -\frac{2\sqrt{17}}{17}\right), \left(\frac{4\sqrt{17}}{17}, \frac{2\sqrt{17}}{17}\right)\right\}$.

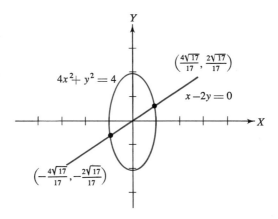

13. $\{(-2, 0), (2, 0)\}$.

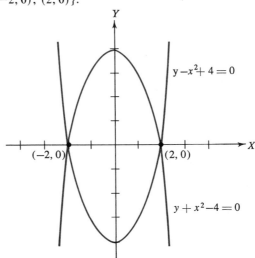

15.

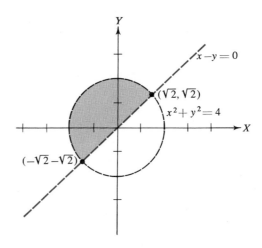

17.

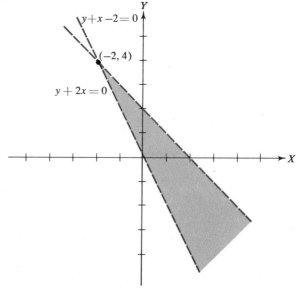

19.

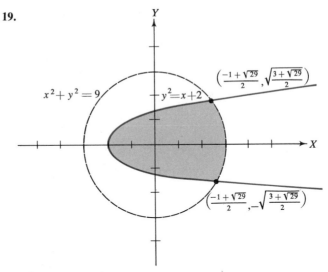

21. $\{-2 + i, 2 - i\}$.

23. $\{-1 + i, -3 - 4i\}$.

Exercise 12.1

1. The set describes a function.

3. The set does not describe a function.

5. The set describes a function.

7. The table does not describe a function.

9. (a) Let $f = \{(1, 5), (2, 6), (3, -2)\}$. Then $D_f = \{1, 2, 3\}$ and $R_f = \{5, 6, -2\}$.

　(b) Let $f = \{(9, 1), (0, 5), (6, 7)\}$. Then $D_f = \{9, 0, 6\}$ and $R_f = \{1, 5, 7\}$.

11. Let f denote $\{(x, y) \mid y = 3x - 1\}$ and suppose $(a, b) \in f \wedge (a, c) \in f$ is true. Show that $b = c$.

$$\big\downarrow \quad (a, b) \in f \ \wedge \ (a, c) \in f,$$
$$(a \in R \wedge b = 3a - 1) \ \wedge \ (a \in R \wedge c = 3a - 1).$$

Therefore, $b = c$.

13. Let f denote $\{(x, y) \mid y = x^2 - 5x + 2\}$ and suppose $(a, b) \in f \wedge (a, c) \in f$ is true. Show that (a, b) and (a, c) are not distinct. That is, show that $b = c$.

$$\big\downarrow \quad (a, b) \in f \ \wedge \ (a, c) \in f,$$
$$(a \in R \wedge b = a^2 - 5a + 2) \ \wedge \ (a \in R \wedge c = a^2 - 5a + 2).$$

Therefore, $b = c$.

15. Let f denote $\{(u, v) \mid v = -u^4 + 2u^2 + 5\}$ and suppose (a, b) $\in f \wedge (a, c) \in f$ is true. Show that (a, b) and (a, c) are not distinct. That is, show that $b = c$.

$$(a, b) \in f \wedge (a, c) \in f,$$

$$(a \in R \wedge b = -a^4 + 2a^2 + 5) \wedge (a \in R \wedge c = -a^4 + 2a^2 + 5).$$

Therefore, $b = c$.

17. $(1, -1)$ and $(1, 1)$ both belong to $\{(x, y) \mid x^2 = y^2\}$.

19. $(0, -1)$ and $(0, 1)$ both belong to $\{(r, s) \mid r = s^2 - 1\}$.

21. The given set is not a set of ordered pairs, since π is one of its elements and is not an ordered pair.

23. $f(-2) = -4$, $f(3) = 11$, $f(0) = 2$, $f(\pi) = 3\pi + 2$.

25. $H(5) = 17$, $H(3) = 11$, $H(5) - H(3) = 6$, $\dfrac{H(5) - H(3)}{2} = 3$.

27. (a) $\dfrac{f(x + h) - f(x)}{h} = 3$, $h \neq 0$.

(b) The secant line through the distinct points $(x, f(x))$, $(x + h, f(x + h))$ on the graph of f has the constant slope of 3.

Exercise 12.2

1. (a) $D_f = R - \{1\}$.

(b) x is the independent variable and y is the dependent variable.

(c) $f = \{(x, y) \mid y = \dfrac{1}{x - 1} \wedge x \in R - \{1\}\}$.

3. (a) $D_f = R - \{2\}$.

(b) x is the independent variable and y is the dependent variable.

(c) $f = \{(x, y) \mid y = \dfrac{x + 1}{(x^2 + 1)(x - 2)} \wedge x \in R - \{2\}\}$.

5. (a) $D_f = R - \{-1, 2\}$.

(b) s is the independent variable and r is the dependent variable.

(c) $f = \{(s, r) \mid r = \dfrac{2s + 7}{(s + 1)(s - 2)} \wedge s \in R - \{-1, 2\}\}$.

7. (a) $D_f = \{u \mid 5 - u \geq 0 \wedge u \in R\}$.

(b) u is the independent variable and z is the dependent variable.

(c) $f = \{(u, z) \mid z = \sqrt{5 - u} \wedge u \in R \wedge 5 - u \geq 0\}$.

9. (a) $D_f = \{u \mid u - 7 > 0 \wedge u \in R\}$.

(b) u is the independent variable and t is the dependent variable.

(c) $f = \left\{ (u, t) \mid t = \dfrac{1}{\sqrt{u-7}} \wedge u \in R \wedge u - 7 > 0 \right\}$.

11. $f(3) = \frac{1}{2}, f(-4) = -\frac{1}{5}$.

13. $f(4) = \frac{5}{34}, f(-1) = 0$.

15. $f(a) = \dfrac{2a+7}{(a+1)(a-2)}, a \in R - \{-1, 2\}$;

$f(a+h) = \dfrac{2a+2h+7}{(a+h+1)(a+h-2)}, a+h \in R - \{-1, 2\}$.

17. $f(1) \cdot f(4) = 2$.

19. (a) $g\left(\dfrac{1}{x}\right) = \dfrac{1}{x} + x$.

(b) $g\left(x - \dfrac{1}{x}\right) = \dfrac{x^2 - 1}{x} + \dfrac{x}{x^2 - 1}, x \in R - \{-1, 0, 1\}$.

(c) $x^2 g(x) - g(1) = x^3 + x - 2, x \in R - \{0\}$.

21. $V = \dfrac{1}{6\sqrt{\pi}} S^{\frac{3}{2}}$.

Exercise 12.3

1.

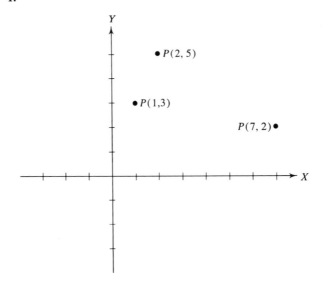

3.

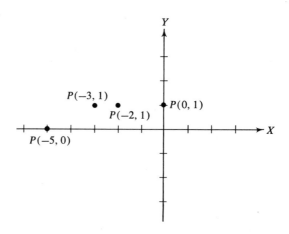

5.

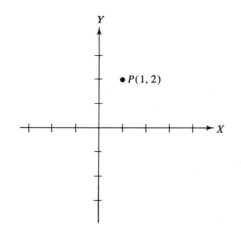

7.

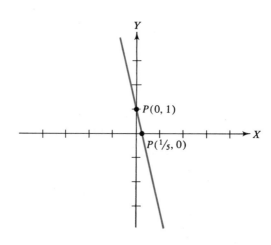

9.

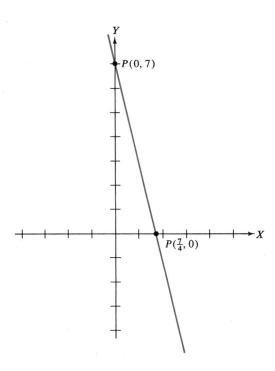

11.

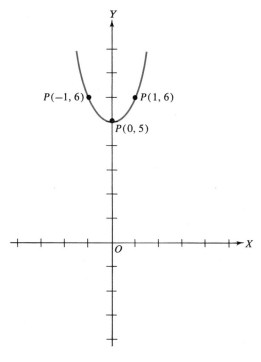

13.

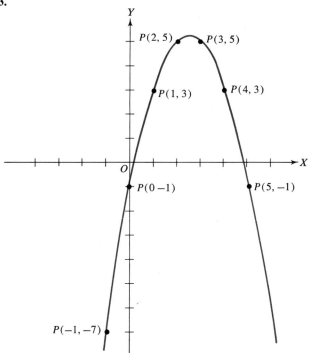

15.

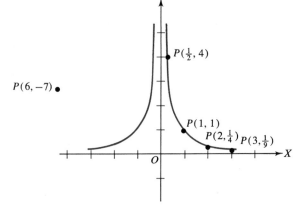

17.

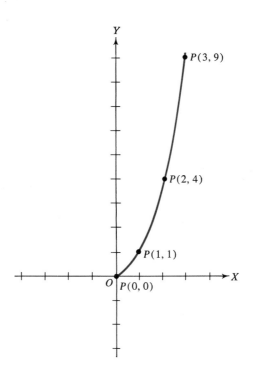

19.

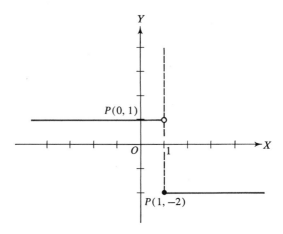

21.

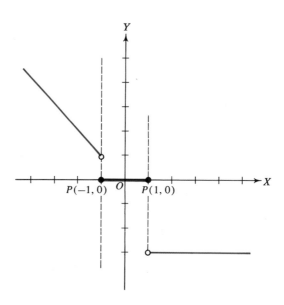

$P(-1, 0)$ $P(1, 0)$

Exercise 12.4

1. $h(x) = x^3 + x^2 - x + 5$ for all $x \in D_f \cap D_g = R$.

3. $h(x) = x^3 + x^2 + x + 6 + \dfrac{1}{x}$ for all $x \in D_f \cap D_g$. $D_f \cap D_g$

$= R - \{0\}$.

5. $h(x) = \dfrac{2x^2 + 5x}{(x + 3)(x^2 - 4)}$ for all $x \in D_f \cap D_g$. $D_f \cap D_g$

$= R - \{-3, -2, 2\}$.

7. $P(x) = x^6 - x^4 - 4x^2 + 4$ for all $x \in D_f \cap D_g = R$ (the set of real numbers).

9. $P(x) = \dfrac{1}{x^2 - 1}$ for all $x \in D_f \cap D_g$. $D_f \cap D_g = R - \{-1, 1\}$.

11. $Q(x) = x - 1$ for all $x \in D_f \cap D_g - \{-1\} = R - \{-1\}$.

13. $Q(x) = \dfrac{1}{x - 2}$ for all $x \in D_f \cap D_g - \{2\} = R - \{1, 2\}$.

15. $Q(x) = \dfrac{x-1}{x^2+x+1}$ for all $x \in D_f \cap D_g = R - \{1\}$.

17. $F(x) = x^2 + 2x - 11$ and $F = \{(x, y) \mid y = x^2 + 2x - 11\}$; $G(x) = x^2 - 8x + 9$ and $G = \{(x, y) \mid y = x^2 - 8x + 9\}$.

19. $F(x) = G(x) = x$ and $F = G = \{(x, y) \mid y = x \wedge x \in R - \{0\}\}$.

21. $f[g(1)] = 1; f[g(2)] = 1; f[g(3)] = 2;$ $g[f(1)] \cdot g[f(2)] = 4; g[f(3)] = 1.$

Exercise 12.5

1. $\quad (a, b) \in f \wedge (c, b) \in f,$

$\qquad b = 3a - 2 \wedge b = 3c - 2,$

$\qquad 3a - 2 = 3c - 2,$

$\qquad 3a = 3c,$

$\qquad a = c.$

Thus, f is one-to-one.

3. $\quad (a, b) \in f \wedge (c, b) \in f,$

$\qquad b = a^3 \wedge b = c^3,$

$\qquad a^3 = c^3,$

$\qquad a = c.$

Thus, f is one-to-one.

5. $\quad (a, b) \in f \wedge (c, b) \in f,$

$\qquad b = a^2 \wedge b = c^2,$

$\qquad a^2 = c^2,$

$\qquad a = c, \qquad$ since $a \leqslant 0 \wedge c \leqslant 0.$

Thus, f is one-to-one.

7. $f^{-1} = \{(x, y) \mid y = -x + 4\}$ and $f^{-1}(x) = -x + 4.$

9. $f^{-1} = \{(x, y) \mid y = \sqrt{x} \wedge x \geqslant 0\}$ and $f^{-1}(x) = \sqrt{x} \wedge x \geqslant 0.$

11.

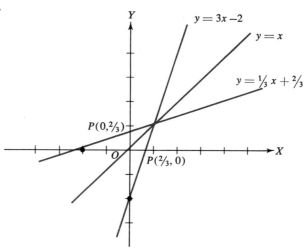

13.

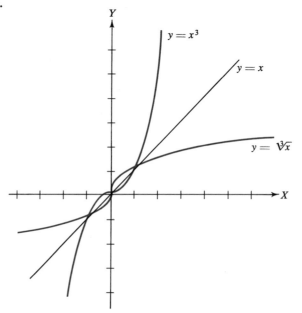

15.

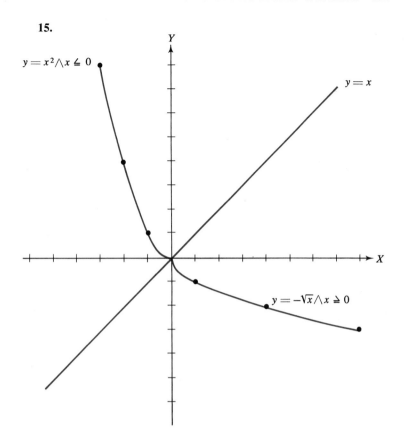

17. $f(2) = 1, f(3) = 0, f(4) = -1, f(5) = -2.$

REVIEW EXERCISES FOR CHAPTER 12

1. Not a function.
3. (a) Function,
   (b) $D_f = \{1, 2, 3\}, R_f = \{1\}.$
5. Not a function.
7. $(a, b) \in f \wedge (a, c) \in f,$

   $a + 2b = 1 \wedge a + 2c = 1,$

   $a + 2b = a + 2c,$

   $2b = 2c,$

   $b = c.$

Thus, f is a function.

$$(a, b) \in f \wedge (c, b) \in f,$$
$$a + 2b = 1 \wedge c + 2b = 1,$$
$$a + 2b = c + 2b,$$
$$a = c.$$

Thus, f is one-to-one.

$$f^{-1} = \{(x, y) \mid 2x + y = 1\}; f^{-1}(2) = -3; f^{-1}(-1) = 3.$$

9. $H(x) = \dfrac{\sqrt{x^2 + 2x + 2}}{|x + 1|}$, $x \in D_g = R - \{-1\}$.

11. First prove that G is a function (see Exercise 7).

$$(a, b) \in G \wedge (c, b) \in G,$$
$$b = \frac{1}{a + 1} \wedge b = \frac{1}{c + 1},$$
$$\frac{1}{a + 1} = \frac{1}{c + 1},$$
$$a = c, \quad a, c \in R - \{-1\}.$$

Thus, G is one-to-one; $G^{-1} = \left\{(x, y) \mid y = \dfrac{1 - x}{x}, x \in R - \{0\}\right\}$.

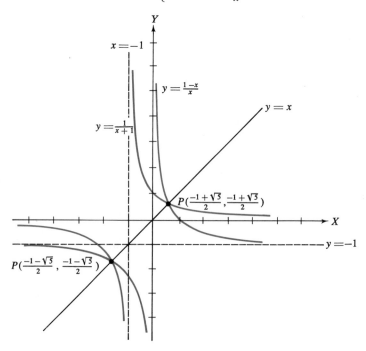

15. (a) $D_f = R$.

(b) x is the independent variable and y is the dependent variable.

(c) $f = \{(x, y) \mid y = 2x^2 + 1 \land x \in R\}$.

17. (a) $D_f = R - \{2, -3\}$.

(b) x is the independent variable and y is the dependent variable.

(c) $f = \left\{(x, y) \mid y = \dfrac{1}{(x - 2)(x + 3)} \land x \in R - \{2, -3\}\right\}$.

19. (a) $D_f = R$.

(b) x is the independent variable and y is the dependent variable.

(c) $f = \{(x, y) \mid y = \sqrt[4]{x + |x|} \land x \in R\}$.

21. $h = \{(1, 1), (2, -2), (3, 1)\}$.

23. $h = \{(1, 2), (2, \frac{1}{3}), (3, \frac{3}{2})\}$.

25. $h = \{(1, 3), (2, 1), (3, 2)\}$.

27. *Definition:* A *function* is a set of ordered pairs, no two distinct members of which have the same first element.

Exercise 13.1

1. Somewhat larger than 1,000,000.

3. Somewhat larger than 10^{15}.

5. Somewhat larger than 10^{27}.

7.

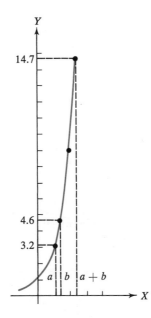

9. 2.

11. 4.

13. −2.

15. 1.

17. 0.

19. 4.

21. 0.

23. {4}.

25. {−2, 2}.

Exercise 13.2

1. 3.

3. −4.

5. 3.

7. 0.9030.

9. 1.7323.

11. 0.07525.

13. 0.8239.

15. 0.1461.

17. $\log_b b = x \leftrightarrow b^x = b \leftrightarrow b^x = b^1 \leftrightarrow x = 1$.

19. Let $\log_b (x) = r$ and $\log_b (y) = s$. Then $x = b^r$ and $y = b^s$. Therefore

$$\frac{x}{y} = \frac{b^r}{b^s} = b^{r-s} \quad \text{and} \quad \log_b \left(\frac{x}{y}\right) = r - s = \log_b (x) - \log_b (y).$$

(The other two parts are proved in a similar way.)

21.

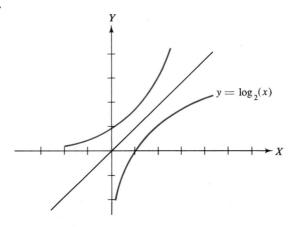

$y = \log_2(x)$

23. Replacement set $= \{x \mid x > 19\}$; solution set $= \{24\}$.
25. Replacement set $= \{x \mid x > 16\}$; solution set $= \{21\}$.
27. Replacement set $= \{x \mid x > 3\}$; solution set $= \{5\}$.
29. Replacement set $= \{x \mid x > \frac{5}{2}\}$; solution set $= \{5\}$.

Exercise 13.3

1. (a) 51.531, (b) 51.5313.
3. (a) 21.313, (b) 21.3131.
5. (a) 121.015, (b) 121.0152.
7. $(1602) \cdot 10^{-6}$.
9. $(1.000561) \cdot 10^9$.
11. $(1.561) \cdot 10^9$.
13. Constant of gravitation, $K = (6.670) \cdot 10^{-8} =$ the attraction in dynes between two gram masses 1 cm apart. Mass of a hydrogen atom $= (1.6731) \cdot 10^{-24}$ g. Planck's constant $= (6.62) \cdot 10^{-27}$ erg-sec.
15. $-4 + 0.7126$.
17. 1.4983.
19. $-7 + 0.9605$.
21. $-4 + 0.9096$.
23. 8130.
25. If $n = a \cdot 10^k$ with $1 \leqslant a < 10$, then the characteristic is k. Thus, if $n > 1$, $k \geqslant 0$, and so on.

Exercise 13.4

1. 0.3696.
3. 0.4917.
5. 0.9852.
7. 2.6181.
9. $-6 + 0.7906$
11. 364.1.
13. 1,032,000.
15. 1035.
17. 0.01496.
19. 0.9588.

Exercise 13.5

1. 33.78.
3. 321,800.

5. 1.66.

7. 0.000005023.

9. 0.9288.

11. 0.4361.

13. 0.8222.

15. 7.937.

17. 0.4771.

19. 51.95.

21. 1600 yr.

23. 2.096.

REVIEW EXERCISES FOR CHAPTER 13

1. 4.

3. −4.

5. 0.

7. $\frac{1}{3}$.

9. 2.

11. $\log \dfrac{a^2 \sqrt{b}}{\sqrt[3]{c}}$.

13. $\log \dfrac{b^5}{c^3}$.

15. $\log c$.

17. 0.3640, 205.1.

19. 0.2342, 51.88.

21. 0.0022, 10.12.

23. 0.1456.

25. 1.156.

27. 474,000.

29. 1.486.

31. Replacement set $= \{x \mid x > 0\}$; solution set $= \{\frac{50}{3}\}$.

Exercise 14.1

1. (a) 4, (b) 10, 14, 17, (c) 3200.

3. (a) $\frac{1}{2}$, (b) 4, $\frac{9}{2}$, 5, (c) 510.

5. (a) 5, (b) 4, 9, 14, (c) 3660.

7. (a) 8, (b) 20, 28, 36, (c) 6400.

9. (a) 8, (b) 9, 17, 25, (c) 5960.

11. (a) −8, (b) −10, −18, −26, (c) −6000.

13. (a) $2 + \pi$, (b) $4 + \pi, 6 + 2\pi, 8 + 3\pi$, (c) $740\pi + 1560$.

15. (a) 1, (b) $\frac{7}{2}, \frac{9}{2}, \frac{11}{2}$, (c) 840.

17. (a) $-\frac{88}{10}$, (b) $-\frac{91}{10}, -\frac{179}{10}, -\frac{267}{10}$, (c) -6524.

19. (a) $-\frac{3}{28}$, (b) $\frac{9}{14}, \frac{15}{28}, \frac{3}{7}$, (c) $-\frac{345}{7}$.

21. (a) $22,560, (b) $36,720.

23. (a) 2, (b) $s_1 = 1$, (c) 15, (d) 64.

Exercise 14.2

1. $r = 4$; 48,192,768.

3. $r = -2$; $-4, 8, -16$.

5. $r = -3/\pi$; $-9/\pi, 27/\pi^2, -81/\pi^3$.

7. $r = -\frac{1}{2}$; $s_1 = -64$, $s_2 = 32$, $s_3 = -16$.

9. $r = \frac{1}{2}$; $s_1 = -256$, $s_2 = -128$, $s_3 = -64$.

11. $r = \frac{1}{4}$ or $r = -\frac{1}{4}$. In case $r = \frac{1}{4}$, $s_1 = 128$, $s_2 = 32$, $s_3 = 8$. In case $r = -\frac{1}{4}$, $s_1 = 128$, $s_2 = -32$, $s_3 = 8$.

13. 0.

15. $\frac{819}{2}$.

17. $\frac{171}{2048}$.

19. 9837.

21. $\frac{1365}{8}$.

23. $\left[24 + \dfrac{32}{3^{13}} (3^{14} - 2^{14}) \right]$ in. (approximately 119.6 in.).

Exercise 14.3

1. (a) $1220.19, (b) $1488.86, (c) $1816.70.

3. (a) $9427.98, (b) $12698.13, (c) $17102.54.

5. $\frac{480}{36} [1.01 + (1.01)^2 + \cdots + (1.01)^{36}] \doteq 580.10$. Compare to $16 \times 36 = 576$.

7. $\frac{110}{12} [1.015 + (1.015)^2 + \cdots + (1.015)^{12}] \doteq 121.34$. Compare to $10 \times 12 = 120$.

9. $\frac{288}{36} [1 + \frac{15}{12} \cdot \frac{1}{100} + \cdots + (1 + \frac{15}{12} \cdot \frac{1}{100})^{36}] \doteq 365.44$. Compare to $10 \times 36 = 360$.

11. $\frac{425}{24} (1.01) [1 + 1.01 + \cdots + (1.01)^{23}] \doteq 482.43$. Compare to $20 \times 24 = 480$.

Exercise 14.4

1. $6003.05.

3. $13180.79.

5. $4959.89.

7. $5189.82.

9. $11118.39.

11. $22204.88.

13. $259 (rounded off to three significant figures).

Exercise 14.5

1. There are only two possible outcomes (heads or tails) and more than two tosses. Hence, the same outcome must occur at least twice.

3. There are only three boys she will date and she wants four dates. Even if she dates a different boy the first three times, on the fourth date she must go out with one of the ones she has already dated.

5. The only possible remainders are 0, 1, 2, 3, 4. There are five possible outcomes but six numbers are divided, hence there are six remainders. If we let each of the numbers 0, 1, 2, 3, 4 correspond to a pigeonhole, and if a remainder corresponds to a pigeon who sleeps in hole i if and only if the remainder is i, there are six pigeons and five holes. Then in at least one of the holes there will be more than one pigeon.

7. If we divide 2^1, 2^2, 2^3, 2^4, 2^5, 2^6 by 5, the same remainder will occur twice. If this happens when 2^m and 2^n are divided by 5, then $2^m = 5 \cdot q_1 + r$ and $2^n = 5 \cdot q_2 + r$. Therefore,

$$2^m - 2^n = (5 \cdot q_1 + r) - (5 \cdot q_2 + r) = 5(q_1 - q_2).$$

Hence, $2^m - 2^n$ is divisible by 5.

9. There are n^2 ordered pairs. Associate with each element of S a pigeonhole and with each ordered pair (a, b) a pigeon. The pigeon sleeps in a hole if and only if its ordered pair is associated with the element of S corresponding to that hole. If $n \geq 2$, $n^2 > n$, hence there are more pigeons than holes. The conclusion follows directly.

11. There are only 17 possible remainders, 0, 1, 2, $\cdots$, 16, at each step of the division. After at most 18 steps, a remainder that appeared before appears again.

13. There are natural numbers m and n such that $b = ma$ and $c = nb$. Therefore $c = n(ma) = (nm) \cdot a$. Hence, a divides c.

15. If m is a prime number, then m divides m and we are through. Otherwise, there is a natural number m_1 such that $1 < m_1 < m$ and m_1 divides m. If m_1 is a prime, we are through. Otherwise, there is a natural number m_2 such that $1 < m_2 < m_1$ and m_2 divides m_1 (and hence divides m). Since a natural number m cannot have more than m distinct divisors, if we could continue the process just described more than m steps, the same divisor would have to occur more than once. Since the divisors are decreasing, $m > m_1 > m_2 > \cdots$, this is impossible and the process must end. Therefore, after k steps, m_k is a prime number and divides m.

REVIEW EXERCISES FOR CHAPTER 14

1. 59; 155.

3. 14.

5. (a) \$18,600,

   (b) \$30,480.

7. $[6 + 36(1 - (\frac{3}{4})^{14})]$ ft (approximately 41.36 ft).

9. $\frac{59,049}{100,000}$; after seven strokes.

11. \$700.16.

13. \$1113.26.

15. \$744.19.

17. Let f be the function whose domain is $D = \{1, 2, \cdots, 100\}$ and range $R = \{1, 2, \cdots, 95\}$. Associate with each member i of R a pigeonhole labeled i and with each member j of D a pigeon labeled j. Now agree that pigeon j sleeps in hole i if and only if $F(j) = i$. Since there are 95 holes and 100 pigeons, we must have at least one hole with more than one pigeon. Hence, there are integers p, q, r such that $f(p) = f(q) = r$.

19. No, because if it were possible, by the pigeonhole principle, there would be at least one member of the domain to which several members of the range would correspond. But this contradicts the definition of a function.

N	0	1	2	3	4	5	6	7	8	9
10	0000	0043	0086	0128	0170	0212	0253	0294	0334	0374
11	0414	0453	0492	0531	0569	0607	0645	0682	0719	0755
12	0792	0828	0864	0899	0934	0969	1004	1038	1072	1106
13	1139	1173	1206	1239	1271	1303	1335	1367	1399	1430
14	1461	1492	1523	1553	1584	1614	1644	1673	1703	1732
15	1761	1790	1818	1847	1875	1903	1931	1959	1987	2014
16	2041	2068	2095	2122	2148	2175	2201	2227	2253	2279
17	2304	2330	2355	2380	2405	2430	2455	2480	2504	2529
18	2553	2577	2601	2625	2648	2672	2695	2718	2742	2765
19	2788	2810	2833	2856	2878	2900	2923	2945	2967	2989
20	3010	3032	3054	3075	3096	3118	3139	3160	3181	3201
21	3222	3243	3263	3284	3304	3324	3345	3365	3385	3404
22	3424	3444	3464	3483	3502	3522	3541	3560	3579	3598
23	3617	3636	3655	3674	3692	3711	3729	3747	3766	3784
24	3802	3820	3838	3856	3874	3892	3909	3927	3945	3962
25	3979	3997	4014	4031	4048	4065	4082	4099	4116	4133
26	4150	4166	4183	4200	4216	4232	4249	4265	4281	4298
27	4314	4330	4346	4362	4378	4393	4409	4425	4440	4456
28	4472	4487	4502	4518	4533	4548	4564	4579	4594	4609
29	4624	4639	4654	4669	4683	4698	4713	4728	4742	4757
30	4771	4786	4800	4814	4829	4843	4857	4871	4886	4900
31	4914	4928	4942	4955	4969	4983	4997	5011	5024	5038
32	5051	5065	5079	5092	5105	5119	5132	5145	5159	5172
33	5185	5198	5211	5224	5237	5250	5263	5276	5289	5302
34	5315	5328	5340	5353	5366	5378	5391	5403	5416	5428
35	5441	5453	5465	5478	5490	5502	5514	5527	5539	5551
36	5563	5575	5587	5599	5611	5623	5635	5647	5658	5670
37	5682	5694	5705	5717	5729	5740	5752	5763	5775	5786
38	5798	5809	5821	5832	5843	5855	5866	5877	5888	5899
39	5911	5922	5933	5944	5955	5966	5977	5988	5999	6010
40	6021	6031	6042	6053	6064	6075	6085	6096	6107	6117
41	6128	6138	6149	6160	6170	6180	6191	6201	6212	6222
42	6232	6243	6253	6263	6274	6284	6294	6304	6314	6325
43	6335	6345	6355	6365	6375	6385	6395	6405	6415	6425
44	6435	6444	6454	6464	6474	6484	6493	6503	6513	6522
45	6532	6542	6551	6561	6571	6580	6590	6599	6609	6618
46	6628	6637	6646	6656	6665	6675	6684	6693	6702	6712
47	6721	6730	6739	6749	6758	6767	6776	6785	6794	6803
48	6812	6821	6830	6839	6848	6857	6866	6875	6884	6893
49	6902	6911	6920	6928	6937	6946	6955	6964	6972	6981
50	6990	6998	7007	7016	7024	7033	7042	7050	7059	7067
51	7076	7084	7093	7101	7110	7118	7126	7135	7143	7152
52	7160	7168	7177	7185	7193	7202	7210	7218	7226	7235
53	7243	7251	7259	7267	7275	7284	7292	7300	7308	7316
54	7324	7332	7340	7348	7356	7364	7372	7380	7388	7396
N	0	1	2	3	4	5	6	7	8	9

N	0	1	2	3	4	5	6	7	8	9
55	7404	7412	7419	7427	7435	7443	7451	7459	7466	7474
56	7482	7490	7497	7505	7513	7520	7528	7536	7543	7551
57	7559	7566	7574	7582	7589	7597	7604	7612	7619	7627
58	7634	7642	7649	7657	7664	7672	7679	7686	7694	7701
59	7709	7716	7723	7731	7738	7745	7752	7760	7767	7774
60	7782	7789	7796	7803	7810	7818	7825	7832	7839	7846
61	7853	7860	7868	7875	7882	7889	7896	7903	7910	7917
62	7924	7931	7938	7945	7952	7959	7966	7973	7980	7987
63	7993	8000	8007	8014	8021	8028	8035	8041	8048	8055
64	8062	8069	8075	8082	8089	8096	8102	8109	8116	8122
65	8129	8136	8142	8149	8156	8162	8169	8176	8182	8189
66	8195	8202	8209	8215	8222	8228	8235	8241	8248	8254
67	8261	8267	8274	8280	8287	8293	8299	8306	8312	8319
68	8325	8331	8338	8344	8351	8357	8363	8370	8376	8382
69	8388	8395	8401	8407	8414	8420	8426	8432	8439	8445
70	8451	8457	8463	8470	8476	8482	8488	8494	8500	8506
71	8513	8519	8525	8531	8537	8543	8549	8555	8561	8567
72	8573	8579	8585	8591	8597	8603	8609	8615	8621	8627
73	8633	8639	8645	8651	8657	8663	8669	8675	8681	8686
74	8692	8698	8704	8710	8716	8722	8727	8733	8739	8745
75	8751	8756	8762	8768	8774	8779	8785	8791	8797	8802
76	8808	8814	8820	8825	8831	8837	8842	8848	8854	8859
77	8865	8871	8876	8882	8887	8893	8899	8904	8910	8915
78	8921	8927	8932	8938	8943	8949	8954	8960	8965	8971
79	8976	8982	8987	8993	8998	9004	9009	9015	9020	9025
80	9031	9036	9042	9047	9053	9058	9063	9069	9074	9079
81	9085	9090	9096	9101	9106	9112	9117	9122	9128	9133
82	9138	9143	9149	9154	9159	9165	9170	9175	9180	9186
83	9191	9196	9201	9206	9212	9217	9222	9227	9232	9238
84	9243	9248	9253	9258	9263	9269	9274	9279	9284	9289
85	9294	9299	9304	9309	9315	9320	9325	9330	9335	9340
86	9345	9350	9355	9360	9365	9370	9375	9380	9385	9390
87	9395	9400	9405	9410	9415	9420	9425	9430	9435	9440
88	9445	9450	9455	9460	9465	9469	9474	9479	9484	9489
89	9494	9499	9504	9509	9513	9518	9523	9528	9533	9538
90	9542	9547	9552	9557	9562	9566	9571	9576	9581	9586
91	9590	9595	9600	9605	9609	9614	9619	9624	9628	9633
92	9638	9643	9647	9652	9657	9661	9666	9671	9675	9680
93	9685	9689	9694	9699	9703	9708	9713	9717	9722	9727
94	9731	9736	9741	9745	9750	9754	9759	9763	9768	9773
95	9777	9782	9786	9791	9795	9800	9805	9809	9814	9818
96	9823	9827	9832	9836	9841	9845	9850	9854	9859	9863
97	9868	9872	9877	9881	9886	9890	9894	9899	9903	9908
98	9912	9917	9921	9926	9930	9934	9939	9943	9948	9952
99	9956	9961	9965	9969	9974	9978	9983	9987	9991	9996
N	0	1	2	3	4	5	6	7	8	9

$$\pi = 3.14159\ 26535\ 89793$$
$$\log_{10} \pi = 0.49714\ 98726\ 94134$$
$$e = 2.71828\ 18284\ 59045$$
$$\log_{10} e = 0.43429\ 44819\ 03252$$
$$\log_e 10 = 2.30258\ 50929\ 94046$$
$$\log_{10} \log_{10} e = 9.63778\ 43113\ 00537$$
$$\log_e 2 = 0.69314\ 71805\ 59945$$

NUMBERS CONTAINING π

	N	$\log_{10} N$		N	$\log_{10} N$
π	3.141 5927	0.497 1499	π^2	9.869 6044	0.994 2997
2π	6.283 1853	0.798 1799	$2\pi^2$	19.739 2088	1.295 3297
3π	9.424 7780	0.974 2711	$4\pi^2$	39.478 4176	1.596 3597
4π	12.566 3706	1.099 2099	$\pi^2/2$	4.934 8022	0.693 2697
$\pi/2$	1.570 7963	0.196 1199	$1/\pi^2$	0.101 3212	$9.005\ 7003 - 10$
$3\pi/2$	4.712 3890	0.673 2411	$1/2\pi^2$	0.050 6606	$8.704\ 6703 - 10$
$\pi/3$	1.047 1976	0.020 0286	$1/4\pi^2$	0.025 3303	$8.403\ 6403 - 10$
$2\pi/3$	2.094 3951	0.321 0586	π^3	31.006 2767	1.491 4496
$4\pi/3$	4.188 7902	0.622 0886	$1/\pi^3$	0.032 2515	$8.508\ 5504 - 10$
$\pi/4$	0.785 3982	$9.895\ 0899 - 10$	$\sqrt{\pi}$	1.772 4539	0.248 5749
$3\pi/4$	2.356 1945	0.372 2111	$\sqrt{2\pi}$	2.506 6283	0.399 0899
$\pi/6$	0.523 5988	$9.718\ 9986 - 10$	$\frac{1}{2}\sqrt{\pi}$	0.886 2269	$9.947\ 5449 - 10$
$1/\pi$	0.318 3099	$9.502\ 8501 - 10$	$\sqrt{\pi/2}$	1.253 3141	0.098 0599
$2/\pi$	0.636 6198	$9.803\ 8801 - 10$	$1/\sqrt{\pi}$	0.564 1896	$9.751\ 4251 - 10$
$3/\pi$	0.954 9297	$9.979\ 9714 - 10$	$1/\sqrt{2\pi}$	0.398 9423	$9.600\ 9101 - 10$
$4/\pi$	1.273 2395	0.104 9101	$\sqrt{2/\pi}$	0.797 8846	$9.901\ 9401 - 10$
$1/2\pi$	0.159 1549	$9.201\ 8201 - 10$	$\sqrt[3]{\pi}$	1.464 5919	0.165 7166
$1/3\pi$	0.106 1033	$9.025\ 7289 - 10$	$\sqrt[3]{\pi/6}$	0.805 9960	$9.906\ 3329 - 10$
$1/4\pi$	0.079 5775	$8.900\ 7901 - 10$	$\sqrt[3]{\pi^2}$	2.145 0294	0.331 4332
$\pi/180$	0.017 4533	$8.241\ 8774 - 10$	$1/\sqrt[3]{\pi}$	0.682 7841	$9.834\ 2834 - 10$
$180/\pi$	57.295 7795	1.758 1226	$\sqrt[3]{3/4\pi}$	0.620 3505	$9.792\ 6371 - 10$

N	Log N!	N	Log N!	N	Log N!	N	Log N!	N	Log N!
1	0.00 000	51	66.19 065	101	159.97 433	151	264.93 587	201	377.20 008
2	0.30 103	52	67.90 665	102	161.98 293	152	267.11 771	202	379.50 544
3	0.77 815	53	69.63 092	103	163.99 576	153	269.30 241	203	381.81 293
4	1.38 021	54	71.36 332	104	166.01 280	154	271.48 993	204	384.12 256
5	2.07 918	55	73.10 368	105	168.03 399	155	273.68 026	205	386.43 432
6	2.85 733	56	74.85 187	106	170.05 929	156	275.87 338	206	388.74 818
7	3.70 243	57	76.60 774	107	172.08 867	157	278.06 928	207	391.06 415
8	4.60 552	58	78.37 117	108	174.12 210	158	280.26 794	208	393.38 222
9	5.55 976	59	80.14 202	109	176.15 952	159	282.46 934	209	395.70 236
10	6.55 976	60	81.92 017	110	178.20 092	160	284.67 346	210	398.02 458
11	7.60 116	61	83.70 550	111	180.24 624	161	286.88 028	211	400.34 887
12	8.68 034	62	85.49 790	112	182.29 546	162	289.08 980	212	402.67 520
13	9.79 428	63	87.29 724	113	184.34 854	163	291.30 198	213	405.00 358
14	10.94 041	64	89.10 342	114	186.40 544	164	293.51 683	214	407.33 399
15	12.11 650	65	90.91 633	115	188.46 614	165	295.73 431	215	409.66 643
16	13.32 062	66	92.73 587	116	190.53 060	166	297.95 442	216	412.00 089
17	14.55 107	67	94.56 195	117	192.59 878	167	300.17 714	217	414.33 735
18	15.80 634	68	96.39 446	118	194.67 067	168	302.40 245	218	416.67 580
19	17.08 509	69	98.23 331	119	196.74 621	169	304.63 033	219	419.01 625
20	18.38 612	70	100.07 841	120	198.82 539	170	306.86 078	220	421.35 867
21	19.70 834	71	101.92 966	121	200.90 818	171	309.09 378	221	423.70 306
22	21.05 077	72	103.78 700	122	202.99 454	172	311.32 931	222	426.04 941
23	22.41 249	73	105.65 032	123	205.08 444	173	313.56 735	223	428.39 772
24	23.79 271	74	107.51 955	124	207.17 787	174	315.80 790	224	430.74 797
25	25.19 065	75	109.39 461	125	209.27 478	175	318.05 094	225	433.10 015
26	26.60 562	76	111.27 543	126	211.37 515	176	320.29 645	226	435.45 426
27	28.03 698	77	113.16 192	127	213.47 895	177	322.54 443	227	437.81 028
28	29.48 414	78	115.05 401	128	215.58 616	178	324.79 485	228	440.16 822
29	30.94 654	79	116.95 164	129	217.69 675	179	327.04 770	229	442.52 805
30	32.42 366	80	118.85 473	130	219.81 069	180	329.30 297	230	444.88 978
31	33.91 502	81	120.76 321	131	221.92 796	181	331.56 065	231	447.25 339
32	35.42 017	82	122.67 703	132	224.04 854	182	333.82 072	232	449.61 888
33	36.93 869	83	124.59 610	133	226.17 239	183	336.08 317	233	451.98 624
34	38.47 016	84	126.52 038	134	228.29 949	184	338.34 799	234	454.35 545
35	40.01 423	85	128.44 980	135	230.42 983	185	340.61 516	235	456.72 652
36	41.57 054	86	130.38 430	136	232.56 337	186	342.88 467	236	459.09 943
37	43.13 874	87	132.32 382	137	234.70 009	187	345.15 652	237	461.47 418
38	44.71 852	88	134.26 830	138	236.83 997	188	347.43 067	238	463.85 076
39	46.30 959	89	136.21 769	139	238.98 298	189	349.70 714	239	466.22 916
40	47.91 165	90	138.17 194	140	241.12 911	190	351.98 589	240	468.60 937
41	49.52 443	91	140.13 098	141	243.27 833	191	354.26 692	241	470.99 139
42	51.14 768	92	142.09 477	142	245.43 062	192	356.55 022	242	473.37 520
43	52.78 115	93	144.06 325	143	247.58 595	193	358.83 578	243	475.76 081
44	54.42 460	94	146.03 638	144	249.74 432	194	361.12 358	244	478.14 820
45	56.07 781	95	148.01 410	145	251.90 568	195	363.41 362	245	480.53 736
46	57.74 057	96	149.99 637	146	254.07 004	196	365.70 587	246	482.92 830
47	59.41 267	97	151.98 314	147	256.23 735	197	368.00 034	247	485.32 100
48	61.09 391	98	153.97 437	148	258.40 762	198	370.29 701	248	487.71 545
49	62.78 410	99	155.97 000	149	260.58 080	199	372.59 586	249	490.11 165
50	64.48 307	100	157.97 000	150	262.75 689	200	374.89 689	250	492.50 959
N	Log N!	N	Log N!	N	Log N!	N	Log N!	N	Log N!

INDEX